Nitroalkanes

Nitroalkanes

Synthesis, Reactivity, and Applications

Roberto Ballini
Alessandro Palmieri

WILEY-VCH

Authors

Prof. Roberto Ballini
University of Camerino
Chemistry Division
Via S. Agostino 1
62032 Camerino
Italy

Prof. Alessandro Palmieri
University of Camerino
Chemistry Division
Via S. Agostino 1
62032 Camerino
Italy

Cover
Cover Image: Courtesy of Ballini

Library of Congress Card No.:
applied for

British Library Cataloguing-in-Publication Data
A catalogue record for this book is available from the British Library.

Bibliographic information published by the Deutsche Nationalbibliothek
The Deutsche Nationalbibliothek lists this publication in the Deutsche Nationalbibliografie; detailed bibliographic data are available on the Internet at <http://dnb.d-nb.de>.

Print ISBN: 978-3-527-34745-2
ePDF ISBN: 978-3-527-82676-6
ePub ISBN: 978-3-527-82677-3
oBook ISBN: 978-3-527-82678-0

Typesetting SPi Global, Chennai, India
Printing and Binding
CPI Group (UK) Ltd, Croydon CR0 4YY
Printed on acid-free paper

10 9 8 7 6 5 4 3 2 1

This book is primarily dedicated to Prof. Goffredo Rosini who was my mentor and who firstly introduced me to the chemistry of nitroalkanes, constantly encouraging and passionate me. Then, I want to dedicate this book to my daughters Federica and Francesca who have been a constant stimulus for me.

Roberto Ballini

Contents

About the Authors

Roberto Ballini is a Full Professor of Organic Chemistry at the University of Camerino (Italy) and Past President of the Italian Organic Chemistry Division. The scientific activity of Prof. Ballini is mainly dedicated to Organic Synthesis, with particular interest on the chemistry of nitroalkanes that, thanks to their highly chemical versatility, can be easily employed for new generation of C—C single and C=C double bond, for the cleavage of C—C bond, for their conversion into other important functionalities, thus as key building blocks in the synthesis, often under sustainable conditions, of a large variety of important "fine chemicals" such as natural products featuring enhanced biologically activity, pharmaceuticals, etc.

The scientific activities have been evidenced by:

(a) More than 250 publications, including more than 20 reviews, mainly dedicated to the chemistry of nitroalkanes.
(b) The role of Editor of the Green Chemistry Books: (i) *Eco-Friendly Processes and Procedures for the Synthesis of Fine Chemicals* (2009) and (ii) *Green Synthetic Processes and Procedures* (2019); both of Green Chemistry Series of Royal Chemical Society, Cambridge.
(c) The preparation, as main author, of four Book chapters on the field of "Nitroalkanes."

Alessandro Palmieri obtained his Laurea degree *cum laude* in Chemistry in 2002 at the University of Camerino (Italy) where, five years later, he received the PhD degree in Chemical Sciences. Then, in the period 2007–2010 he held a postdoctoral fellowship and in 2008 he moved, as visiting postdoctoral fellow, to the ITC laboratory at the University of Cambridge (Prof. Steven V. Ley). After the experience as assistant professor (2010–2013), in 2014 he was appointed associate professor in Organic Chemistry at the University of Camerino. Currently, his research interests involve (i) the chemistry of aliphatic nitro compounds, (ii) the realization of new one-pot protocols for generating and derivatizing heterocyclic systems, (iii) the preparation and use of solid supported reagents, (iv) the development of new sustainable processes, and (v) flow chemical protocols.

Acknowledgments

We wish to express our gratitude to all persons who contributed directly or indirectly to the preparation of this book by their researches and publications. We would like to offer a special thanks to Prof. Dieter Seebach and Prof. Noboru Ono for their historical contributions in this field. Finally, we wish to thank our Camerino's colleagues for their high helps in many aspects of the chemistry of nitroalkanes.

Roberto Ballini and Alessandro Palmieri

Introduction

Why the choice of a book dedicated to "nitroalkanes"? Usually a generic molecule containing the nitro group is called "nitro compound"; however there are three different typologies of nitro compounds, with distinct chemical behaviors (Figure 1): (i) aromatic nitro compounds, subjected to electrophilic substitution, (ii) conjugated nitroalkenes that, due to the presence of an electrophilic carbon in β-position, are exposed to nucleophilic addition, and (iii) nitroalkanes that, under basic conditions, act as nucleophiles due to the presence of an acidic proton germinal to the nitro group, providing the formation of C—C bond with electrophiles.

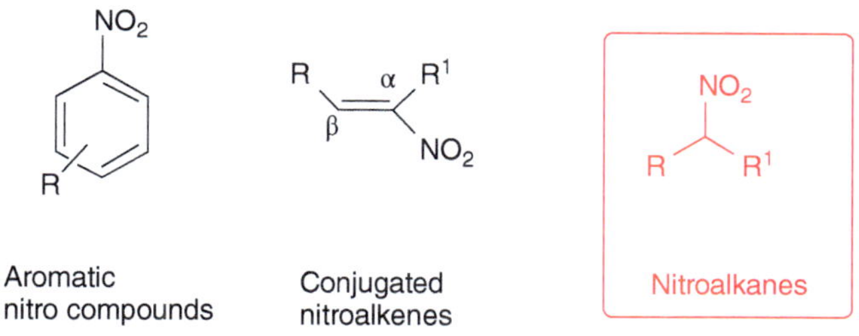

Figure 1 Three different typologies of nitro compounds.

Thus, since the studies dedicated to the chemistry of nitroalkanes and their application in organic synthesis are growing year after year in the last two decades, this category of molecules has proven to be one of the most valuable and versatile collection of substances in organic synthesis. Consequently, this book is dedicated exclusively to this class of molecules.

However, as the chemistry of nitroalkanes is constantly expanded, this book should not be considered as "the book" of nitroalkanes but just "the first" (after the pioneering Ono's book [1]) of a series dedicated to this class of molecules.

Reference

1 Ono, N. (2001). *The Nitro Group in Organic Synthesis*. Wiley-VCH. ISBN 0-471-31611-3.

List of Abbreviations

AAPOs	5-alkyl-5-aryl-1-pyrroline *N*-oxides
AB	ammonia borane
ACE	angiotensin converting enzyme
AIBN	azoisobutyronitrile
ALA	aminolevulinic acid
AuNPs	gold nanoparticles
BSA	*O*,*N*-bis-trimethylsilyl acetamide
BSDA	bis(sulfonamide)-diamine
BTSP	bis(trimethylsilyl) peroxide
CAN	cerium ammonium nitrate
CBOA	chiral bisoxazolidine
CFL	compact fluorescent lamp
m-CPBA	*m*-chloroperbenzoic acid
CTACl	cetyltrimethylammonium chloride
CTABr	cetyltrimethylammonium bromide
CTAOH	cetyltrimethylammonium hydroxide
CIDT	crystallization-induced diastereoselective transformations
CPME	cyclopentyl methyl ether
DABCO	1,4-diazabicyclo[2.2.2]octane
2,6-DMBQ	2,6-dimethylbenzoquinone
DBU	1,8-diazabicyclo[5.4.0]undec-7-ene
DEAD	diethyl azodicarboxylate
DIBAL	diisobutylaluminium hydride
DIPEA	diisopropylethylamine
DMD	dimethyldioxirane
DME	dimethoxyethane
DMF	dimethylformamide
DMO	dimethyldioxane
DMTHP	dimethyl-1,3,4,5-tetrahydropyrimidine
DZC	dinuclear zing catalysis
EGB	electrogenerated base
$Fe(acac)_3$	iron acetyl acetate

FUD-ED	Fudan University-ethylenediamine
γ-AA	γ-amino acid
HMPA	hexamethylphosphoramide
HMPT	hexamethylphosphoric triamine
HFIP	hexafluoroisopropanol
KI@Fe_3O_4NPs	KI loaded Fe_3O_4 nanoparticles
LBC	lanthanum binaphthol complex
LDA	lithium diisopropylamide
LiHMDS	lithium hexamethyldisiloxane
MCPBA	*meta*-chloroperbenzoic acid
MA	Michael asymmetric
MH	Michael heterogeneous
MG	Michael general
MIL	Michael in ionic liquids
MOM	methoxymethyl
MS	molecular sieves
MTBE	methyl *t*-butyl ether
MVK	methyl vinyl ketone
NAE	nitrous acid elimination
NAPI	*N*-acetoxyphtalimide
NI-HAp	nickel hydroxyapatite
NMO	*N*-methylmorpholine *N*-oxide
NMP	*N*-methylpiperidine
NP	nano particles
NSBP	non-symmetric bis-pyrrolidine
PDC	potassium dichromate
PEI	polyethylenimine
PPA	polyphosphoric acid
PPTS	pyridinium *p*-toluenesulfonate
PTM	phase transfer method
PTR	phase transfer reaction
PTT	phase transfer technique
SFC	solvent-free conditions
SG-MNP-NH_2	silica gel magnetic nanoparticles NH_2
NHPI	*N*-hydroxyphthalimide
PCC	pyridinium chlorochromate
$Pd_2(dba)_3$	tris(dibenzylideneacetone) dipalladium
PMHS	polymethylhydrosiloxane
PPA	polyphosphoric acid
PPI	poly(propyleneimide)
PS-DMAP	4-(dimethylamino)pyridine, polymer-bound
PS-BEMP	2-*tert*-butylimino-2-diethylamino-1,3-dimethyl-perhydro-1,3,2-diazaphosphorine on polystyrene
SPB	sodium perborate
TBAN	tetrabutylammonium nitrite

TBD	1,5,7-triazabicyclo[4.4.0]dec-5-ene
TBHP	*tert*-butyl hydroperoxide
TBAHS	tetra-*n*-butylammonium hydrogen sulfate
TBDMS	*tert*-butyldimethylsilyl
TBME	terbutyl methyl ether
TBO	tridentate bis-oxazoline
TCAA	trichloroacetic acid
TFA	trifluoroacetic acid
TFAN	trifluoroacetyl nitrate
TEMPO	2,2,6,6-tetramethyl-1-piperidinyloxy
THBA	tetrahexylammonium bromine
THP	tetrahydro pyranyl
TMEDA	tetramethylethylenediamine
TMG	tetramethylguanidine
TNM	tetranitromethane
TPAP	tetrahydropropylammonium perrutenate
UHP	urea-hydrogen peroxide
Triton B	benzyltrimethylammonium hydroxide
VOS	volatile organic solvent
VS	very small

1 Synthesis of Nitroalkanes

1.1 Synthesis of Nitroalkanes

Simple nitroalkanes ($CH_3(CH_2)_{0-5}NO_2$, c-$C_{5,6}H_{9-11}NO_2$, i-$PrNO_2$, etc.) are commercially available, while the more complex ones need to be prepared following a variety of synthetic procedure and, in this sense, there are a variety of methods to obtain a large collection of nitroalkanes.

1.1.1 Displacement of Alkyl Halides

Treatment of alkyl halides with metal nitrite is certainly the best known and the most used method for the preparation of nitroalkanes, mainly the primary ones (R^1 = H, Scheme 1.1) [1].

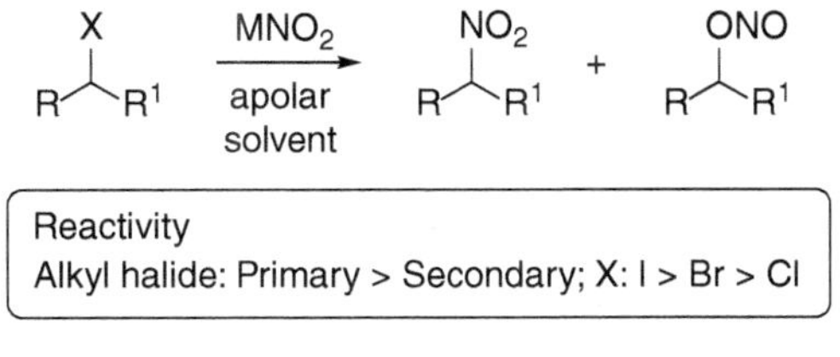

Scheme 1.1 Displacement (nitration) of alkyl halides.

In literature a variety of efficient procedures have been reported using sodium or silver nitrite (M = Ag, Victor-Meyer reaction [2]; M = K or Na Kornblum reaction [3]), anyway the formation of nitroalkanes is accomplished by the formation of alkyl nitrite as by-product, removed via distillation or chromatography. The nitrite is formed because of the ambident nature of the nitrite anion, which bears two different nucleophilic centers. The reaction is typically performed at room temperature adding the halide to a stirred mixture of dimethyl formamide (DMF) and metal nitrite. The use of DMF is due to the ability of this solvent to dissolve the nitrite and to accelerate the reaction of the alkali nitrite with the halide, while minimizing any side reaction such as those reported in Scheme 1.1.

An interesting application of the nitration of alkyl halides has been reported [4] for the preparation of fatty nitroalkanes, mainly the polyunsaturated ones, from

Nitroalkanes: Synthesis, Reactivity, and Applications, First Edition.
Roberto Ballini and Alessandro Palmieri.

naturally alcohols (Scheme 1.2) through their successive conversion into (i) bromo alkanes, (ii) iodo alkanes (more reactive vs bromo derivatives), and (iii) nitroalkanes with silver nitrite, in Et_2O and at room temperature for three days.

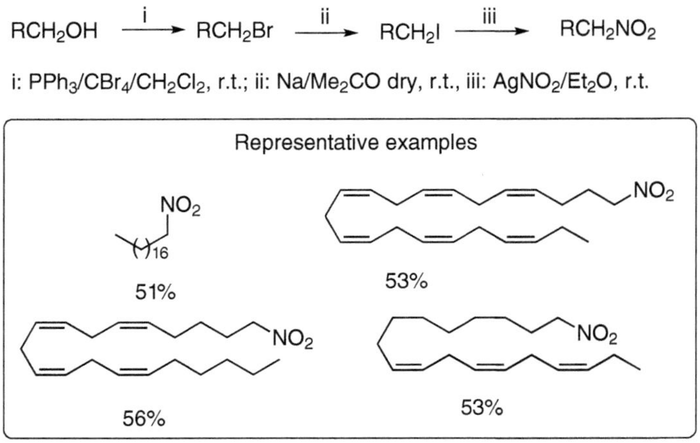

Scheme 1.2 Synthesis of fatty nitroalkanes.

It is important to point out that the need to convert bromo alkanes into the iodo alkanes is due to the fact that, contrary to short-chain bromo alkanes, the longer ones are inert toward the reaction with $AgNO_2$.

Benzylic nitro compounds (aryl nitromethanes) are an important class of nitroalkanes and their preparation shows some drawbacks due to the lack of selectivity, with consequent formation of an amount of nitrite ester. Guillaumet and coworkers [5] reported an improved approach (based on the Victor Meyer reaction) for the synthesis of aryl nitromethanes, minimizing the complex purification process on the basis of the difference in pK_a values between aryl nitromethane and benzyl nitrite ester (side product). In fact, the nitro compound is quite a strong acid ($pK_a = 8–12$) due to the high stability of the formed nitronate. Indeed, the negative charge is delocalized onto the nitro and the aryl group, while the nitrite ester does not bear any acidic proton because the conjugate base is no longer stabilized in any nitro group. This significant acidic difference has been the ideal property to develop a practical purification method usable on large scales.

In fact (Scheme 1.3), adding a methanolic solution of sodium methoxide into the filtrate of the reaction, it is possible to observe a precipitate of the formed sodium nitronate that could be simply isolated by filtration. The nitronate, treated with acetic acid and urea, converts into the *aci*-nitro structure through an O-protonation. The latter quickly precipitates in the acidic medium and isomerizes into the target aryl nitromethane in satisfactory yields (>50%) and with the survival of a variety of other functional groups.

However, the nitration of alkyl halides with $AgNO_2$ (Victor Meyer procedure) shows important drawbacks because usually the need of long reaction times and the use of toxic solvents produce important amounts of by-products and require

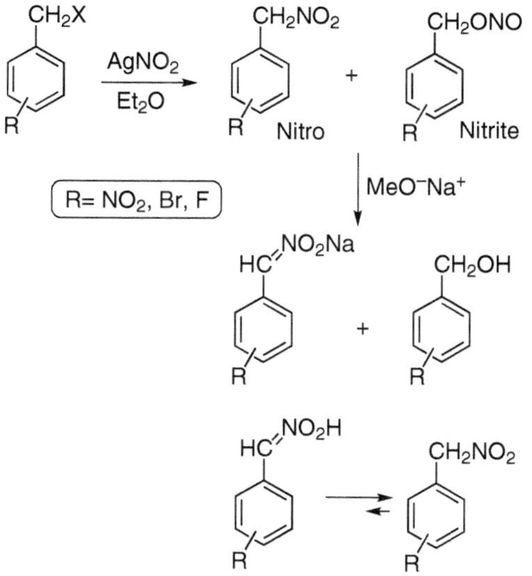

Scheme 1.3 Aryl nitromethanes synthesis under minimized purification step.

a tedious work-up. Thus, an improved procedure has been developed, in aqueous medium, treating primary alkyl halides with 4 equiv of silver nitrite (r.t. or 60 °C) [6]. The method allows satisfactory to good yields of a variety of primary nitroalkanes, in short reaction times (0.5–1.25 hours, Scheme 1.4). The need of 4 equiv of silver nitrite is due to the fast of the reaction that can minimize the competitive formation of the corresponding alcohol. In addition, the formation of alkyl nitrite is strongly depressed.

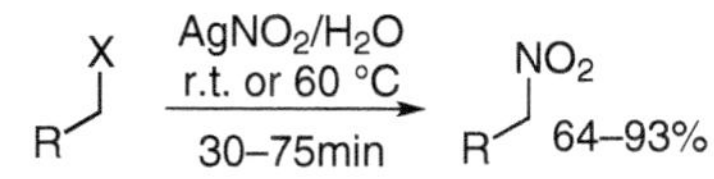

Selected examples:
R = n-C_nH_{2n+1}, $CH_2{=}CH(CH_2)_8$, $AcO(CH_2)_3$, $PhCOCH_2$
X = Br, I.

Scheme 1.4 Nitration of halides with silver nitrite.

Although the method works well with both alkyl bromides and alkyl iodides, the latter often shows higher reactivity, while it fails with secondary halo derivatives. Furthermore, due to the mildness of the reaction conditions, other functionalities such as C=C double bond, ester, imide, and ketones are preserved.

An important application of the above method is the synthesis of α,ω-dinitroalkanes from α,ω-diiodo structures (Scheme 1.5).

Thus, this procedure can be considered as a significant progress in the sustainability of the nitration of halo alkanes. However, a possible drawback consists of the need for a large excess of expensive reagents such as silver nitrite and the formation of the corresponding alcohol due to the competitive action of water vs nitrite.

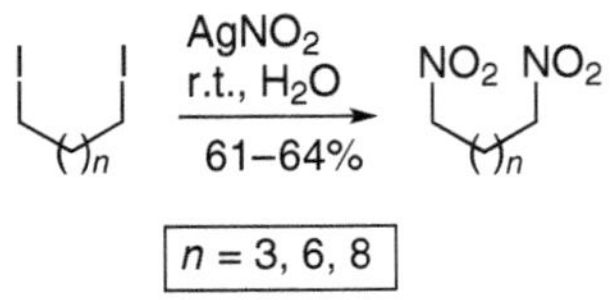

Scheme 1.5 Nitration of halides with silver nitrite (selected examples).

Then, few years later, our group developed an improved eco-friendly nitration of alkyl halides [7], using sodium nitrite in polyethylene glycol (PEG 400) as green reaction medium [8, 9]. Thus, treatment of primary alkyl halides with 1.5–3 equiv of sodium nitrite, at room temperature, in PEG 400 as solvent, allows the chemoselective (other functionalities such as esters, hydroxyl group, C=C double bond, ethers, aromatics, and amides are preserved) formation of primary nitroalkanes (Scheme 1.6). It is important to point out that only traces of alkyl nitrite are observed. In addition, it is relevant that PEG 400, after extraction with cyclohexane, can be recycled at least twice for the same reaction before its disposal as waste.

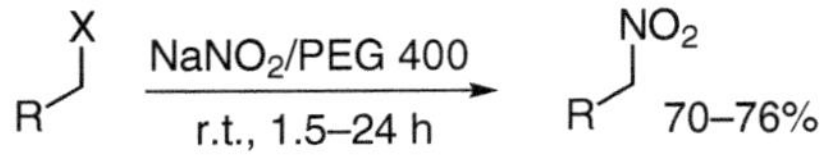

Selected examples:
n-$C_{11}H_{23}$, CH_2=$CH(CH_2)_8$, $Ph(CH_2)_2$,$Cl(CH_2)_3$
X = Br, I

Scheme 1.6 Nitration of halides under PEG solvent.

Although the procedure works well with both alkyl iodides and bromides, the latter needs larger amounts of sodium nitrite (3 equiv vs 1.5 equiv) and longer reaction times (*ratio* > 3).

With this procedure, the conversion of secondary halo derivatives fails, and this result can be applied to the chemoselective nitration of primary iodides or bromides in the presence of a secondary one (Scheme 1.7).

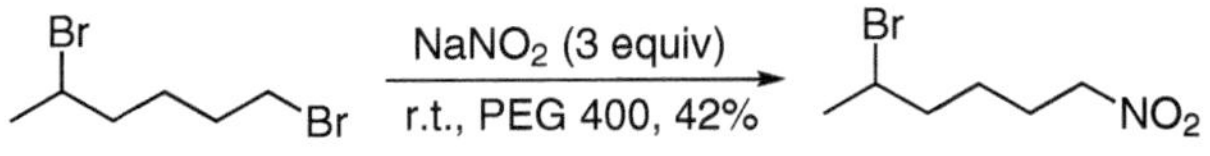

Scheme 1.7 Chemoselective nitration of alkyl halides.

1.2 Nitration of Mesylates and Tosylates

The displacement of alkyl halides with metal nitrite, thanks to the large number of available halides, is the most used procedure for the preparation of nitroalkanes. However, despite the advantages introduced with these methodologies, the search for alternative sources is surely welcomed. In this regard, alcohols are a broad class of molecules easily available from both commercial and nature sources. The conversion of alcohol into nitroalkanes firstly requires the conversion into the corresponding mesylates or tosylates (usually in quantitative yields) that can be easily converted (Scheme 1.8) into the corresponding nitroalkanes using tetrabutylammonium nitrite (TBAN) in toluene for few hours (2.5 hours for primary compounds and 28–40 hours for the secondary ones) [10].

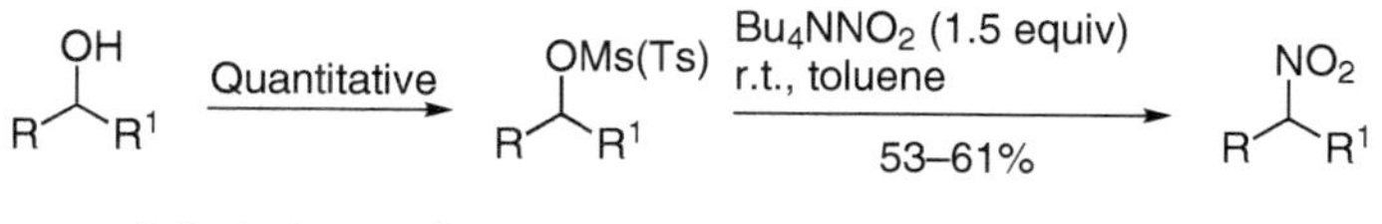

Scheme 1.8 Nitration of mesylates and tosylates.

The method was realized at room temperature and, after a simple work up, a variety of primary and secondary nitroalkanes can be easily obtained is 41–61% overall yields.

1.3 Oxidation of Nitrogen Derivatives

The oxidation of nitrogen derivatives is a useful alternative to furnish nitroalkanes that are difficult to prepare by the direct nitration of hindered halo derivatives.

1.3.1 Oxidation of Amines

The direct oxidation of primary amines into the corresponding nitro derivatives is very useful for fundamental and industrial applications because it provides targets that may be difficult to obtain via nitration methods.

The pioneering oxidation of primary amines has been performed with various reagents such as ozone [11], $NaMnO_4$ [12], $KMnO_4$ [13], peracetic acid [14], *meta*-chloroperbenzoic acid (MCPBA) [15], and dimethyldioxirane (DMO) [16].

The reactions proceed (Scheme 1.9) via a few intermediates and an efficient oxidant is preferred for a rapid conversion to nitroalkanes. In fact, incomplete oxidation was sometimes observed, with consequent formation of by-products such as nitroso and nitroso-dimers.

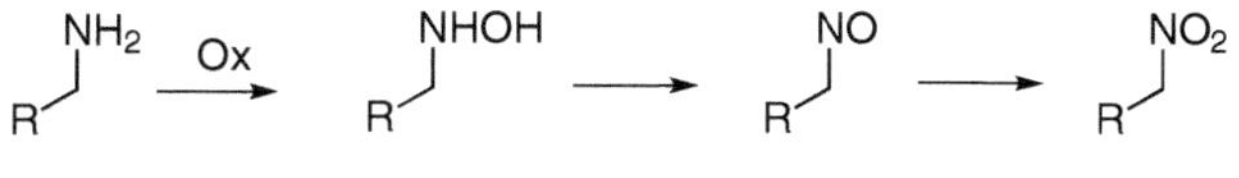

Scheme 1.9 Oxidation of primary amines.

From an environmental point of view, ozonization is also a benign and atom-efficient oxidation process, but an important drawback deals with the safety since it can lead the formation of unstable and potentially explosive ozonides and peroxides. Moreover, the subsequent quenching step is generally exothermic, thus requiring an efficient temperature control. To overcome the aforementioned disadvantages, few years ago Kappe and coworkers [17] proposed a continuous flow procedure as a valuable alternative to batch protocols for the oxidative conversion of important functionalities, including some primary amines to the corresponding nitroalkanes.

Later, the oxidation of amines has been developed using an alternative and improved oxygen transfer reagent, such as "HOF·CH_3CN complex," simply obtained passing fluorine through acetonitrile. The method (Scheme 1.10) [18] proceeds by dissolving the amine in CH_2Cl_2, then the mixture was cooled to −15 °C and added to the glass reactor containing NaF and HOF·CH_3CN (3 equiv) in cold (−15°C) aqueous CH_3CN.

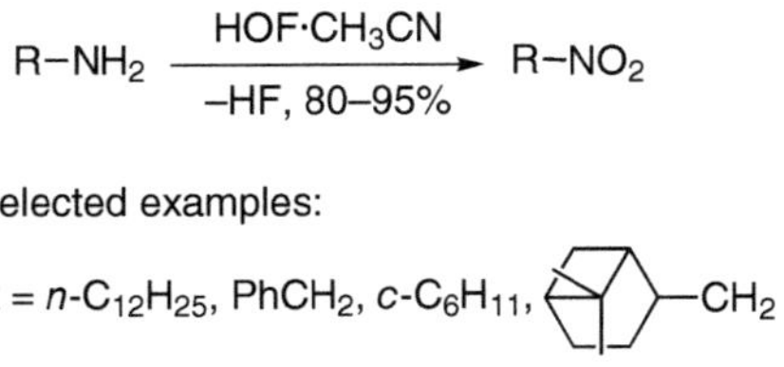

Scheme 1.10 Oxidation of amines using HOF·CH_3CN.

The reaction proceeds for few minutes allowing nitroalkanes in excellent yields. Considering that HOF·CH_3CN cannot be used in a large-scale process due to its instability (half-life = four hours at room temperature), it cannot be packaged and shipped but must be prepared and used immediately. In addition, the oxidation reaction has been performed in batch and all the problems of using concentrated oxidizing media remain. In this context, Sandford and coworkers [19] developed a flow process to solve the above drawbacks. However, for an efficient and selective oxidation, any fluorine gas that enters into the flow channel must be completely consumed before the organic substrate could be added to prevent fluorination and substrate degradation. In addition, the oxidation needs 3 equiv of HOF·CH_3CN to perform the efficient conversion of the ammines to nitroalkanes. Thus, a flow reactor designed specifically for gas–liquid reactions was adapted in order to perform the sequential formation and reaction of HOF·CH_3CN.

Some years before, Krohn and Küpke reported [20] an efficient procedure to convert aliphatic amines into nitro compounds based on a zirconium-catalyzed oxidation of primary amines with *tert*-butyl hydroperoxide (TBHP), in the presence of $Zr(O\text{-}t\text{-}Bu)_4$ as catalyst (Scheme 1.11).

In fact, a solution of amine (10 mmol), in dry CH_2Cl_2, is treated with freshly activated powered molecular sieves (3 Å) and $Zr(O\text{-}t\text{-}Bu)_4$ (1 mmol). After stirring for 30 minutes, a solution of TBHP (59 mmol) in CH_2Cl_2 is added. After complete consumption of the starting material and work-up, good yields of nitroalkanes are obtained. The method works well with a variety of substrates and it is mild enough to preserve other functional groups as acetals and esters.

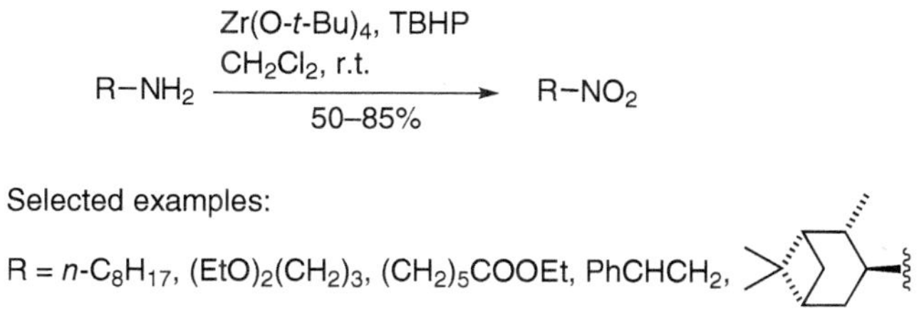

Scheme 1.11 Oxidation of amines using Zr(O-*t*-Bu)$_4$.

1.3.2 Oxidation of Oximes

Conversion of carbonyl function to nitro group (retro-Nef reaction) [21] is an important method for the preparation of nitro compounds. Such conversion is generally effected via oximes using strong oxidants (Scheme 1.12).

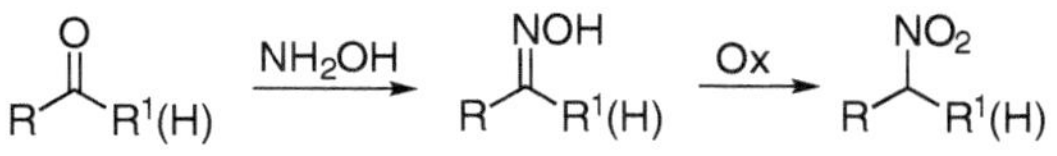

Scheme 1.12 Retro-Nef reaction via oximes.

Following a pioneering procedure performed with CF_3CO_3H [22], there was reported a revised and improved procedure (Scheme 1.13) [23] using peroxytrifluoroacetic acid obtained through reaction of urea–hydrogen peroxide complex (UHP) with trifluoroacetic anhydride (TFAA) in acetonitrile at (0 °C).

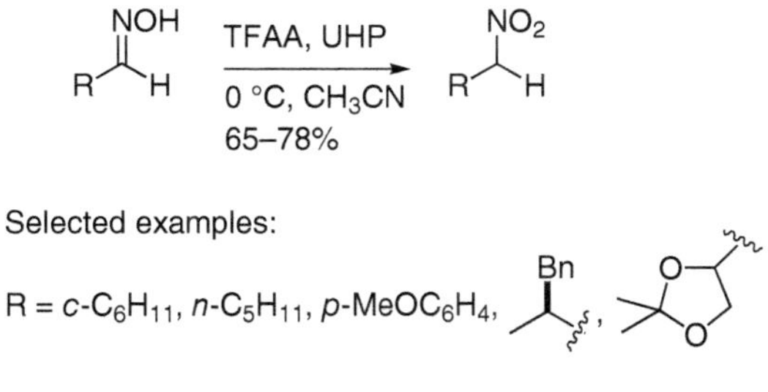

Scheme 1.13 Oxidation of oximes with TFAA/UHP.

The procedure allows good yields with aldoximes, while fails to react with ketoximes. On the other hand, ketoximes can be converted into secondary nitroalkanes, following the Olah method [24], oxidizing ketoximes with sodium perborate in glacial acetic acid. However, this procedure failed with aldoximes.

An interesting conversion of both aldoximes and ketoximes to the corresponding nitroalkanes has been realized by a complementary synthetic route of the UHP method. In fact, the oxidation achieved with "Benz-Mo," the Mo(VI) oxidiperoxo complex $[\text{Benz-MoO}(O_2)_2]^-(t\text{-Bu})_4N^+$ in acetonitrile, affords good yields of both primary and secondary nitroalkanes (Scheme 1.14) [25].

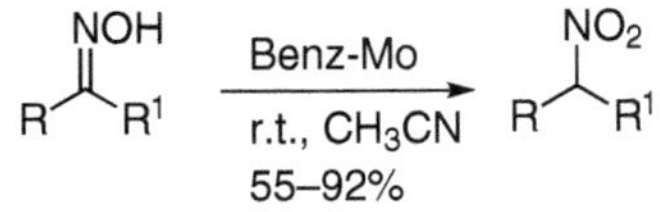

Selected examples:
R= c-$C_{12}H_{22}$, Ph; R^1= H, Me, Ph

Scheme 1.14 Oxidation of oximes using Benz-Mo.

1.3.3 Oxidation of Azides

Recently azides have been identified as a new precursor of primary nitroalkanes by oxidation. In this context, Rozen et al. [26] found that the carbon-bonded nitrogen of azides would be nucleophilic enough to interact with the oxygen atom of $HOF{\cdot}CH_3CN$. Thus, thanks to the efficiency of this oxidant, a variety of azides, easily obtainable from alkyl bromides and sodium azide, have been converted into primary nitroalkanes in quantitative or very good yields and in short reaction times (Scheme 1.15).

$$R{-}N_3 \xrightarrow[80\text{–}98\%]{HOF{\cdot}CH_3CN,\ 0\,°C} R{-}NO_2$$

Selected examples:
R= c-C_6H_{11}, c-C_5H_9, n-$C_{10}H_{21}$, $PhCH_2$, $AcO(CH_2)_5$

Scheme 1.15 Oxidation of azides using $HOF{\cdot}CH_3CN$.

Concerning the use of fluorine, a very corrosive material, it should be used only with the appropriate vacuum line techniques. However, for the occasional user, various premixed mixtures of F_2 in inert gases are commercially available.

Moreover, as for the oxidation of primary amines, a flow process has been developed [19] and, following this procedure, quantitative yields (<90%) of primary and secondary nitroalkanes were obtained.

1.4 Reduction of Conjugate Nitroalkenes

Conjugated nitroalkenes are a very important class of molecules easily prepared by nitroaldol condensation or direct nitration of the corresponding olefins. Their reduction provides a convenient route to a variety of different functionalities including oximes, carbonyl compounds, hydoxylamines, alkylamines, and nitroalkanes [27, 28].

1.4.1 Reduction of Nitroalkenes into Nitroalkanes

The reduction of conjugate nitroalkenes into nitroalkanes (Scheme 1.16) has been achieved under several distinct protocols; however, often there is a loss of product

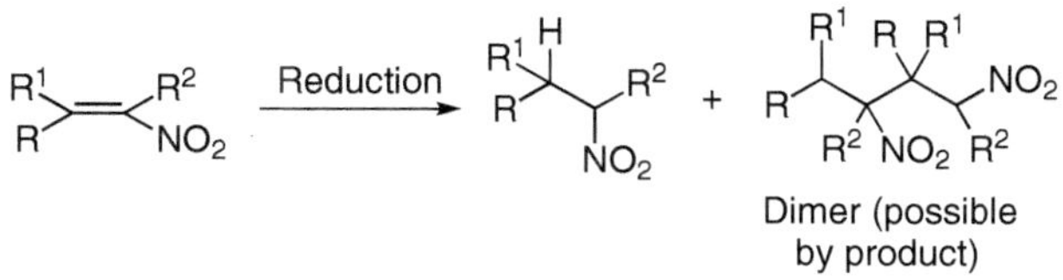

Scheme 1.16 Reduction of nitroalkenes.

due to dimerization caused by the competitive Michael addition of the formed nitronate to another molecule of nitroalkene.

However, because this by-product may be suppressed at reduced pH, the main theme of several reports is the selective reduction of the double bond avoiding the formation of the Michael by-product. Thus, a plethora of reductive methods have been reported. A modified reduction with sodium borohydride in the presence of both silica gel and a mixture of chloroform-propanol as solvent system achieves good results of nitroalkanes (Table 1.1) [29].

Successively, Kabalka et al. [28] proposed three different reductive agents (Table 1.2): (i) trialkylborohydride, (ii) $NaBH_4$ in a mixed MeOH-THF solvents, and (iii) $NaBH_4$ supported on an ion exchange resin, for an effective conversion of nitroalkenes to nitroalkanes.

Table 1.1 $NaBH_4/SiO_2$ reduction of nitroalkenes (selected examples).

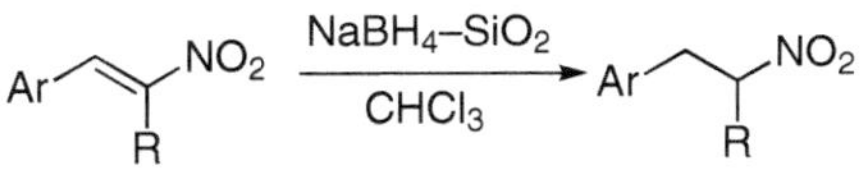

Nitroalkene	Nitroalkane	Yield (%)
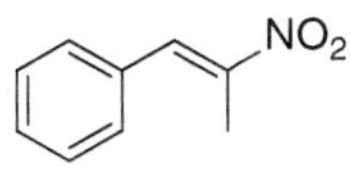	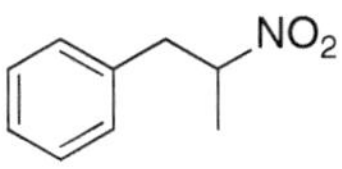	93
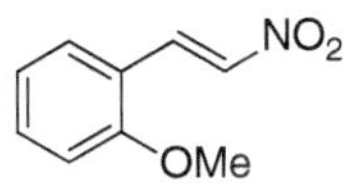	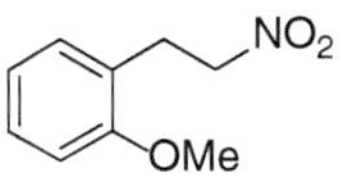	92
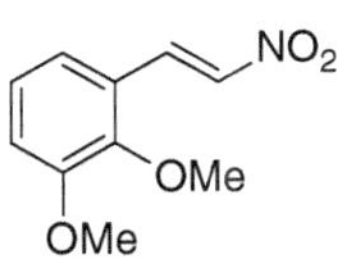	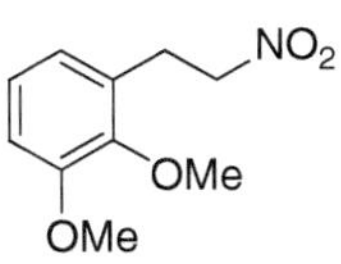	94
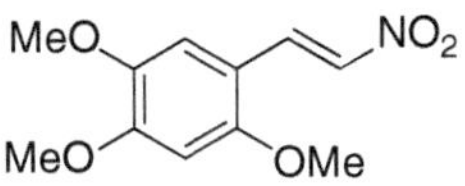	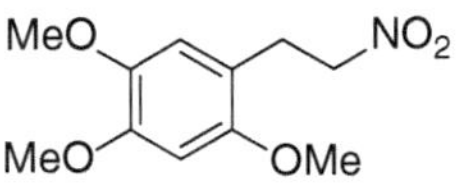	94
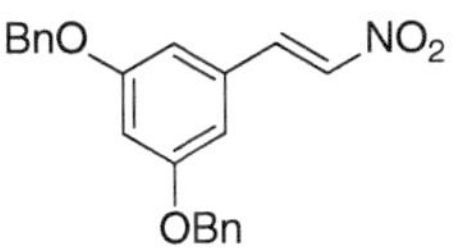	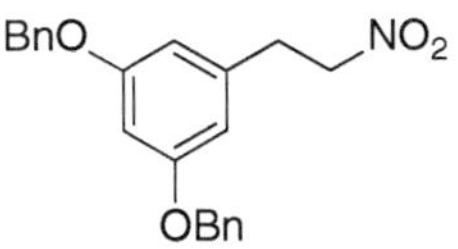	98

Table 1.2 Comparative reduction of nitroalkenes with methods (i)–(iii) (selected examples).

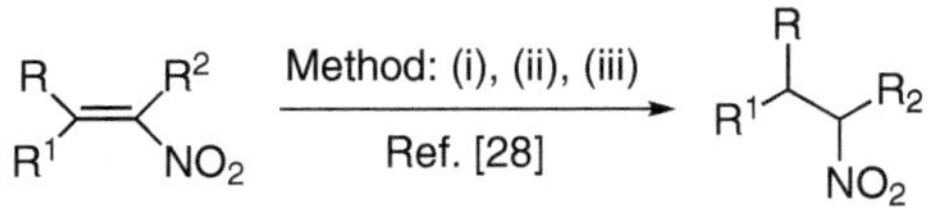

Nitroalkene	Nitroalkane	Yield (%) (i)	Yield (%) (ii)	Yield (%) (iii)
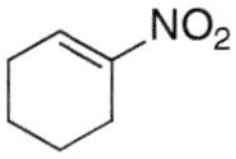	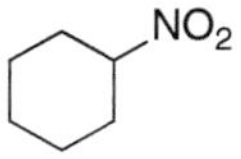	69	62	80
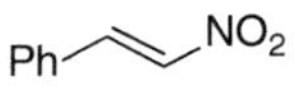	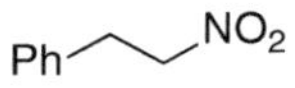	78	64	—
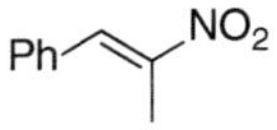	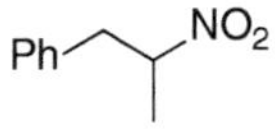	81	82	80
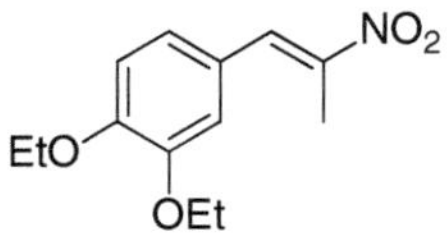	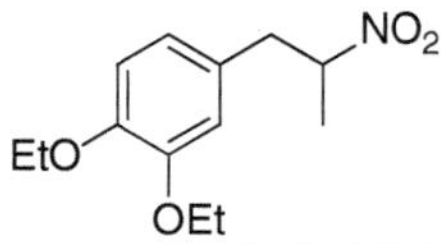	69	75	83

Later, Vankar and coworkers [30] developed an efficient, chemoselective procedure for the reduction of conjugated nitroalkenes with the use of $NaCNBH_3$ (1 equiv) in methanol (Table 1.3), by the help of Zeolite (H-ZSM-5, [Si:Al = 35 : 1], 0.5 equiv), and a careful control of the pH.

1.4.2 Stereoselective Reduction of Conjugated Nitroalkenes

The stereoselective reduction of conjugated nitroalkenes is an important goal and a variety of efficient procedures have been reported [31–36]; however, they seem to be restricted to the reduction of nitrostyrene derivatives. Thus, a survey of the most representative examples is reported in Table 1.4.

1.4.3 Aldehyde Reductive Nitromethylation

Several decades ago, Wollemberg and Miller developed a useful procedure for the preparation of primary nitroalkanes with an extra atom beginning from aldehydes [37]. The starting point is the nitroaldol (Henry) reaction (Scheme 1.17) of an aldehyde with nitromethane, catalyzed with KF and in the presence of *i*-PrOH as solvent. The formed nitroalkanol is acetylated (acetic anhydride in the presence of 4-dimethylaminopridine as catalyst) and treated with sodium borohydride affording the desired nitroalkane via "one-pot" acetic acid-elimination and C=C double bond reduction.

So, this procedure offers the opportunity to increase the chain length of the starting aldehyde.

Table 1.3 Reduction of nitroalkenes with $NaCNBH_3$ (selected examples).

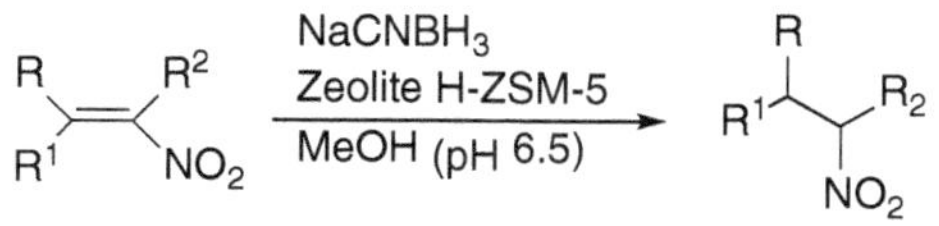

Nitroalkene	Nitroalkane	Yield (%)
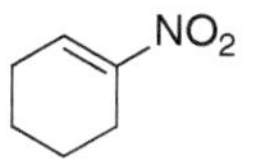	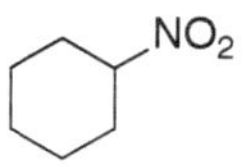	79
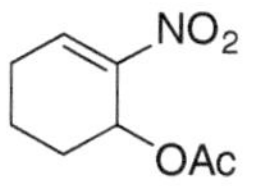	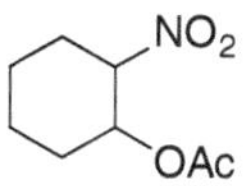	74
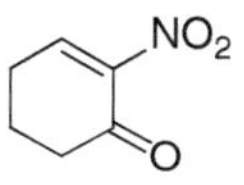	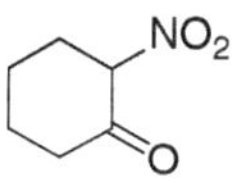	70
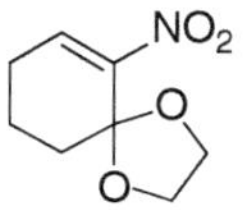	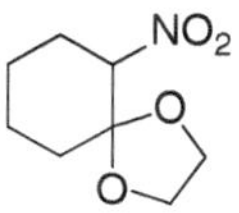	78
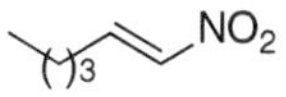	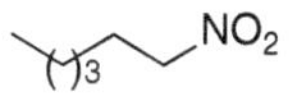	78
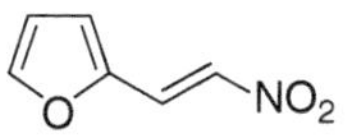	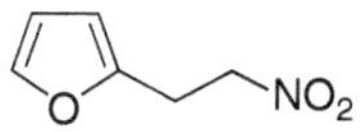	69

1.5 Nitration of Alkanes

In contrast with the nitration of aromatic hydrocarbons that can easily performed using nitric acid in the presence of sulfuric acid, the selective nitration of aliphatic hydrocarbons is very difficult due to the exceeding low reactivity of the latter.

Currently, nitration reactions of aliphatic hydrocarbons are carried out at fairly high temperature using nitrogen dioxide or nitric acid, thanks to a free radical process, involving C—H bond homolysis [38]. Often, under such temperature conditions higher alkanes undergo also cleavage of the C–C skeleton. Thus, Ishii and coworkers [39] developed a milder method for the nitration of light alkanes and alkyl side-chain aromatic compounds with NO_2 and HNO_3 under *N*-hydroxyphthalimide (NHPI) or *N*-acetoxyphtalimide (NAPI) catalysis (Table 1.5).

1.6 Metal-Catalyzed Alkylation or Arylation of Nitroalkanes

Commercially available nitroalkanes can be used as precursors of more complex structures through their alkylation or arylation, performed under metal catalysis.

Table 1.4 Stereoselective reduction of nitroalkenes (selected examples).

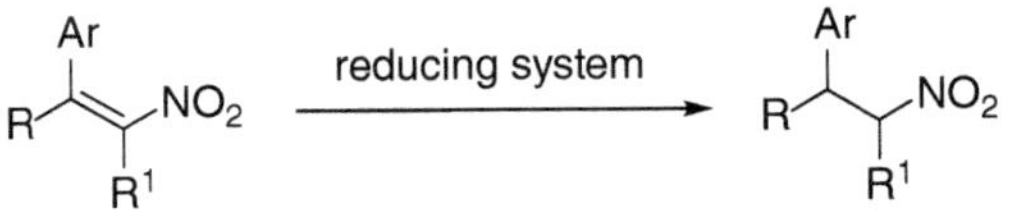

Substrate	Reducing system	Yield (%), (ee)
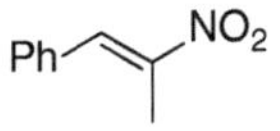	Baker's yeast, EtOH·H_2O [33]	91 (*S*, 12%)
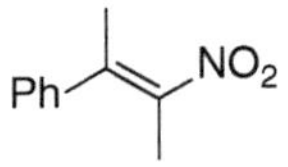	Baker's yeast, EtOH·H_2O [33]	72 (*S*, 12%)
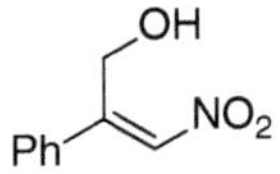	$CuF_2/PhSiH_3$, (*R*)(*S*)-JOSIPHOS, PHMS/PhMe·H_2O [35]	52 (*R*, 90%)
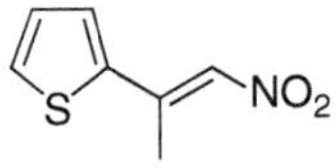	$CuF_2/PhSiH_3$, (*R*)(*S*)-JOSIPHOS, PHMS/PhMe·H_2O [34]	88 (*R* 92%)
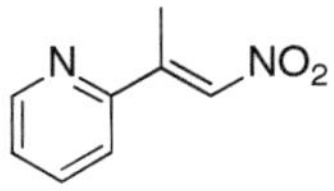	CuO-*t*-Bu/$PhSiH_3$, (*S*)(*R*)-JOSIPHOS, PMHS/PhMe·H_2O [34]	55 (*S*, 72%)
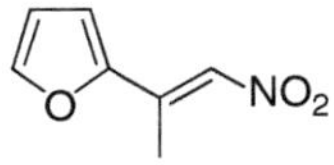	CuO-*t*-Bu/$PhSiH_3$, (*S*)(*R*)-JOSIPHOS, PMHS/PhMe·H_2O [34]	72 (*S*, 90%)
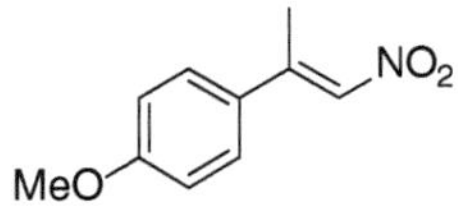	Iridium complex, HCOOH/H_2O, pH 2 [34]	94 (*R*, 92%)
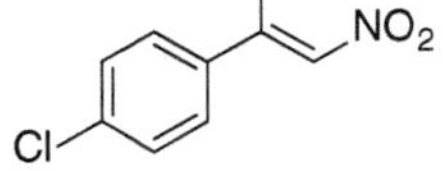	Iridium complex, HCOOH/H_2O, pH 2 [34]	92 (*R*, 90%)

PMHS, polymethylhydrosiloxane.

1.6.1 Nitroalkylation of Aryl Halides

Aryl halides can be converted into "aryl nitromethanes" via a cross-coupling reaction with nitromethane, catalyzed by $Pd_2(dba)_3$, tris(dibenzylideneacetone)dipalladium, as reported in Scheme 1.18 [40].

A more general protocol provides the arylation of primary nitroalkanes (different from nitromethane), via Pd-catalyzed reaction with aryl bromide or chlorides (Scheme 1.19) [41].

More detailed applications are reported in Chapter 6.

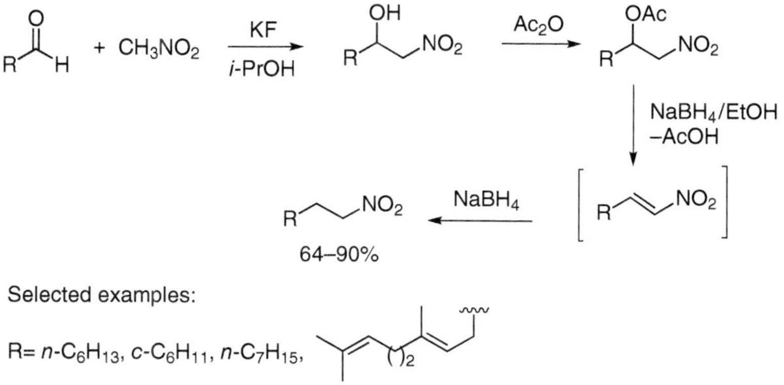

Scheme 1.17 Reductive nitromethylation of aldehydes.

Table 1.5 Nitration of alkanes (selected examples).

Substrate	Nitrating agent	Nitroalkane	Yield (%)
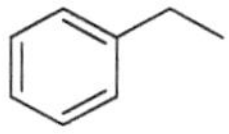	NHPI (NO_2)	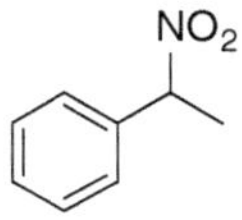	63
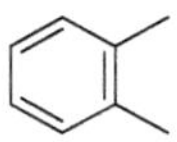	NAPI (HNO_3)	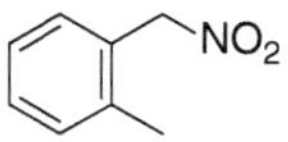	69
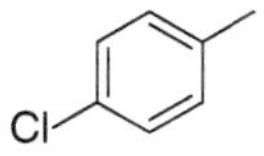	NAPI (NO_2)	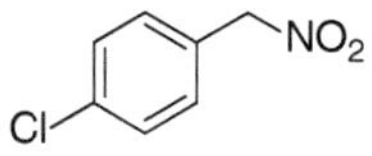	69
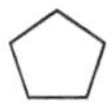	NHPI (HNO_3)	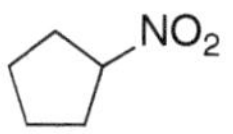	70
	NHPI (NO_2)	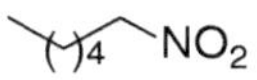	54
	NHPI (NO_2)	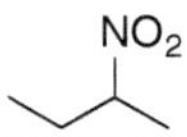	65

1.6.2 Nitroalkylation of Allylic Esters

Pd catalyzes even the asymmetric allylic alkylation of cyclic allyl esters [42] in the presence of *O*,*N*-bis-trimethylsilylacetamide (BSA) as base and a chiral biphosphine ligand (Table 1.6). The method allows a practical way to primary chiral nitroalkanes with an extra carbon atom.

1.6.3 Nitroalkylation of Allylic Alcohols

Unactivated nitroalkenes (mainly nitromethane) can be added to vinyl epoxides, under Pd(0) catalysis, affording nitroalkenols (Scheme 1.20).

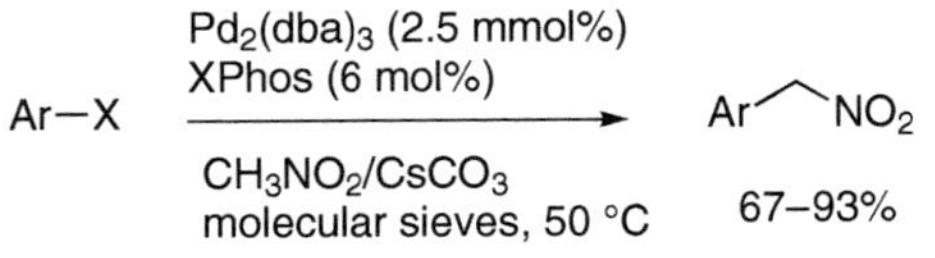

Selected examples:
Ar = Ph, p-$MeOC_6H_4$, p-$CF_3C_6H_4$, o-BrC_6H_4, o-$NO_2C_6H_4$

Scheme 1.18 Nitromethylation of aryl halides.

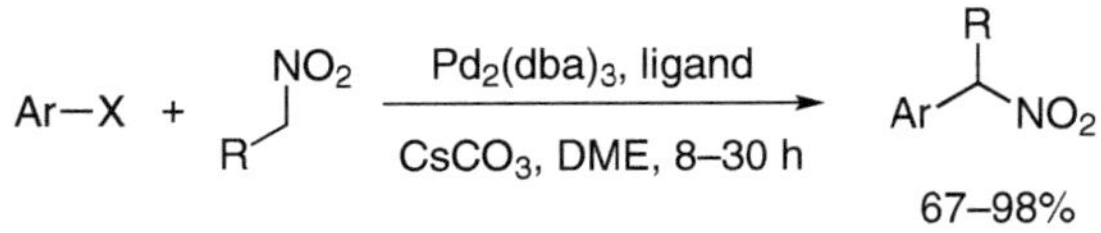

Selected examples:
Ar= Ph, p-$MeOOCC_6H_4$, p-$MeOC_6H_4$, p-$Me_2NC_6H_4$, o-$NO_2C_6H_4$
R= Me, Et, $MeOOC(CH_2)_2$
X= Cl, Br

Scheme 1.19 Nitroalkylation of aryl halides.

Table 1.6 Asymmetric nitromethylation of cyclic allyl esters (selected examples).

Substrate	Reaction conditions	Nitroalkane	Yield (%), (ee)
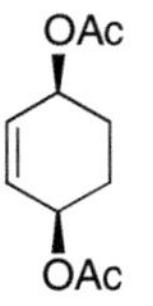	CH_3NO_2/CH_2Cl_2, $Pd_2(dba)_3$, ligand, BSA	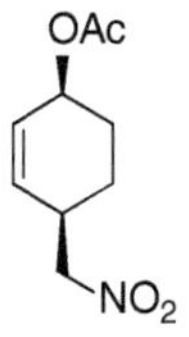	82 (99%)
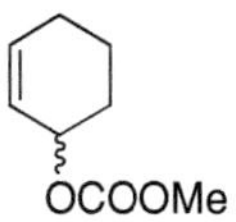	CH_3NO_2/CH_2Cl_2, $Pd_2(dba)_3$, ligand, Bu_4NCl/BSA	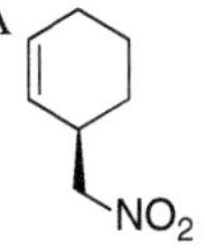	99 (99%)
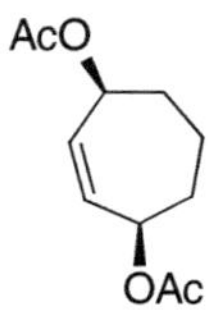	CH_3NO_2/CH_2Cl_2, $Pd_2(dba)_3$, ligand, BSA	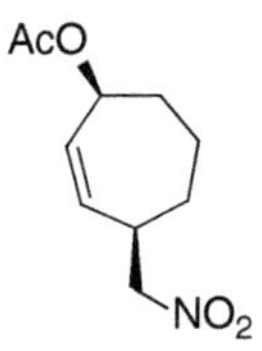	84 (99%)
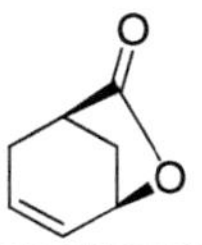	CH_3NO_2/CH_2Cl_2, $Pd_2(dba)_3$, ligand, Bu_4NCl/BSA	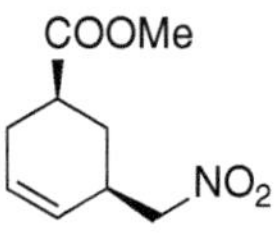	74 (99%)

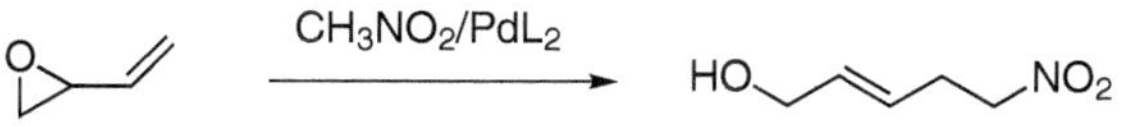

Scheme 1.20 Nitroalkanes addition to vinylepoxides under Pd(0) catalysis.

Table 1.7 Two-step process of nitro alkylation of allylic alcohols (selected examples).

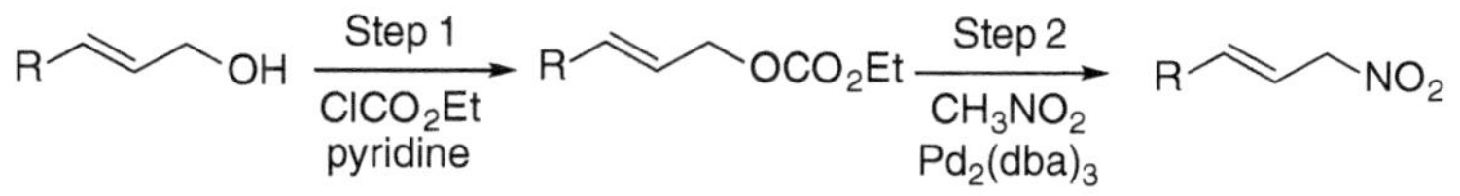

Allylic alcohol	Nitroalkane	Yield (%)
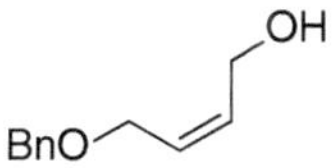	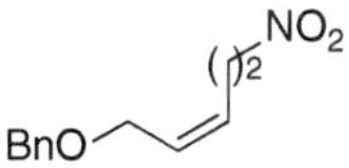	62
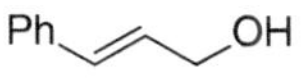	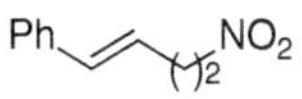	74
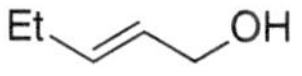	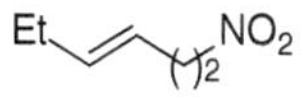	70
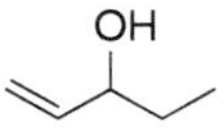	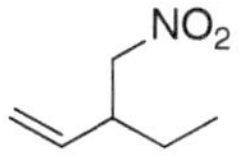	71
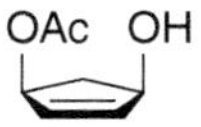	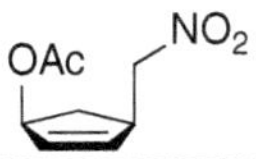	63

Developing this result, Deardorff et al. [43] reported a two-step process for nitromethylation of allylic alcohols that include (i) the preparation of the corresponding allylic carbonate and (ii) the addition of nitromethane, under $Pd_2(dba)_3$ catalysis.

So, this procedure represents a useful method for the synthesis of functionalized nitroalkanes and some of the most representative examples are reported in Table 1.7.

1.6.4 Two-Carbon Homologation of Vinyl Triflates and Bromides

Palladium catalyzes a two-carbon homologation of vinyl bromides (or triflates) via reaction with nitromethane and in the presence of Cs_2CO_3 as base, giving access to homoallylic nitrocompounds (Table 1.8) [44].

The process exploits the anion stabilizing effect and the leaving properties of the nitro group to generate the homoallylic nitro products via a tandem cross-coupling/π-allylation sequence of the alkenyl derivatives with nitromethane. Following this method a variety of homoallylic nitro compounds can be easily obtained.

Table 1.8 Two carbon homologation of vinyl bromides or triflates (selected examples).

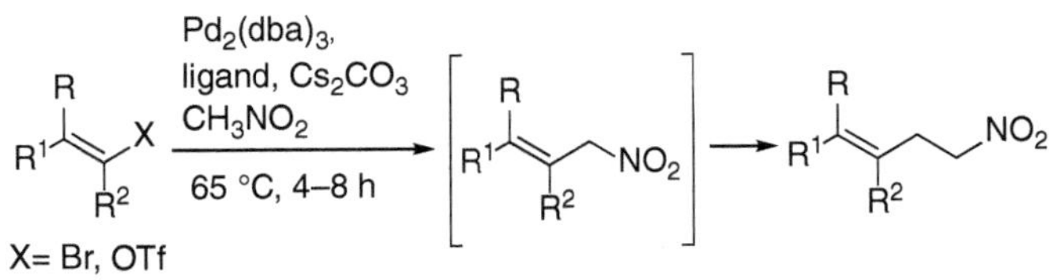

Substrate	Nitroalkane	Yield (%)
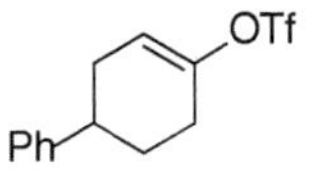	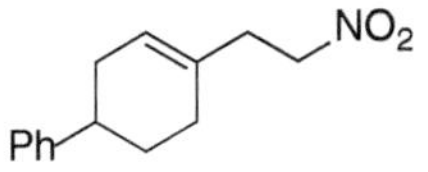	76
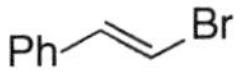	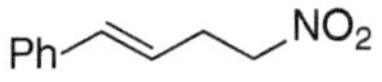	69
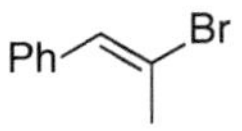	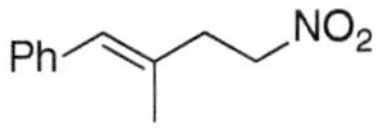	69
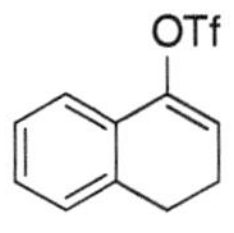	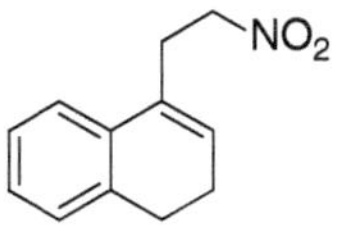	74
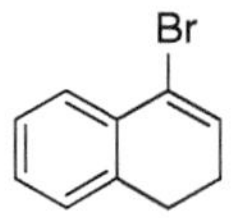	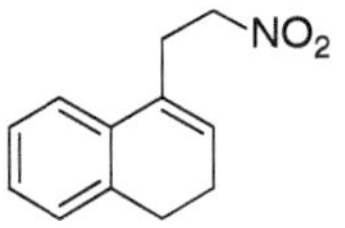	83
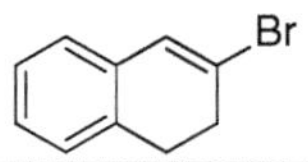	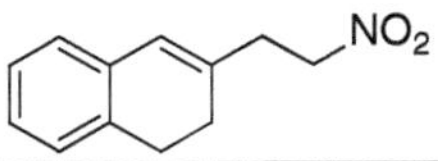	93

References

1 Ballini, R. and Palmieri, A. (2018). *Adv. Synth. Catal.* 360: 2240–2266.
2 Meyer, V. and Stuber, O. (1872). *Ber. Dtsch. Chem. Ges.* 5: 203–205.
3 Kornblum, N., Larson, H.O., Blackwood, R.K. et al. (1956). *J. Am. Chem. Soc.* 78: 1497–1501.
4 Easton, C.J., Xia, L., Pitt, M.J. et al. (2001). *Synthesis*: 451–458.
5 Alaime, T., Delots, A., Pasquinet, E. et al. (2016). *Tetrahedron* 72: 1337–1341.
6 Ballini, R., Barboni, L., and Giarlo, G. (2004). *J. Org. Chem.* 69: 6907–6908.
7 Ballini, R., Barboni, L., and Palmieri, A. (2008). *Green Chem.* 10: 1004–1006.
8 Sheldon, R.A. (2005). *Green Chem.* 7: 267–278.
9 Chen, J., Spear, S.K., Huddleston, J.G., and Rogers, R.D. (2005). *Green Chem.* 7: 64–82.
10 Palmieri, A., Gabrielli, S., and Ballini, R. (2013). *Beilstein J. Org. Chem.* 9: 533–536.
11 Keinan, E. and Mazur, Y. (1977). *J. Org. Chem.* 42: 844–847.

12 Menger, F.M. and Lee, C. (1981). *Tetrahedron Lett.* 22: 1655–1656.
13 Kornblum, N., Clutter, R.J., and Jones, W.J. (1956). *J. Am. Chem. Soc.* 78: 4003–4004.
14 Emmons, W.D. (1957). *J. Am. Chem. Soc.* 79: 5528–5530.
15 Gilber, K.E. and Border, W.T. (1979). *J. Org. Chem.* 44: 659–661.
16 Murray, R.W., Rajadhaksha, S.N., and Mohan, L. (1989). *J. Org. Chem.* 54: 5783–5788.
17 Irfan, M., Glsnov, T.N., and Kappe, C.O. (2011). *Org. Lett.* 13: 984–987.
18 Harel, T. and Rozen, S. (2007). *J. Org. Chem.* 72: 6500–6503.
19 McPake, C.B., Murray, C.B., and Sandford, G. (2012). *ChemSusChem* 5: 312–319.
20 Krohn, K. and Küpke, J. (1998). *Eur. J. Org. Chem.*: 679–682.
21 Nef, J.U. (1894). *Liebigs Ann. Chem.* 280: 264–268.
22 Emmons, W.D. and Pagano, A.S. (1955). *J. Am. Chem. Soc.* 77: 4557–4559.
23 Ballini, R., Marcantoni, E., and Petrini, M. (1992). *Tetrahedron Lett.* 33: 4835–4838.
24 Olah, G.A., Ramaiah, P., Lee, C.S., and Prakash, G.K.S. (1992). *Synlett*: 337–339.
25 Ballistreri, F.P., Barbuzzi, E., Tomaselli, G.A., and Toscano, R.M. (1996). *Synlett*: 1093–1094.
26 Rozen, S. and Carmeli, M. (2003). *J. Am. Chem. Soc.* 125: 8118–8119.
27 Barret, G.M. and Graboski, G.G. (1986). *Chem. Rev.* 86: 751–762.
28 Kabalka, G.W., Guind, L.H.M., and Varma, R.S. (1990). *Tetrahedron* 46: 7443–7457.
29 Sinhababu, A.K. and Borchard, R.T. (1983). *Tetrahedron Lett.* 24: 227–230.
30 Gupta, A., Haque, A., and Vankar, Y.D. (1996). *Chem. Commun.*: 1653–1654.
31 Kawai, Y., Inaba, Y., Hayashi, M., and Tokitoh, N. (2001). *Tetrahedron Lett.* 42: 3367–3368.
32 Kawai, Y., Inaba, Y., Hayashi, M., and Tokitoh, N. (2001). *Tetrahedron: Asymmetry* 12: 309–318.
33 Czekelius, C. and Carreira, E.M. (2003). *Angew. Chem. Int. Ed.* 42: 4793–4795.
34 Czekelius, C. and Carreira, E.M. (2004). *Org. Lett.* 6: 4575–4577.
35 Soltani, O., Ariger, M.A., and Carreira, E.M. (2009). *Org. Lett.* 11: 4196–4198.
36 Fryszkowska, A., Fisher, K., Gardiner, J.M., and Stephens, G.M. (2008). *J. Org. Chem.* 73: 4295–4298.
37 Wollenberg, R.H. and Miller, S.J. (1978). *Tetrahedron Lett.* 19: 3219–3222.
38 Markofsky, S.B. (1999). *Ullmann's Encyclopedia of Industrial Organic Chemicals*, vol. 6 (eds. B. Elvers and S. Hawkins), 3487–3501. Weinheim: Wiley-VCH.
39 Sakagichi, S., Nishiwali, Y., Kitamura, T., and Ishii, Y. (2001). *Angew. Chem. Int. Ed.* 40: 222–224.
40 Walvoord, R.R., Berritt, S., and Kozlowski, M.C. (2012). *Org. Lett.* 14: 4086–4089.
41 Vogl, E.M. and Buchwald, S.L. (2002). *J. Org. Chem.* 67: 106–111.
42 Trost, B.M. and Surivet, J.-P. (2000). *Angew. Chem. Int. Ed.* 39: 3122–3124.
43 Deardoff, D.R., Savin, K.A., Justman, C.J. et al. (1996). *J. Org. Chem.* 61: 361–362.
44 Padilla-Salinas, R., Walvoord, R.R., Tcyrulnikov, S., and Kozlowski, M.C. (2013). *Org. Lett.* 15: 3966–3969.

2

Reduction of the Nitro Group into Amines

The nitro group bearing to nitroalkanes is amenable of further transformations into a plethora of other functionalities, such as oximes [1], hydroxylamines [2], thiols (just from tertiary nitroalkanes) [3], amides (R^1 = H) [4], nitrones [5], silyl nitronates [6], nitriles [7], nitrile oxides [8], carbamates [9], amines [10], carbonyls (Nef Reaction) [11], etc. (Figure 2.1).

Anyhow, the reduction of the nitro group to amines (this Chapter) or its conversion to a carbonyl (Nef reaction, Chapter 3) is considered the most used and strategic transformations for the preparation of a large variety of targets.

The reduction of the nitro groups into the corresponding amine is of fundamental importance because it represents the most used gate to primary amines [12a] that, on the other hand, are strategic for the construction of structurally defined targets of practical interest [12b]. As consequence, a large number of procedures for this transformation have already been reported in literature some decade ago (Figure 2.2, Methods a–j).

However, in the last few years a variety of improved procedures for the reduction of nitroalkanes into the corresponding amines have been reported with the aim to develop their selectivity and sustainability (Figure 2.3, Methods a^1–i^1).

Method a^1: Very recently, an interesting hydrogenation of the nitro group has been described by one-pot tandem catalysis over Pd@MIL-101 (combined with palladium nanoparticles, NPs) coupled with NH_3BH_3 (AB) dehydrogenation [13]. The ammines were obtained, at room temperature, in quantitative yields and in very short reaction times (1.5–5 minutes). Moreover, the activity of the catalyst is well retained during six consecutive runs, demonstrating the recyclability and reusability of the catalyst.

Method b^1: Reduction of the nitro group to primary amine has been developed at room temperature, by transfer hydrogenation with $NaBH_4$, as hydrogen source, and under aluminum oxy-hydroxide supported Pd nanoparticles catalysis [14]. The method is highly efficient (quantitative yields of the products) and the catalyst can be efficiently reused, at least, ten additional times.

Method c^1: An interesting reduction of the nitro group has been realized via tandem dehydrogenation of ammonia borane (AB) and hydrogenation of nitro group, under graphene-supported NiPd (G-NiPd) nanoparticles catalysis [15].

Nitroalkanes: Synthesis, Reactivity, and Applications, First Edition.
Roberto Ballini and Alessandro Palmieri.

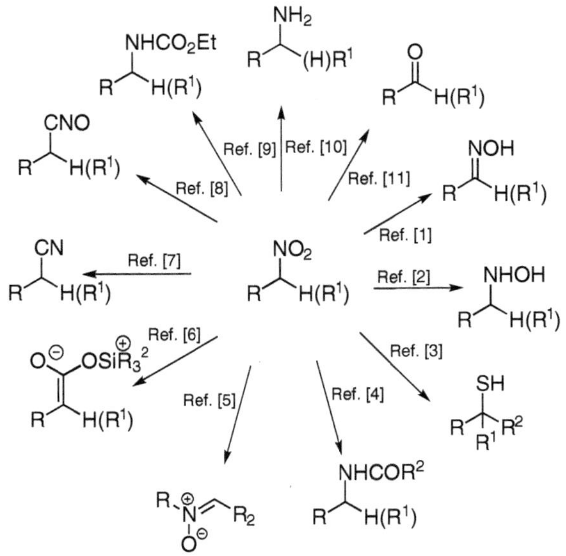

Figure 2.1 Conversion of nitroalkanes into other functionalities.

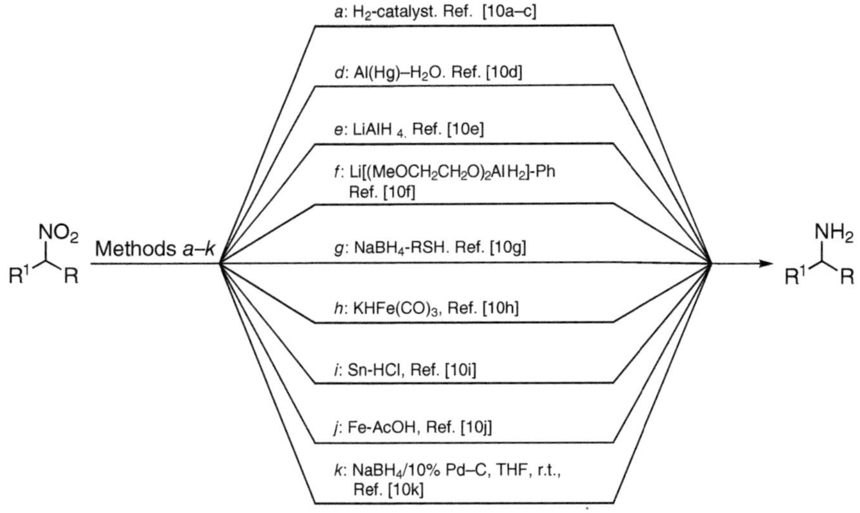

Figure 2.2 Common historical procedures for the reduction of nitroalkanes to amines.

The corresponding amines are obtained in few minutes and in quantitative yields. It is important to observe that the tandem reaction is performed in an environmentally friendly and safe manner, while the catalyst is both reusable and cost-effective.

Method d[1]: Hexafluoro-2-propanol (HFIP), in combination with iron powder and 2 N HCl [16], has been found to be the promoter for the easy nitro-reduction at room temperature. The method is mainly devoted to the reduction of aromatic nitro compounds; however 1-nitrooctane converts into 1-aminooctane in very good yield (82%). Moreover, the method shows chemoselectivity since a variety

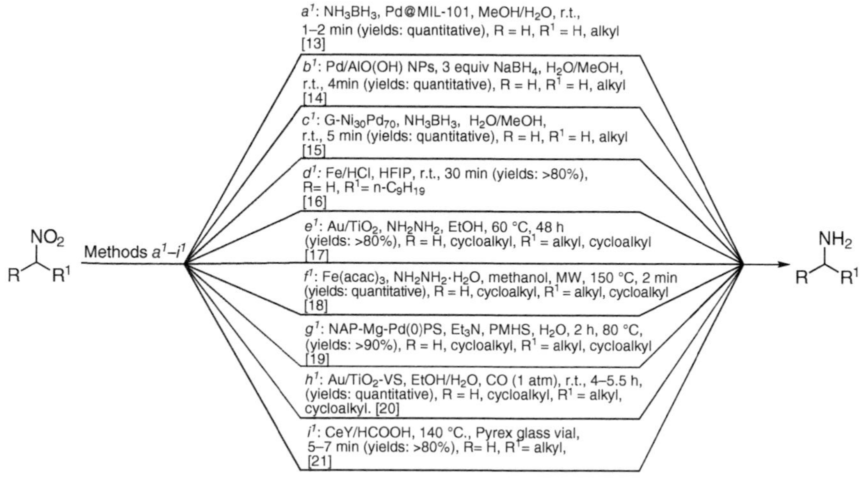

Figure 2.3 Recent procedures for the reduction of nitroalkanes to amines.

of other functionalities (aldehyde, ketone, nitrile, amide, ketal, etc.) can be preserved under the reported reaction conditions.

Method e^1: Few years ago it has been discovered that supported gold nanoparticles (Au NPs) catalyze the activation of hydrazine (NH_2NH_2) as transfer hydrogenation agent for the chemoselective reduction of nitro group [17]. The method, performed at 60 °C for 48 hours, shows highly chemoselectivity since other reducible functionalities (carboxylates, carbonyl, cyano, etc.) are maintained and the yields of the formed amines are >80%.

Method f^1: As earlier, nitroalkanes have been reduced to amines using hydrazine as hydrogen source. The reaction is catalyzed using magnetic iron oxide nanocrystal, generated by $Fe(acac)_3$ in methanol, and heating at 150 °C [18]. This protocol eliminates the use of precious metal catalyst and of hydrogen gas, moreover, the catalyst can be easily retrieved from the crude reaction mixture and reused, isolated, and stored for several weeks. In addition, the procedure is quick (two minutes) and works well (quantitative yields) with primary, secondary, and tertiary nitroalkanes, including nitrocycloalkanes.

Method g^1: Chemoselective and regioselective reduction of nitroalkanes (and aromatic nitro compounds) can be achieved via polymethylhydrosiloxane (PMHS) (derived palladium nanoparticles) as environmentally friendly hydrogenating agent. The Pd nanoparticles are highly stable and were prepared reducing NAP-Mg-$PdCl_4$ with PMHS. The procedure tolerates the presence of other functionalities such as alkenes, alkynes, cyano, carboxylic acid, esters, and amides. The selectivity simplifies the product separation and avoids unwanted waste making the process environmentally friendly.

Method h^1: Supported gold nanoparticles have emerged as active and selective catalysis able to catalyze the reduction of nitroalkanes with CO and H_2O, as hydrogen source, that circumvents H_2 activation to enable the rapid and chemoselective reduction a wide range of organic nitro compounds under mild conditions. Of the various gold catalysts tested, very small (VS) Au NP supported on TiO_2 showed the highest activity (this catalyst system is denoted as Au/TiO_2-VS) performing the reaction under ambient conditions (25 °C, 1 atm). These mild conditions significantly improve the economic and environmental impact.

The method is applicable to non-activated aliphatic nitro compounds and the corresponding amines are obtained in almost quantitative yields. Moreover, it is important to note the chemoselectivity of the procedure since other functionalities such as aldehyde, ester, or C=C double bond can be preserved, in addition the catalyst can be reused without loss of the activity.

Method i[1]: The reduction of the nitro group can be performed with CeY Zeolite instead of expensive Ni, Pt, Pd, etc. In fact, a heterogeneous procedure has been achieved under microwave (MW) in the presence of CeY Zeolite catalysis and HCOOH as reducing system. The catalyst (CeY) can contribute to decreasing the cost of amine synthesis due to its long catalyst lifetime and good stability as heterogeneous solid base catalyst.

The reaction is performed in a Pyrex glass vial and grants high yields (>80%) in very short reaction times (five to seven minutes). Moreover, the catalyst maintains its activity even after being used for 20 cycles.

A sustainable conversion of a variety of nitroalkanes, possessing heteroaromatic structures, to the corresponding amines has been developed [22] under sonication (seven hours) of a mixture of sodium hypophosphite monohydrate, hypophosphorous acid (50% in water) associated with Pd/C (5%), and under 2-MeTHF/H_2O as miscible and eco-friendly solvents [23] (Table 2.1). The method is highly chemoselective and efficient (time and yield) and important targets, such as tryptamine, can be prepared in high yield (94%, Table 2.1, entry 1).

If reactive functionalities are introduced (such as activated methylene of carbonyl derivatives), the formed amine reacts with the carbonyl moiety of malonate, affording the one-pot formation of lactams in good yields (Table 2.1, entries 4–6).

Al-SBA-15 is a mesoporous material largely employed as support for Ni-based catalysts because possesses excellent properties, including high surface area, large pore volume, and suitable acidity. In this context, Ni/Al-SBA-15 lamellar structure (prepared via the deposition–precipitation [DP] method using Al-SBA-15 as the support and nickel nitrate as the metallic precursor) has demonstrated to be a strategic catalyst for the amine formation, in near quantitative yields, by hydrogenation (H_2 at 0.5 MPa) of nitroalkanes [24].

As reported in Table 2.2, amines are obtained in near quantitative yields in all cases. Chain length and steric effect did not have much impact on the formation of amines. Moreover, the catalyst exhibits excellent selectivity since the nitro group can be reduced while the C=C double bond is preserved (entry 6) and dechlorination is minimal (entry 5).

Under the same reaction conditions, the authors achieved even the one-pot synthesis of imines. In fact, in Scheme 2.1, reduction of nitroalkanes (Ni/Al-SBA-15, H_2, EtOH) in the presence of an aldehyde allows the one-pot formation of imines via simultaneous catalytic reduction-condensation process.

Table 2.1 Reduction of nitroalkanes possessing heteroaromatic structures (selected examples).

Entry	Substrate	Product	Yield (%)[a]
1	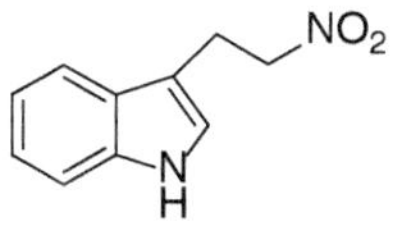	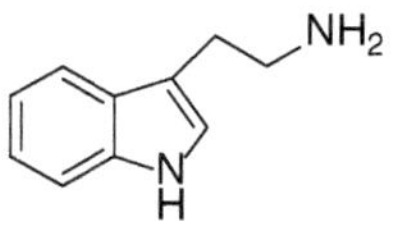	94
2	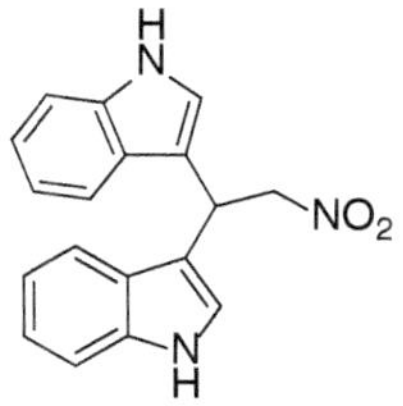	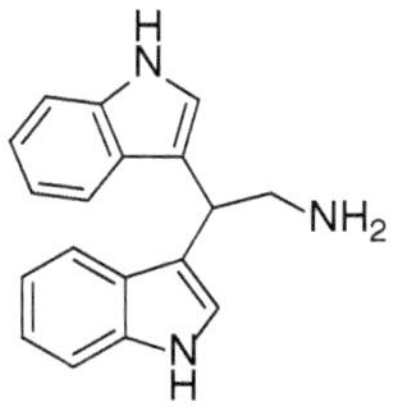	76
3	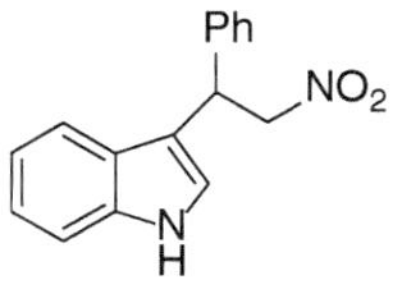	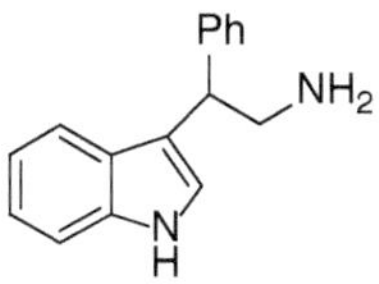	85
4	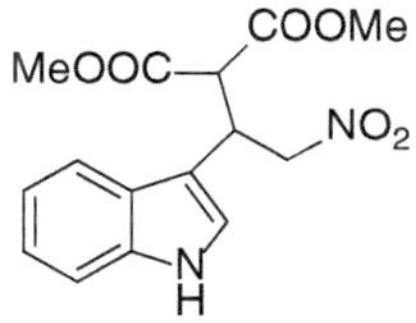	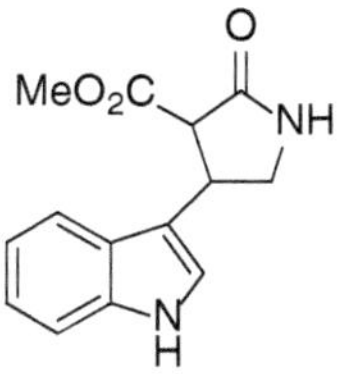	78
5	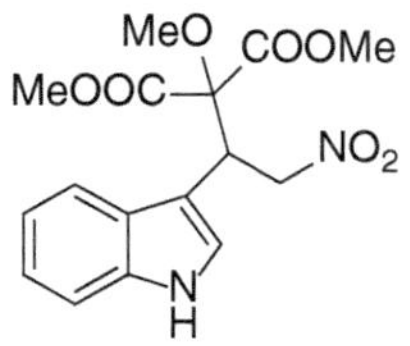	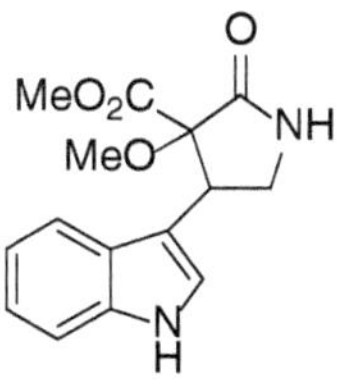	62
6	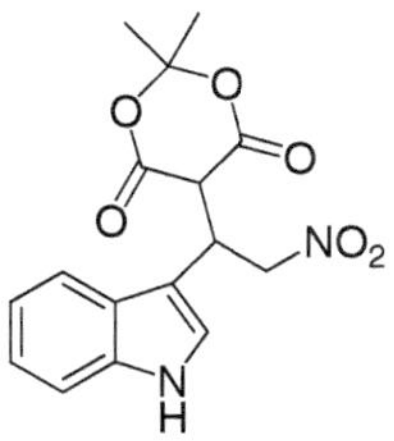	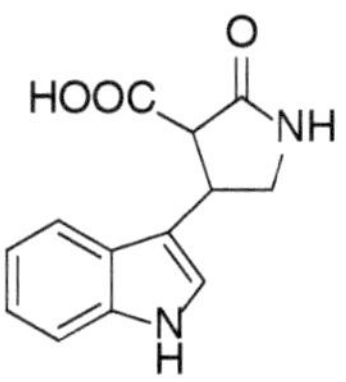	63

a) Reaction conditions: Pd/C 5% (2.5 mol%), H_3PO_2/NaH_2PO_2, sonication, seven hours, H_2O/2-MeTHF.

Table 2.2 Reduction of nitroalkanes by H_2/Ni/Al-SBA-15 (selected examples).

Entry	Substrate	Product	Yield (%)[a]
1	$EtNO_2$	$EtNH_2$	99
2	*n*-$PrNO_2$	*n*-$PrNH_2$	99
3	*n*-$BuNO_2$	*n*-$BuNH_2$	99
4	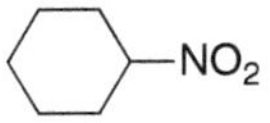	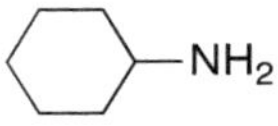	99
5		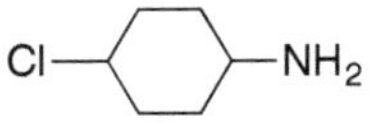	97
6	$MeCH{=}CHNO_2$	$MeCH{=}CHNH_2$	99

a) Reaction conditions: Ni/Al-SBA-15, 2.5 MPa H_2, EtOH, 7.5 hours.

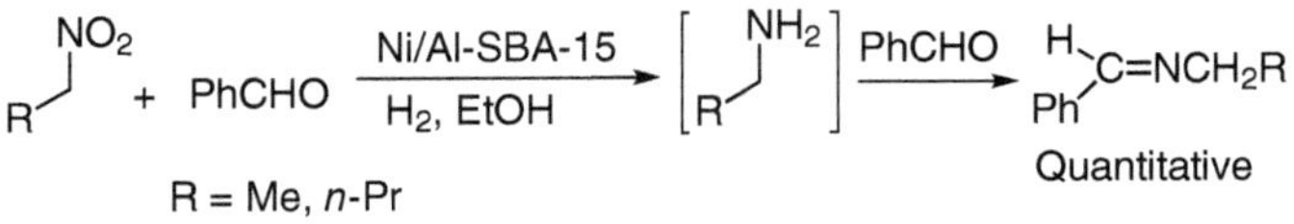

Scheme 2.1 One-pot synthesis of imines.

It is important to note that the catalyst maintains high catalytic effect after multiple recycles and can easily isolated and regenerated. The latter concept is very important for industrial applications and large scale "green production" of amines and imines.

2.1 Representative Synthetic Applications of Nitroalkane Reductions

Most reactions involving nitroalkanes lead to new carbon–carbon bonds formations together with added functional groups such as alcohols or carbonyls, depending from the nature of the chemical process used (mainly nitroaldol-Henry or Michael reaction). Although these processes will be, specifically, the topic of Chapters 4 and 5, we wish to report here some the most representative synthetic procedures in which the reduction of the nitro to the amino group plays a strategic role.

2.1.1 Reduction of Nitroalkanes Obtained via Nitroaldol Reaction

The nitroaldol reaction consists in the nucleophilic attack of a nitroalkane to a carbonyl, mainly under basic conditions, with the formation of β-nitroalkanols.

(*S*)-Propranolol is a potent β-blocker for the treatment of cardiovascular diseases. This target can be easily prepared (Scheme 2.2) starting from the enantioselective nitroaldol reaction of $MeNO_2$ with the aldehyde **1**, in the presence of lanthanum binaphthol complex (LBC) as catalyst.

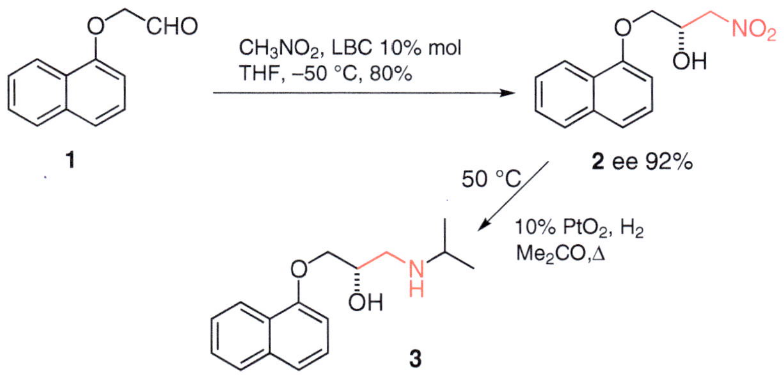

Scheme 2.2 Synthesis of Propranolol **3**.

The obtained nitroalcohol **2** is submitted to reduction under 10% PtO_2/H_2 and alkylated, in a one-pot process, leading to optically active (*S*)-Propranolol **3** [25].

Anticancer drug *threo*-Dihydrosphingosine **7** has been synthesized with a nitroaldol reaction of palmitaldehyde **4** with nitroethanol **5** (Scheme 2.3) [26]. The nitroalkanol **6** is obtained with *syn* selectivity and high ee due to the mildness of the catalytic system that prevents any epimerization.

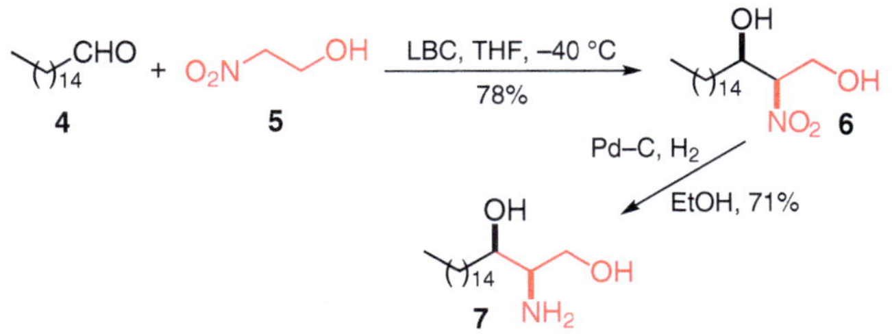

Scheme 2.3 Synthesis of *threo*-Dihydrosphingosine **7**.

Finally, reduction of the nitro group with Pd–C/H_2 allows the target compound **7** without any loss of enantiopurity.

Denopamine **13** is an important β-adrenoceptor agonist that has been synthesized starting from nitroaldol reaction of nitromethane and silyloxy benzaldehyde (Scheme 2.4) [27].

The reaction is catalyzed using dinuclear zing catalyst (DZC), in order to obtain the nitroalkanol **9** in both good yield (88%) and enantioselectivity (ee 90%). Successive reduction ($LiAlH_4$) of the latter gives the corresponding amino alcohol **10** that is strategic for the coupling with the molecular structure **11** that produce the amide **12** in high yield (88%). Further reduction and removal of the silyl protecting group ($SiMe_2t$-Bu) complete the preparation to the target molecule **13**.

Nitromethane has demonstrated to be a key starting compound for the synthesis of piperidine system [28]. If fact (Scheme 2.5), asymmetric nitroaldol reaction of 4-oxobutanoate **14** with $MeNO_2$, catalyzed by chiral bisoxazolidine (CBOA), affords

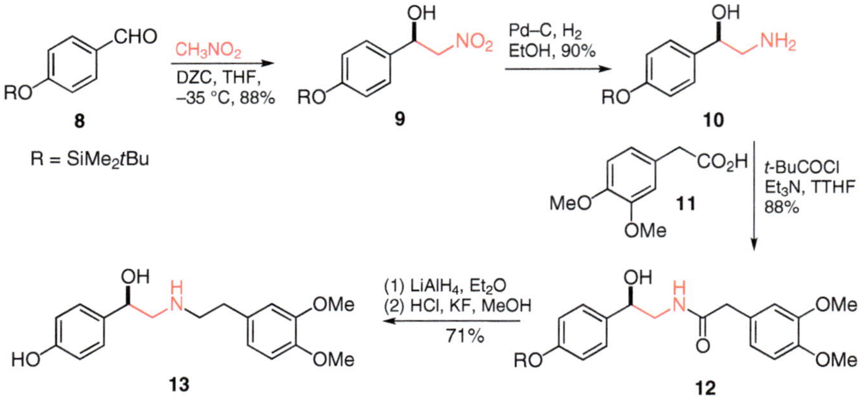

Scheme 2.4 Synthesis of (−)-Denopamine **13**.

15. Although the ee produced in the formation of nitro alcohol **15** is not very high, reduction of the nitro group results to concomitant ring closure to piperidin-2-one **16** that, after recrystallization, results almost enantiomerically pure.

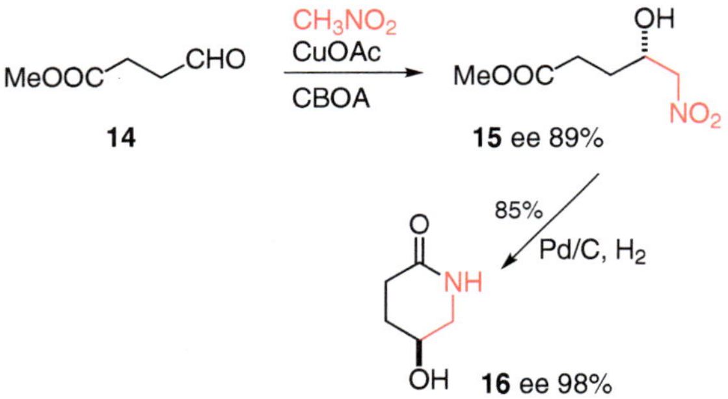

Scheme 2.5 Enantioselective synthesis of piperidin-2-one

2.1.2 Reduction of Nitroalkanes Obtained via Michael Reaction

The nucleophilic attack of nitroalkanes, under basic conditions, to electron poor alkenes gives access to a variety of more complex structures having at least two functionalities, including the nitro group. Chiral thiourea derivative **20** is an important molecular structure due to its *anti*-HIV activities and nitroalkanes have demonstrated to be strategic for their synthesis. As example Scheme 2.6 reports a possible procedure for the preparation of the structure **20** [29].

The synthesis starts with the asymmetric Michael addition of nitroethane to *p*-methylnitrostirene **17**, at room temperature, and catalyzed with tridentate bis-oxazoline (TBO). Under these conditions the 1,3-dinitro compound **18** is obtained with outstanding diastereo- and enantioselectivity. A dinuclear zinc catalyst could be formed via interaction of diethylzinc with TBO, although activation of the organozinc reagent with $Ti(O^iPrO)_4$ is mandatory for optimal results.

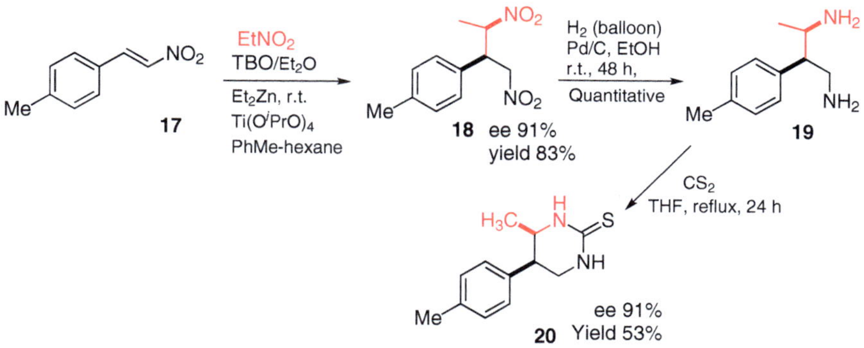

Scheme 2.6 Synthesis of thiourea derivative **20**.

The obtained *syn*-dinitro compounds **18** are then reduced leading to the diamino structure **19**. The latter, under reaction with carbon disulfide, allows the target thiourea derivative **20**.

Bicyclic-γ-lactam **24** is an important inhibitor of penicillin binding protein [30] and its synthetic precursor is the pyrrolidinones **23** (Scheme 2.7). The latter can be easily prepared from diethylgluconate **21** through a Michael reaction with nitromethane and in presence of DBU as base, followed by reduction of the formed nitro adduct **22**. It is important to observe that hydrolytic resolution of racemic **22**, using two enantiocomplementary enzymes, allows the preparation of optically active **23** in both enantiomeric forms.

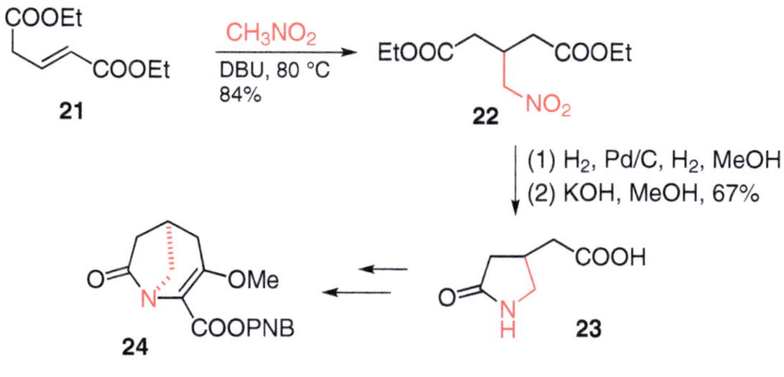

Scheme 2.7 Synthesis of bicycle-γ-lactam **24**.

Rolipram is a type IV inhibitor of cyclic AMP phosphodiesterase clinically used as antidepressant. Its enantioselective synthesis (Scheme 2.8) involves [31] the Michael reaction of nitromethane with enal **25**, under chiral pyrrolidine (CP) catalysis and in the presence of water as solvent. The subsequent step is dedicated to the oxidation of the aldehyde **26**, esterification (trimethylsilyl diazomethane) of the obtained carboxylic acid and formation of the nitro ester **27**.

Reduction of the nitro group occurs with the concomitant cyclization yielding, directly, (*S*)-Rolipram **28** in good yield and almost enantiopure form.

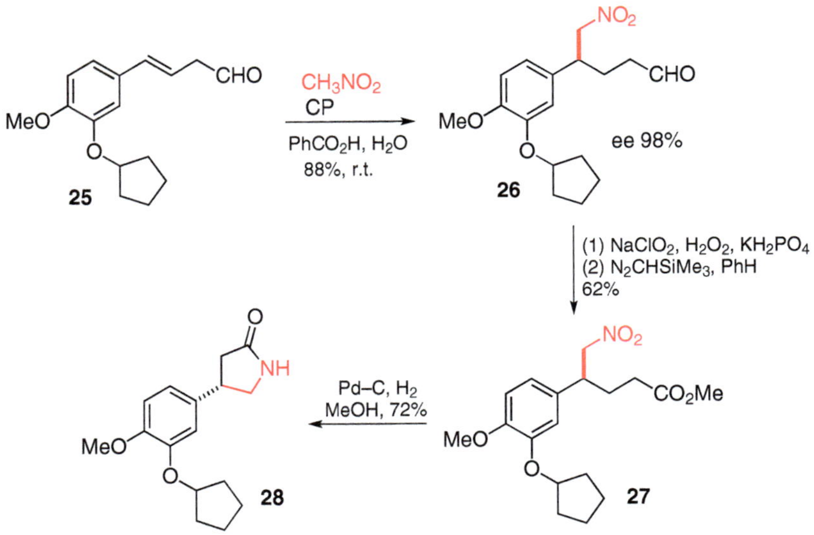

Scheme 2.8 Synthesis of (*S*)-Rolipram **28**.

References

1 (a) Wehrli, P.A. and Schaer, B. (1977). *Synthesis*: 649–650. (b) Saito, J., Takami, M., and Matsuura, T. (1975). *Tetrahedron Lett.* 16: 3155–3158. (c) Moles, P.J. and Reid, S.T. (1976). *Tetrahedron Lett.* 17: 2283–2286. (d) Hwu, J.R., Tseng, W.N., Patel, H.V. et al. (1999). *J. Org. Chem.* 64: 2211–2218. (e) Pham-Huu, D.-P., Petrušová, M., BeMiller, J.N., and Petruš, L. (1998). *Synlett*: 1319–1320.

2 (a) Shin, C., Narukava, H., Yamaura, M., and Yoshimura, J. (1977). *Tetrahedron Lett.* 18: 2147–2150. (b) Lee, T.D. and Keana, J.F.W. (1976). *J. Org. Chem.* 41: 3237–3241. (c) Stermitz, F.R. and Neiswander, D.W. (1975). *Tetrahedron* 31: 655–656.

3 Kornblum, N. and Widmer, J. (1978). *J. Am. Chem. Soc.* 100: 7086–7088.

4 (a) Yamashita, M., Mizushima, K., Watanabe, Y. et al. (1976). *J. Chem. Soc., Chem. Commun.*: 670–670. (b) Yamashita, M., Watanabe, Y., Mitsudo, T., and Takegami, Y. (1976). *Tetrahedron Lett.* 17: 1585–1586.

5 Murahashi, S.-I. and Imada, Y. (2019). *Chem. Rev.* 119: 4684–4716.

6 (a) Colvin, E.W. and Seebach, D. (1978). *J. Chem. Soc., Chem. Commun.*: 689–691. (b) Torssell, K. and Zeuthen, O. (1978). *Acta Chem. Scand.* B32: 118–124.

7 (a) Tu, Z., Jang, Y., Lin, C. et al. (2005). *Tetrahedron* 61: 10541–10551. (b) Czekelius, C. and Carreira, E.M. (2005). *Angew. Chem. Int. Ed.* 44: 612–615. (c) Temelli, B. and Unaleroglu, C. (2014). *Synthesis* 46: 1407–1412.

8 (a) Barco, A., Benetti, S., Pollini, G.P. et al. (1978). *Synth. Commun.* 8: 219–226. (b) Giacomelli, G., De Luca, L., and Porcheddu, A. (2003). *Tetrahedron* 59: 5437–5440.

9 Chandrasekhar, S., Narsihmulu, C., and Jagadeshwar, V. (2002). *Synlett* 5: 771–772.

10 (a) Knifton, J.F. (1975). *J. Org. Chem.* 40: 519–520. (b) Hill, R.K., Joule, J.A., and Loeffler, L.J. (1962). *J. Am. Chem. Soc.* 84: 4951–4956. (c) Joseph, M., Leigh, T., and Swain, M.L. (1977). *Synthesis*: 459–461. (d) McDonald, E. and Martin, R.T. (1977). *Tetrahedron Lett.* 18: 1317–1320. (e) Wassermann, H.H., Hearn, M.J., Haveauz, B., and Thyes, M. (1976). *J. Org. Chem.* 41: 153–155. (f) Butterick, J.R. and Unrau, A.M. (1974). *J. Chem. Soc., Chem. Commun.*: 307–308. (g) Maki, Y., Sugiyama, H., Kikuchi, K., and Seto, S. (1975). *Chem. Lett.* 4: 1093–1094. (h) Watanabe, Y., Mitsudo, T., Yamashita, M., and Takegami, Y. (1975). *Bull. Chem. Soc. Jpn.* 48: 1478–1479. (i) Baldwin, J.E., Haber, S.B., Hoskins, C., and Kruse, L.I. (1977). *J. Org. Chem.* 42: 1239–1241. (j) Kornblum, N. and Fishbein, I. (1955). *J. Am. Chem. Soc.* 77: 6266–6269. (k) Petrini, M., Ballini, R., and Rosini, G. (1987). *Synthesis*: 713–714.

11 (a) Ballini, R. and Petrini, M. (2004). *Tetrahedron* 60: 1017–1047. (b) Ballini, R. and Petrini, M. (2015). *Adv. Synth. Catal.* 357: 2371–2402.

12 (a) Nishimura, S. (2001). *Handbook of Heterogeneous Catalytic Hydrogenation for Organic Synthesis*, 315–387. Chichester: Wiley. (b) Ballini, R., Gabrielli, S., Palmieri, A., and Petrini, M. (2011). *Curr. Org. Chem.* 15: 1482–1506.

13 Yang, Q., Chen, Y.-Z., Wang, Z.U. et al. (2015). *Chem. Commun.* 51: 10419–10422.

14 Göksu, H. (2015). *New J. Chem.* 39: 8498–8504.

15 Göksu, H., Ho, S.F., Metin, Ö. et al. (2014). *ACS Catal.* 4: 1777–1782.

16 Chen, X.-L., Ai, B.-R., Dong, Y. et al. (2017). *Tetrahedron Lett.* 58: 3646–3649.

17 Gkizis, P., Stratakis, M., and Lykakis, I.N. (2013). *Catal. Commun.* 36: 48–51.

18 Cantillo, D., Mirhosseini, M., and Kappe, O. (2013). *J. Org. Chem.* 78: 4530–4542.

19 Damodara, D., Arundhathi, R., Babu, T.V.R. et al. (2014). *RSC Adv.* 4: 22567–22574.

20 He, L., Wang, L.-C., Sun, H. et al. (2009). *Angew. Chem. Int. Ed.* 48: 9538–9541.

21 Arya, K. and Dandia, A. (2010). *J. Korean Chem. Soc.* 54: 55–58.

22 Baron, M., Métay, E., Lemaire, M., and Popowycz, F. (2013). *Green Chem.* 15: 1006–1015.

23 Henderson, R.K., Jiménez-Gonzáles, C., Constable, D.J.C. et al. (2011). *Green Chem.* 13: 854–862.

24 Ren, R. and Ma, J. (2015). *RSC Adv.* 5: 74802–74810.

25 Sasai, H., Itoh, N., Suzuki, T., and Shibasaki, M. (1993). *Tetrahedron Lett.* 34: 855–858.

26 Sasai, K., Tokunaga, T., Watanabe, S. et al. (1995). *J. Org. Chem.* 60: 7388–7389.

27 Trost, B.M. and Yeh, V.S.C. (2002). *Angew. Chem. Int. Ed.* 4: 2621–2623.

28 Spangler, K.Y. and Wolf, C. (2009). *Org. Lett.* 11: 4724–4727.

29 Lu, S.-F., Du, D.-M., Xu, J., and Zhang, S.-W. (2006). *J. Am. Chem. Soc.* 128: 7418–7418.

30 Aszodi, J., Rowlands, D.A., Mauvais, P. et al. (2004). *Bioorg. Med. Chem. Lett.* 14: 2489–2492.

31 Palomo, C., Landa, A., Mielgo, A. et al. (2007). *Angew. Chem. Int. Ed.* 46: 8431–8435.

3

Nitro Group to Carbonyl (Nef Reaction)

More than one century ago (1894) [1], Nef discovered that primary and secondary nitroalkanes can be converted into carbonyls and the synthetic importance of this result was growing year by year. It is important to observe that in this transformation the carbon having the nitro group reverses the polarity of the neighboring carbon from nucleophilic to electrophilic, thus a wide range of synthetic applications can be carried out.

The original Nef reaction (Scheme 3.1) was performed (Step 1) under basic conditions, in order to obtain the corresponding nitronate form, followed by hydrolysis of nitronic acid generated from the corresponding nitronate anion (Step 2) under strong acidic conditions.

It is obvious that the acidic conditions of Step 2 are incompatible with the presence of other sensitive functionalities and this why a variety of alternative methods have been, successively, developed. Thus, as alternative procedure, the cleavage of the C=N bonds has been proposed under oxidative (mainly), reductive or basic conditions.

3.1 Nef Reaction under Oxidative Conditions

3.1.1 Method Ox1

A pioneering oxidative procedure was reported via ozonolysis (at −78 °C) of the nitronate form obtained under sodium methoxide (Table 3.1) [2]. The method works well with both primary and secondary nitroalkanes giving access to aldehydes and ketones.

3.1.2 Method Ox2

Few years later, Olah and Gupta proposed Ceric Ammonium Nitrate (CAN) as oxidant, following two possible ways (Scheme 3.2) [3]. Nitro compounds can be transformed (way a) into carbonyls, in high yields, treating with CAN at 50–60 °C, and in the presence of Et_3N as mild base. The authors also tested the oxidation with CAN of the silyl nitronates (way b) prepared from the nitroalkanes. The latter method needs of milder reaction conditions (temperature and reaction time) allowing better yields.

Nitroalkanes: Synthesis, Reactivity, and Applications, First Edition.
Roberto Ballini and Alessandro Palmieri.

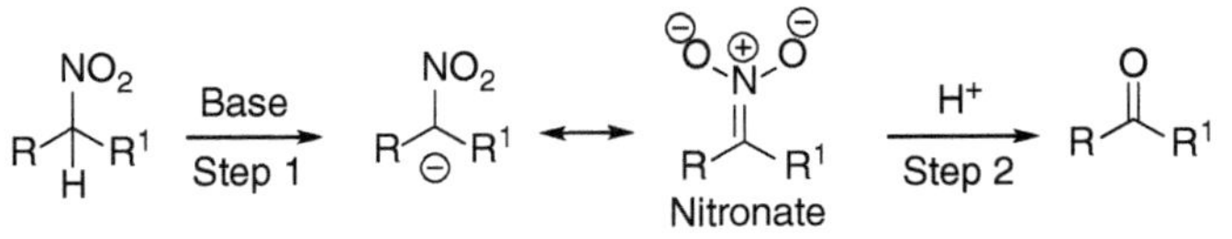

Scheme 3.1 Nef original reaction.

Table 3.1 Nef reaction by ozonolysis of nitronate (selected examples).

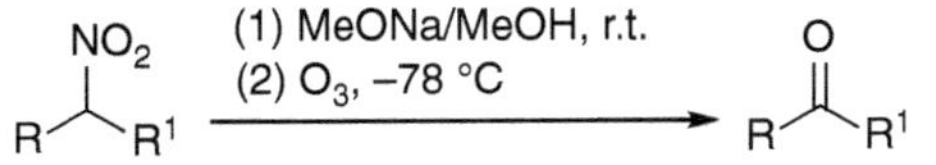

Entry	Nitroalkane	Product	Yield (%)
1	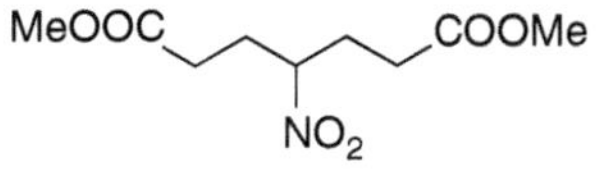	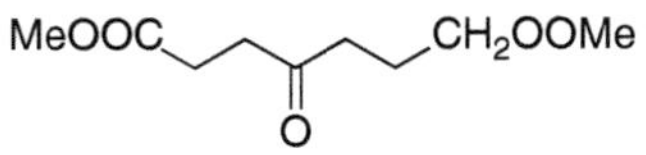	88
2	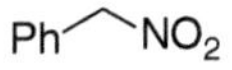	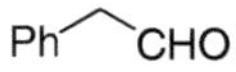	68
3	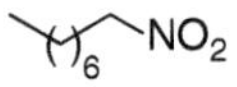	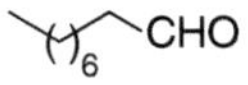	65

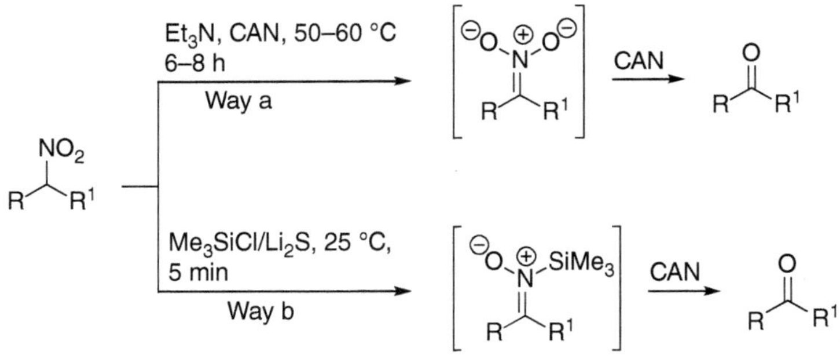

Scheme 3.2 Nef reaction by CAN.

However, good to excellent yields are obtained following both procedures (Table 3.2).

3.1.3 Method Ox3

The use of potassium permanganate provides a simple means to convert nitronates, obtained by treatment of nitroalkanes with NaH, into aldehydes and ketones in high yields (>82%) [4]. Under these conditions it is possible to achieve the transformation in excellent yields, including the preparation of the quaternary aldehydes (entries 2 and 6). Some representative examples of aldehydes and ketones prepared are reported in Table 3.3.

Table 3.2 Nef reaction by CAN (selected examples).

			Yield (%) (reaction time)	
Entry	**Nitroalkane**	**Product**	**Way a**	**Way b**
1	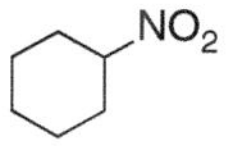	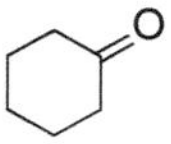	80 (2 h)	92 (5 min)
2	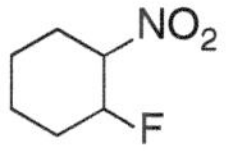	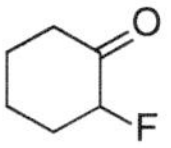	80 (2.5 h)	90 (5 min)
3	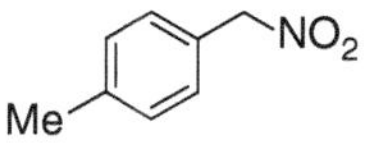	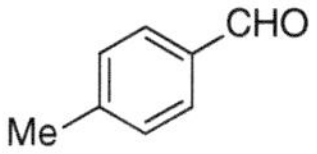	85 (1 h)	
4	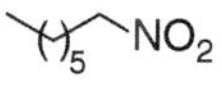	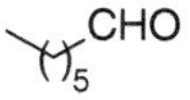	67 (1 h)	

Table 3.3 Nef reaction with NaH/$KMnO_4$.

Entry	Nitroalkane	Product	Yield (%)
1		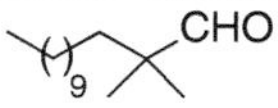	96
2	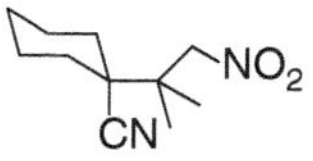	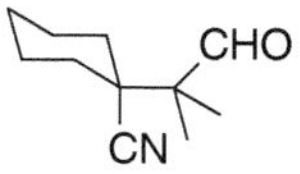	81
3	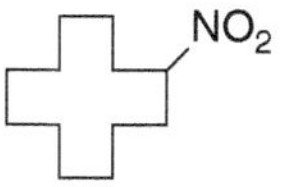	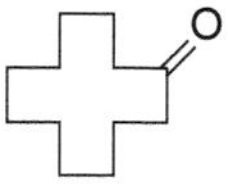	91
4	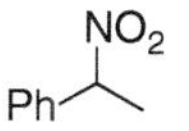	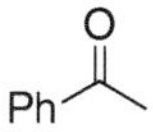	90
5	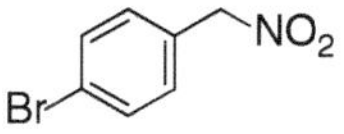	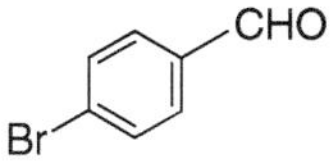	90
6	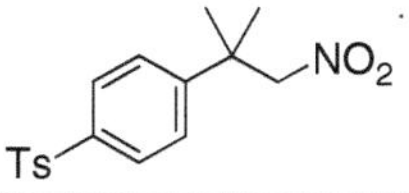	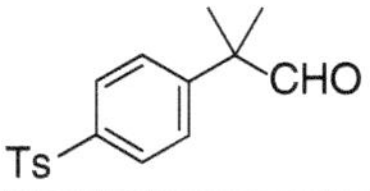	91

3.1.4 Method Ox4

In 1985, Steliou and Poupart reported a more selective method for the Nef reaction under permanganate oxidation [5]. In fact, they found that treatment of primary and secondary nitro alkanes with methanolic potassium hydroxide, at 0 °C, gives a stable and soluble form of the corresponding potassium nitronate. The in situ oxidation of this salt can be readily achieved with concentrated aqueous solution of $KMnO_4$ and $MgSO_4$. Some selected examples of the results obtained are reported in Table 3.4.

Table 3.4 Nef reaction with KOH/$KMnO_4$–$MgSO_4$.

Entry	Nitroalkane	Product	Yield (%)
1	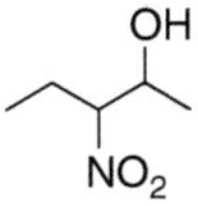	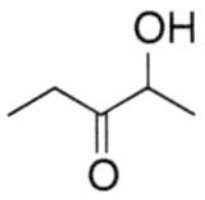	83
2	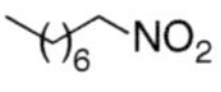	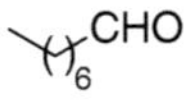	73
3	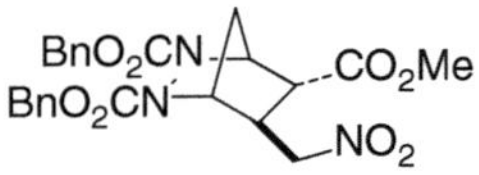	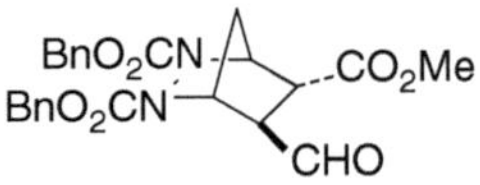	95
4	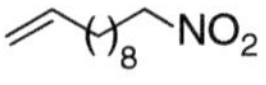	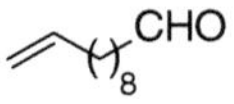	50

3.1.5 Method Ox5

Palomo and coworkers found that trialkyl nitronates, derived from secondary nitroalkanes, by reaction with trialkylsilyl chloride in the presence of DBU (1,8-diazabicyclo[5.4.0]undec-7-ene) as base, can be oxidized into ketone using *m*-chloroperbenzoic acid (MCPA)[6]. The method grants high yields of the products and it is mild enough for the survival of other functionalities such as ether, nitrile, and C=C double bond (Table 3.5). The silylation step is crucial in order to obtain ketone, in fact other nitronate anions, diverse from silyl nitronate, are unreactive. On the other hand, other base diverse from DBU furnishes bad results in the silylation step.

Of interest is the Nef transformation of entry 5 in which the obtained azetidin-2-one is considered as strategic intermediate in the synthesis of bicyclic β-lactams.

3.1.6 Method Ox6

Nitroalkanes are converted into the corresponding carbonyl derivatives with sodium chloride under phase transfer catalysis conditions [7]. In a typical procedure, nitroalkane is dissolved in CH_2Cl_2 and tetra-*n*-butylammonium hydrogen sulfate (TBAHS). After cooling with ice bath, an aqueous solution of 1 N NaOH and solid sodium chloride ($NaClO_2$) was added. After work up, the reaction allows the formation of linear and cyclic ketones in 65–85% yield (4–27 hours), and aldehydes in 65–79% yields (0.5–1 hours).

3.1.7 Method Ox7

Direct transformation of primary nitroalkanes into alkanoic acids has been reported, under mild conditions with buffered (pH = 11) potassium permanganate [8]. The

Table 3.5 Nef reaction promoted by MCPA (selected examples).

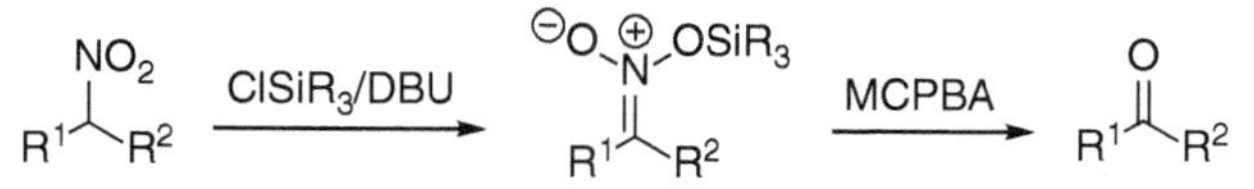

Entry	Nitroalkane	Product	Yield (%)
1	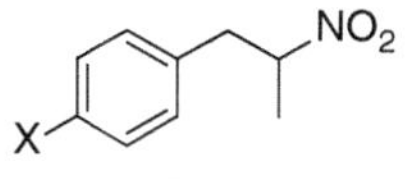 X = H, Cl, Me, OMe, CN	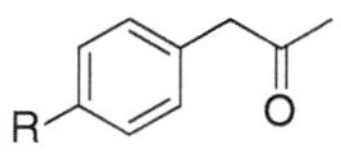	92–99
2	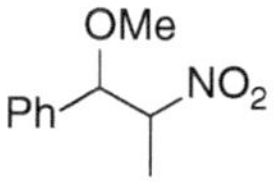	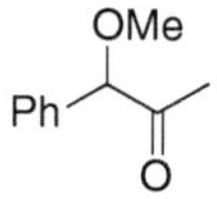	95
3	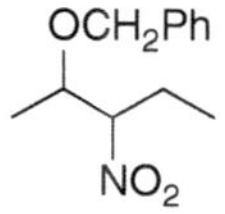	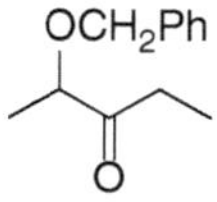	95
4	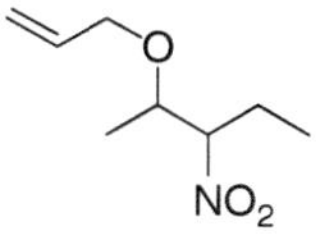	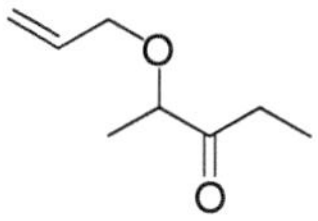	81
5	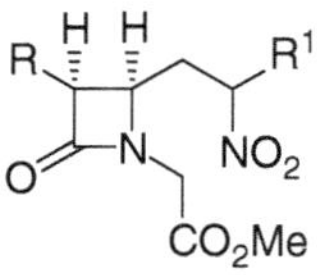	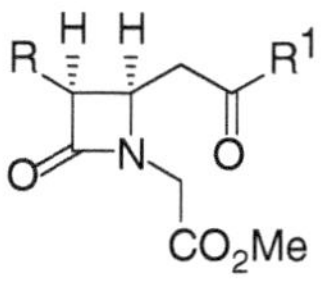	Not reported

procedure consists in the addition, at r.t., of KOH and K_2HPO_4, to the nitroalkane in *t*-BuOH. After stirring two minutes, aqueous $KMnO_4$ is added and the target carboxylic acids are obtained after one hour.

Under the aforementioned conditions a variety of other functionalities, such as ester, amide, alcohol, and tetrahydropyranyl (THP)-ether groups, can be preserved (Table 3.6).

3.1.8 Method Ox8

The reaction of nitroalkanes with sodium percarbonate ($Na_2CO_3{\cdot}3/2H_2O_2$) and silica gel in dimethylformamide–H_2O (DMF-H_2O), at 40 °C for one hour, produces good yields (65–81%) of the corresponding carbonyl compounds (aldehydes and ketones) [9]. The exact role of silica gel is not clear, but in its absence the reaction is very slow (>12 hours).

Table 3.6 Primary nitroalkanes to carboxylic acid by basic phosphate buffer $KMnO_4$.

Entry	Nitroalkane	Product	Yield (%)
1	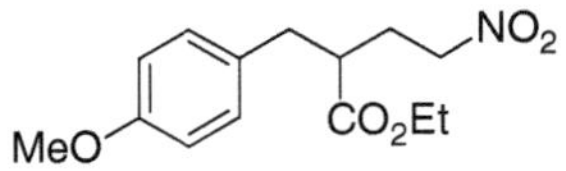	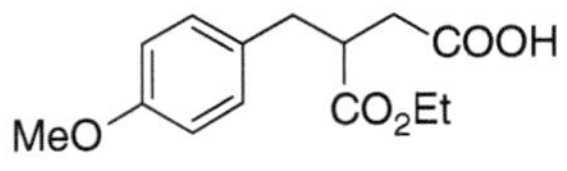	99
2	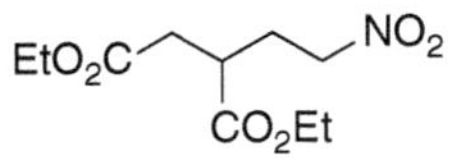	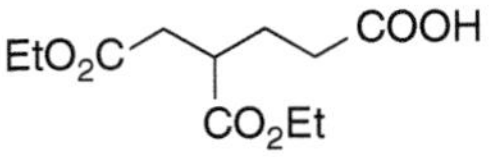	99
3	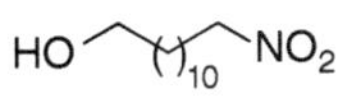	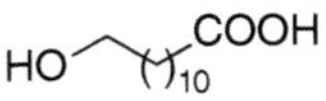	99
4	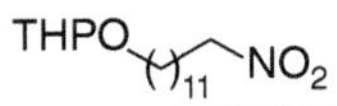	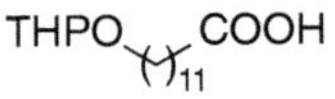	90

3.1.9 Method Ox9

Primary and secondary nitroalkanes can be easily converted into aldehydes or ketones with dioxygen, catalyzed by metallic copper, in the presence of *N*-ligand such as *N*,*N*,*N'*,*N'*-tetramethylethylenediamine [10]. In this transformation copper reacts with the *aci*-form of the nitro compounds to give copper nitronate complexes, prone to be oxidized with dioxygen. Although the method seems of interest, the authors reported just few examples of this procedure.

Table 3.7 Nef reaction by TPAP/NMO/MS.

Entry	Nitroalkane	Product	Yield (%)
1	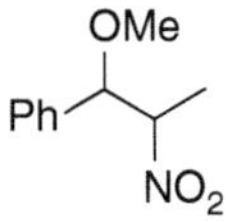	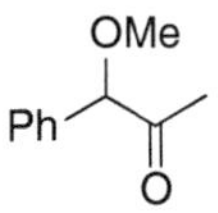	52
2	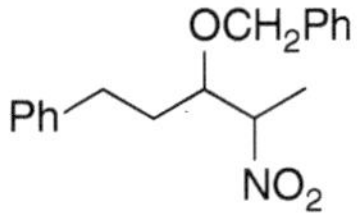	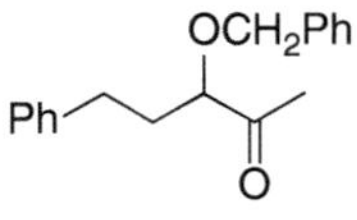	76
3	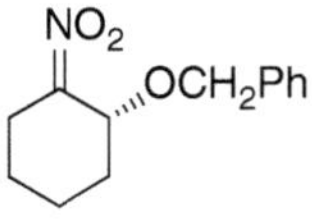	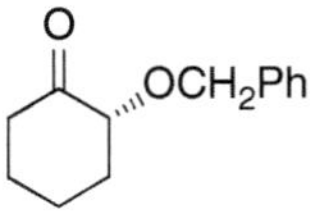	62
4	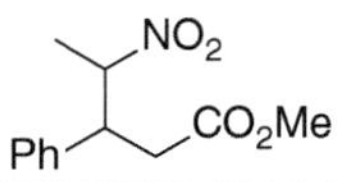	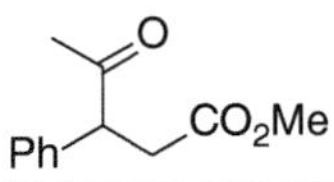	89

3.1.10 Method Ox10

Oxidation of secondary nitroalkanes with catalytic amount of tetrapropylammonium perruthenate (TPAP), in the presence of *N*-methylmorpholine *N*-oxide (NMO), silver(I) acetate, potassium permanganate, and 4 Å molecular sieves (MS), provides the corresponding ketones in moderate to good yields [11]. The treatment of primary nitroalkanes under the same reaction conditions yielded an intractable mixture.

Some representative examples are reported in Table 3.7.

3.1.11 Method Ox11

An interesting oxidative Nef reaction, using Oxone® (potassium hydrogen persulfate), was developed by Marcotullio and coworkers two decades ago [12]. Nitroalkane is treated with a hydroalcoholic solution of Na_2HPO_4/NaOH and the obtained nitronate form is oxidized with Oxone. Secondary nitroalkanes give the corresponding ketones, while the primary ones are converted, directly, into carboxylic acids. Some representative examples are reported in Table 3.8, where it is possible to observe the chemoselectivity of the method since important functionalities such as hydroxyl, ester, and *tert*-butyldimethylsilyl (TBDMS) protecting hydroxyl can be preserved.

Table 3.8 Nef reaction by Oxone (selected examples).

Entry	Nitroalkane	Product	Yield (%)
1	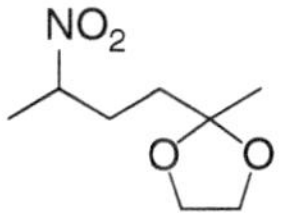	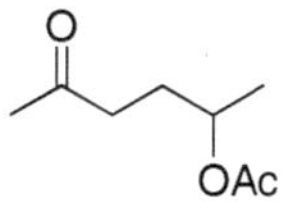	87
2	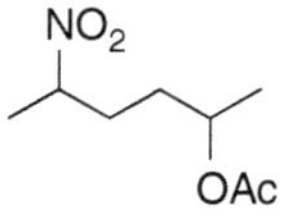	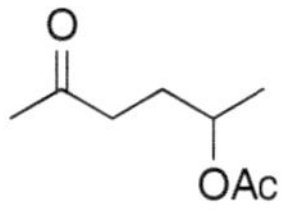	93
3	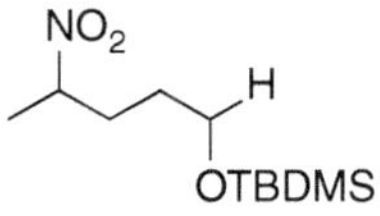	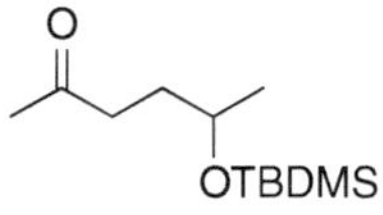	88
4	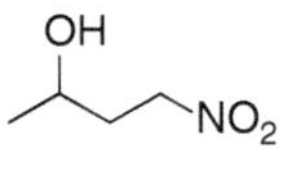	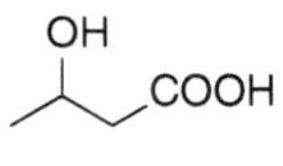	70
5	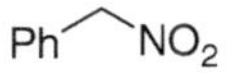	PhCOOH	91

Table 3.9 Nef reaction promoted by DMD (selected examples).

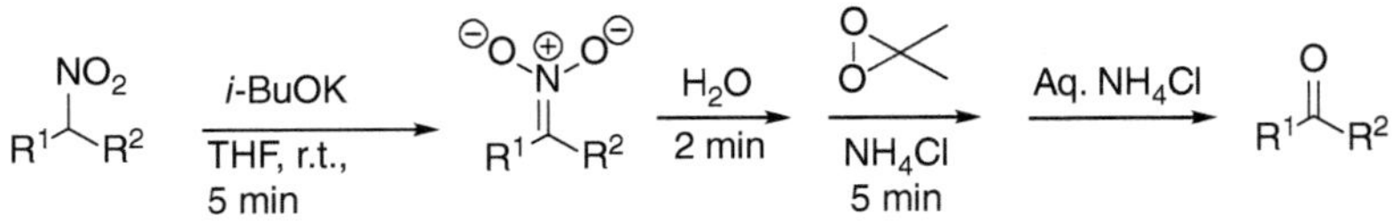

Entry	Nitroalkane	Product	Yield (%)
1	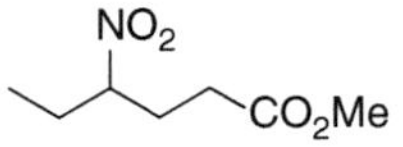	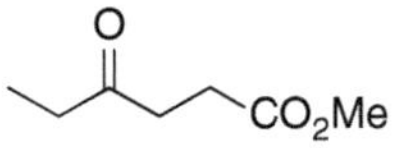	90
2	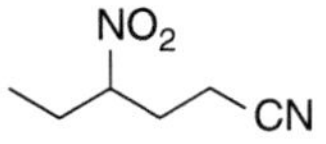	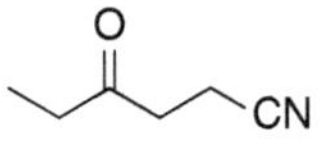	86
3	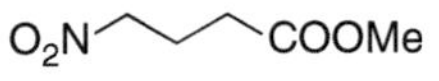	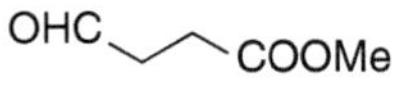	73
4	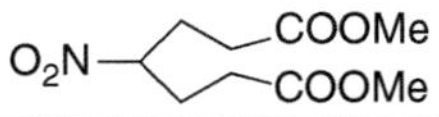	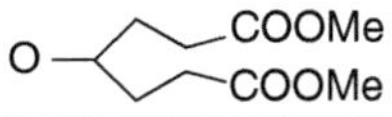	83

3.1.12 Method Ox12

Dimethyldioxirane (DMD) has been tested as mild and efficient oxidant of nitronates, obtained treating nitroalkanes with ^{t}BuOK, for the formation of ketones or aldehydes [13]. The nitroalkane, under ^{t}BnOK, converts into its nitronate, the successive addition of water and (after two minutes) of DMD yields, selectively, the corresponding ketone or aldehyde. Authors reported that the addition of water before DMD is very important. In Table 3.9 it describes a possible mechanism of the reaction and some of the most representative results obtained.

3.1.13 Method Ox13

Secondary nitroalkanes, upon treatment with bis(trimethylsilyl) peroxide (BTSP), in the presence of NaH as base, convert into ketones in satisfactory to good yields (65–84%) [14]. Primary nitroalkanes are unreactive under these reaction conditions, just benzylic primary nitroalkanes convert into the corresponding aldehydes.

3.1.14 Method Ox14

Aryl nitroaldols (Table 3.10) can be converted, directly, into α-ketoacids by treatment with $Cu(OAc)_2$ as oxidizing agent and in a solution of 30% AcOH/MeOH [15]. The presence of the aryl group is fundamental since the reaction failed for alkyl nitroaldols.

Moreover, it is interesting to note that the presence of an electron-withdrawing groups in the aromatic ring, such as NO_2 (entry 5), stops the reaction at the keto-aldehyde without undergoing further oxidation to the corresponding acid.

Table 3.10 Nef reaction promoted by $Cu(OAc)_2$ (selected examples).

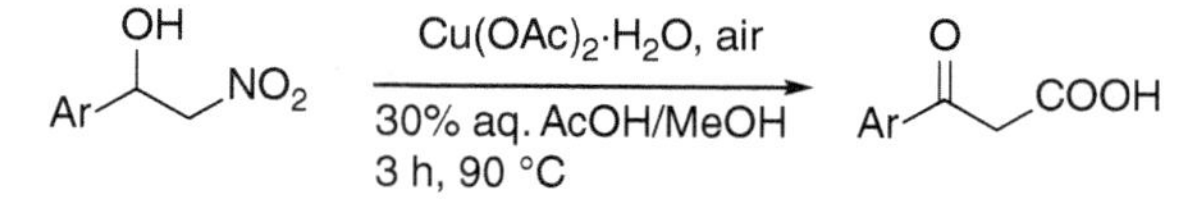

Entry	Nitroalkane	Product	Yield (%)
1	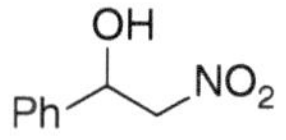	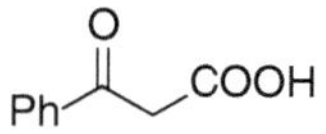	61
2	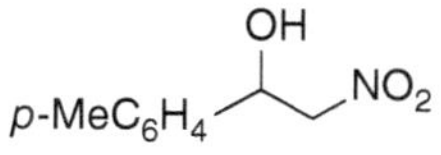	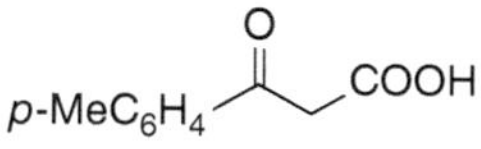	88
3	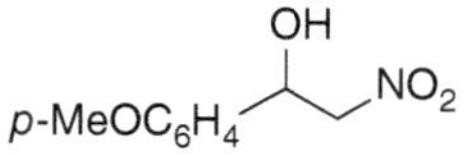	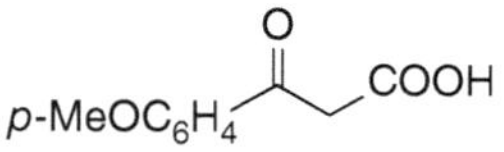	87
4	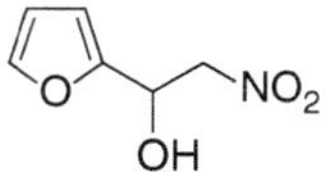	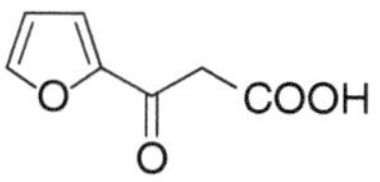	72
5	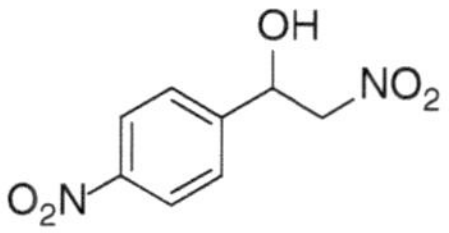	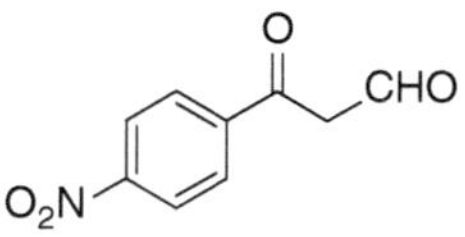	50

3.1.15 Method Ox15

1-Aryl-2-nitroalkanes are a class of secondary nitrocompounds in which the nitro group can be transformed into ketone, in high yields (80–95%), by oxidative hydrolysis under Fe–HCl mixture (Scheme 3.3) [16].

It is well known that reduction of nitro groups to amines take place through the intermediary of the nitroso derivatives of the oximes. The authors suppose that the

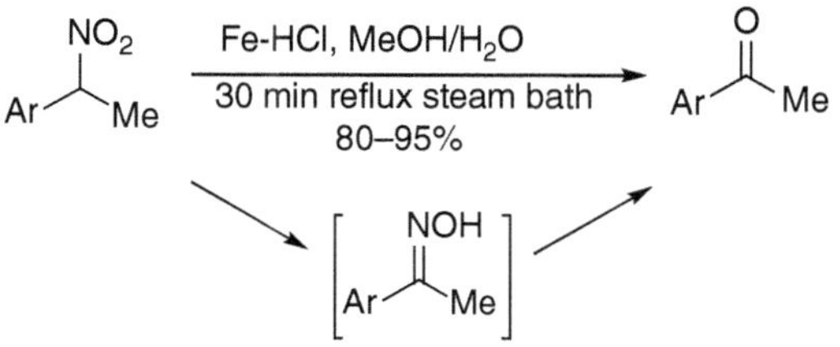

Selectected examples:

Ar = Ph, *p*-$MeOC_6H_4$, *p*-HOC_6H_4, 2-furyl, 3-indolyl, 3,4,5-$(MeO)_3C_6H_2$

Scheme 3.3 Nef reaction by Fe–HCl.

Table 3.11 Nef reaction with IL/SPB (representative examples).

Entry	Nitroalkane	Product	Yield (%)
1	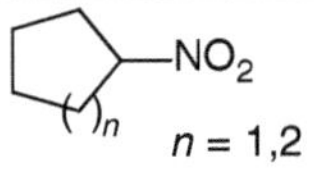	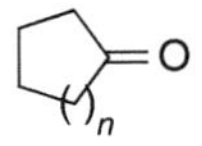	95–96
2	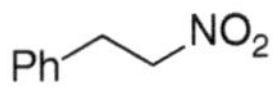	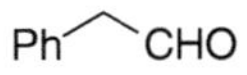	60
3	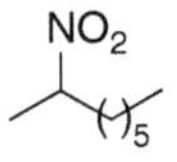	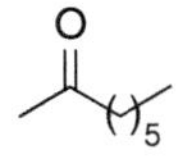	94
4	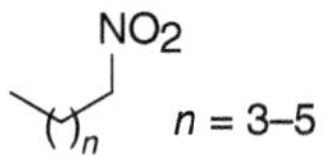	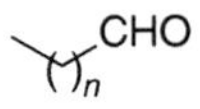	60–78

presence of a methyl group in α-position to NO_2 favors the formation of oximes that subsequently undergo oxidative hydrolysis to the corresponding ketones.

3.1.16 Method Ox16

Mild oxidative conversion of primary and secondary nitroalkanes to the corresponding aldehydes and ketones has been performed at 70 °C, by sodium perborate (SPB), as oxidant, in ionic liquids bmim(OTf) (bmim = 1-butyl-3-methyl imidazolium; $OTf^- = CF_3SO_3^-$) following the green chemistry approaches (Table 3.11) [17]. SPB is a solid peroxygen with great storage stability that functions as a source of moderately alkaline H_2O_2. In addition, SPB is cheap, safe, easy to handle, and readily available.

One criticism of this method is the need of volatile organic solvent (VOS) during the work-up and/or the purification step. However, a single extraction allows the quantitative recovery of both the product and the IL. The latter can be reused for further transformations.

The authors observed a decrease of the substrate conversion increasing its lipophilicity; anyway, the method offers important advantages such as low cost, good yields, easy work-up, and the possibility to reuse the IL.

3.1.17 Method Ox17

Continuous flow process is an emerging field of green chemistry that has been proposed for the Nef reaction under $KMnO_4$ oxidative conditions (Table 3.12) [18].

The choice of $KMnO_4$ is due because is a cheap reagent and the manganese dioxide by-product can be easily isolated and reused in other oxidative processes. The conversion of primary nitro group into aldehyde or carboxylic acid is strongly dependent from the ratio nitroalkane/$KMnO_4$. Thus, two different procedures (a and b)

Table 3.12 Nef reaction by $KMnO_4$ under flow process (representative examples).

Entry	Nitroalkane	Product (yield (%))	
		Method a	Method b
1	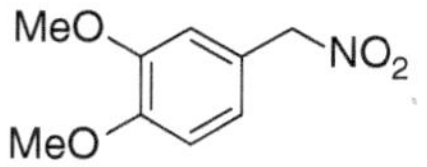	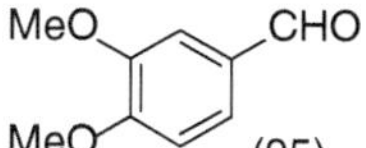 (95)	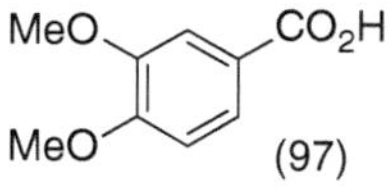 (97)
2	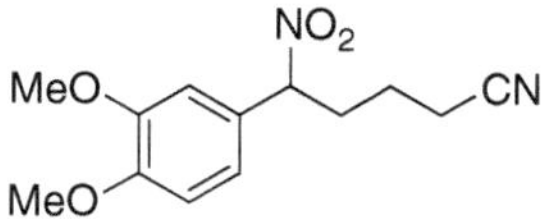	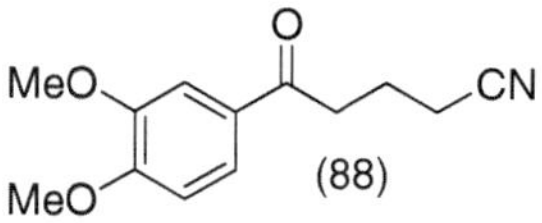 (88)	
3	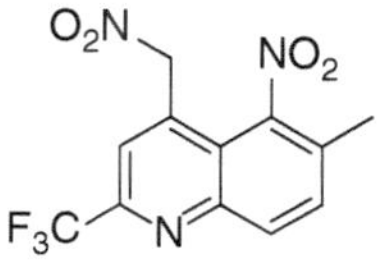	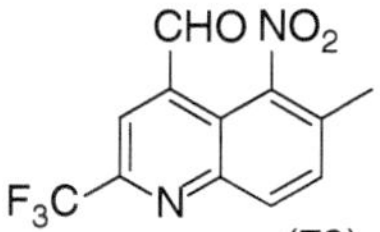 (73)	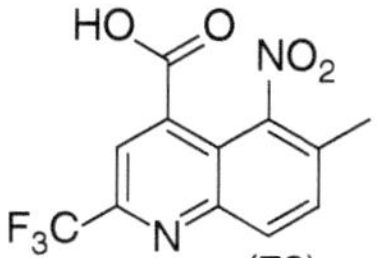 (78)
4	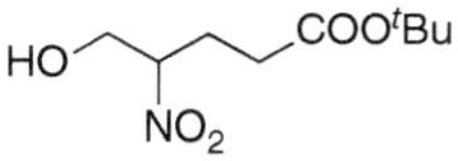	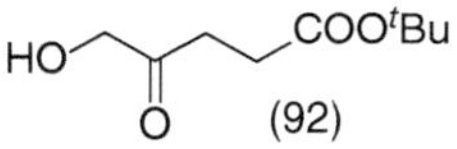 (92)	

have been proposed in order to obtain CHO or COOH from the same primary nitro group.

Method a (primary nitroalkanes into aldehydes): 1 equiv of nitroalkane and KOH (1.2 equiv) in methanol is injected in the flow apparatus, together with 0.8 equiv of aqueous $KMnO_4$. After few minutes the conversion nitro to carbonyl has taken place, while other functionalities such as esters, alcohols, and nitrile can be preserved.

Method b (primary nitroalkanes into carboxylic acids): 1 equiv of primary nitroalkane and KOH (1.2 equiv) in methanol with buffered solution of $KMnO_4$ (2 equiv) and Na_2HPO_4 in water were added to the flow apparatus at room temperature. The excess of permanganate was reduced with sodium thiosulfate during the work up with HCl. Under these reaction conditions, the formed MnO_2 was solubilized by a simple extractive work-up.

3.2 Nef Reaction Under Reductive Conditions

3.2.1 Method Red1

In any reductive method nitroalkanes or their nitronate ions are expected to allow oximes (the main tautomer of the nitroso structures) as intermediates, prone to convert into imines that, during the aqueous work up, may be hydrolyzed to carbonyls [19].

Table 3.13 Nef reaction promoted by $TiCl_3-NH_4OAc$ (representative examples).

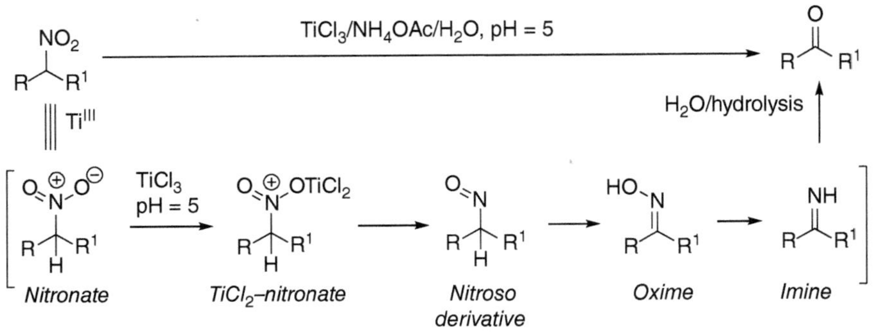

Entry	Nitroalkane	Product	Yield (%)
1	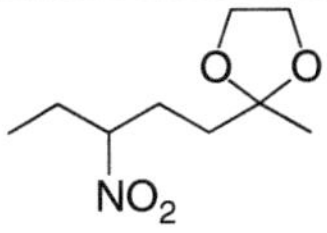	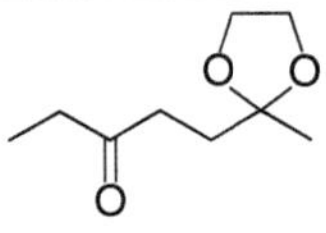	70
2	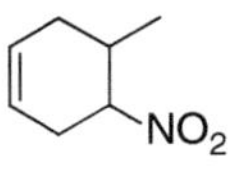	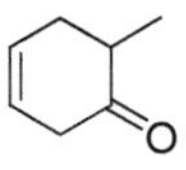	30
3	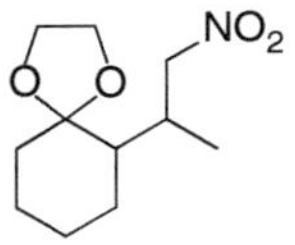	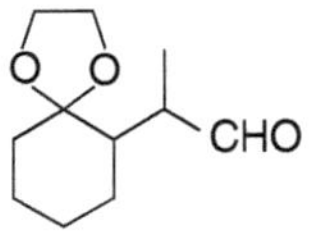	70
4	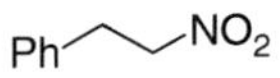	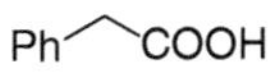	70

Thus, treatment of nitroalkanes (primary and secondary) with a buffered aqueous solution of $TiCl_3$, prepared adding NH_4OAc/H_2O to 20% aqueous $TiCl_3$ (pH = 5), under nitrogen and at room temperature, gives after the appropriate reaction times (mainly 12 hours) and after the work up, the formation of aldehydes or ketones in satisfactory yields (Table 3.13).

A possible mechanism consists in the tautomerization of the nitro compound to its *aci*-form, helped be the presence of $TiCl_3$ as Lewis acid. The nitronate reacts with $TiCl_3$ yielding a $TiCl_2$-nitronate structure, followed by its transformation into an oxime via tautomerization of a nitroso-structure. Finally, the oxime is reduced to imine and the latter hydrolyses to carbonyls.

3.2.2 Method Red2

Aqueous vanadium(II) chloride is an effective reducing agent for the Nef reaction. In fact, adding aqueous VCl_2 (75 mmol) to a nitroalkane (15 mmol) dissolved in a mixture $H_2O/HCl/DMF$, under nitrogen atmosphere and at room temperature, allows

Table 3.14 Nef reaction promoted by VCl_2 (representative examples).

Entry	Nitroalkane	Product	Yield (%)
1	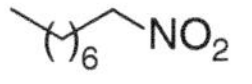	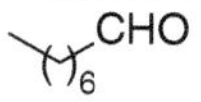	24
2	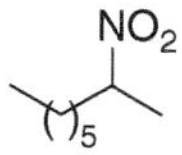	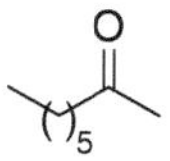	71
3	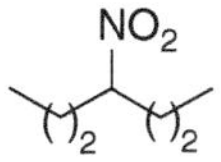	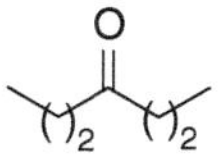	63
4	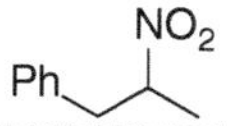	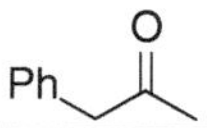	65

the formation of aldehyde and ketone in short reaction times (20–30 minutes) [20]. However, the achieved yields are moderate (Table 3.14).

3.2.3 Method Red3

A decade ago, Vilarrasa and Burés tested trialkylphosphine, as reducing agent, for the conversion of secondary nitroalkanes into ketones [21]. After an exhaustive search the authors found the combination of trimethyl phosphine (PMe_3), as the most effective reducing agent, with 4,4′-bis-^{t}buthyldiphenyl disulfide ($^tBuC_6H_4SSC_6H_4{}^tBu$, **A**) or *N*-(phenylselenyl) phathalimide (PhthN-SePh, **B**) as additives, the role of which is to shorten the reaction times. Thus, the reaction is carried out, at room temperature, in THF/PMe_3 (2.2 mmol) solution, under N_2. Nitroalkane (1 mmol) and the additive (**A** or **B**, 0.5 mmol) are added, without any solvent, to the aforementioned solution. The reaction times depended from the nature of **A** or **B** (Table 3.15), while the yields are near to quantitative with both the additives.

The procedure is mild enough to preserve other important functionalities and the configuration of chiral carbons (entries 2–5).

3.2.4 Method Red4

An original and green Nef procedure has been reported thanks to the bioreduction of secondary nitroalkanes, catalyzed by flavoproteins (from the old-yellow-enzyme family), with the formation of the corresponding ketones [22]. The method proceeds with the classical steps-sequence (similar to those reported in Scheme 3.9): (i) reduction of the nitro group to nitroso derivative that spontaneously (ii) tautomerizes to yield the more stable oxime, the latter (iii) is enzymatically reduced to imine that, spontaneously, (iv) hydrolyzed to carbonyl.

Although this is the first example of Nef reaction under bioreduction, poor yields of ketones are achieved.

Table 3.15 Nef reaction promoted by PMe_3/additive **A** or **B** (selected examples).

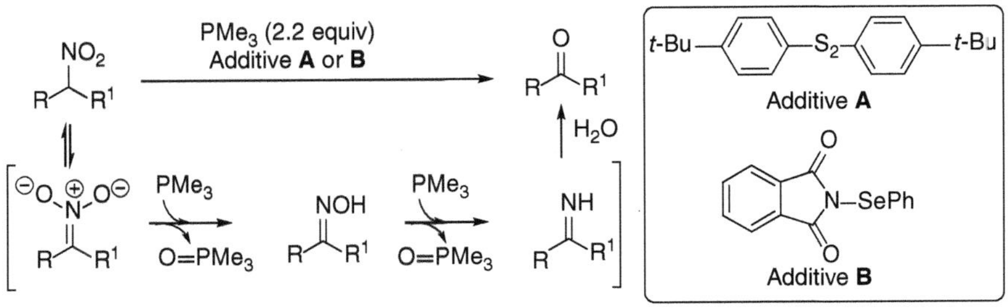

Entry	Nitroalkane	Product	Yield (%) (reaction time) Additive A	Yield (%) (reaction time) Additive A
1	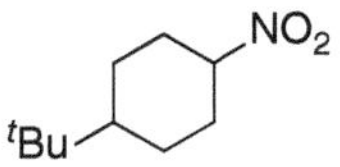	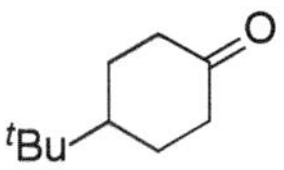	97 (15 min)	96 (2 h)
2	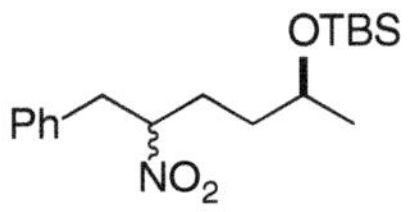	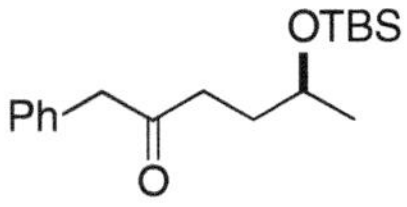	98 (30 min)	98 (3 h)
3	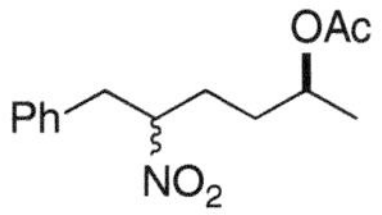	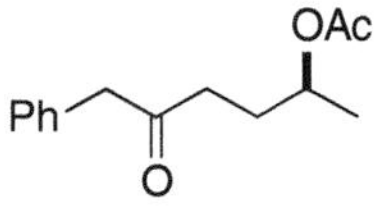	95 (30 min)	91 (3 h)
4	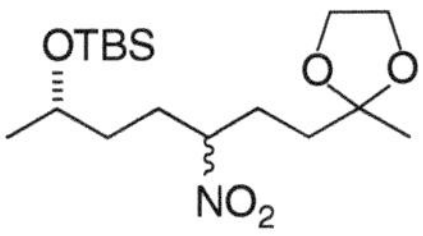	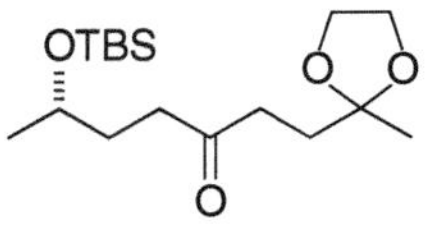	94 (30 min)	96 (3 h)
5	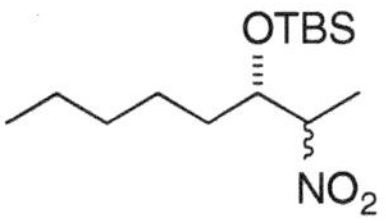	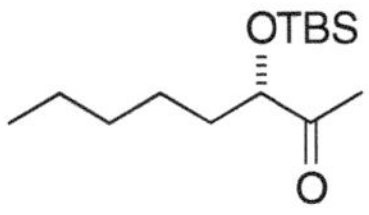	84 (12 h)	90 (12 h)

3.3 Nef Reaction Under Basic Conditions

3.3.1 Method Base1

An original procedure for the Nef reaction under basic conditions has been developed using basic silica gel, readily prepared mixing chromatographic grade silica

Table 3.16 Nef reaction by basic silica gel (representative examples).

Entry	Nitroalkane	Product	Yield (%)
1	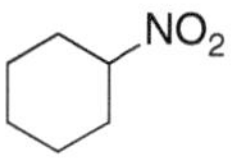	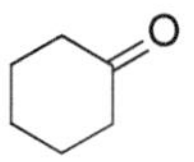	99
2	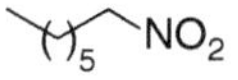	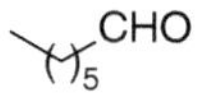	87
3	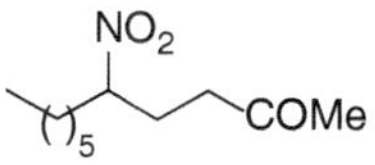	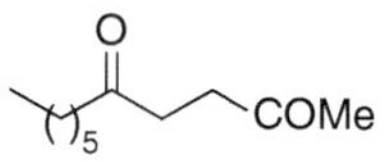	88
4	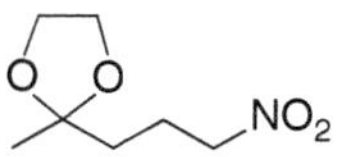	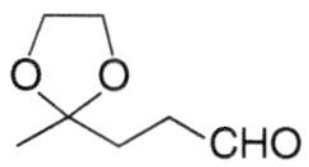	81

Figure 3.1 Intermediate of Nef reaction with basic silica gel.

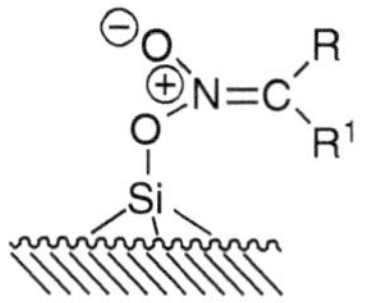

gel with methanolic solution of NaOMe, followed by evaporation and heating at 400 °C for several hours [23]. The method is very simple and consists in a one-pot procedure that involves the embedding the nitroalkane into activated basic silica gel and elution of the resulting carbonyl derivatives. Both primary and secondary nitroalkanes are converted, in excellent yields, into the corresponding aldehydes or ketones (Table 3.16).

The authors supposed that the first step is the formation of the nitronate ion that reacts with the siloxane function of the silica gel, resulting in the structure of Figure 3.1.

This intermediate is then attacked by adsorbed hydroxide anion, resulting in the carbonyl derivative and nitrogen oxide.

3.3.2 Method Base2

DBU, a tertiary amidine base, has been used for the region- and chemoselective conversion of secondary nitro compounds into ketones [24]. In fact, treating 1 equiv of nitroalkane with 2 equiv of DBU, in MeCN, and heating at 60 °C for one to five days, converts secondary nitroalkanes into ketones in satisfactory to good yields, while the primary ones are unreacted. A plausible mechanism, and some selected results, is reported in Table 3.17.

Table 3.17 Nef reaction promoted by DBU (representative examples).

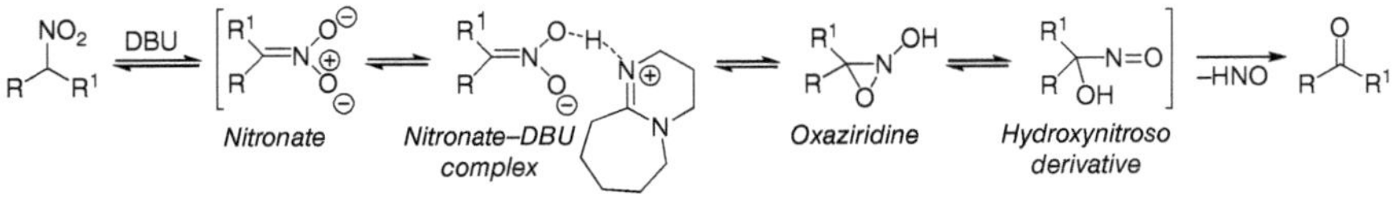

Entry	R	R^1	Yield (%)
1	$Ph(CH_2)_3$	$CH_3(CH_2)_2$	60
2	$MeO_2C(CH_2)_2$	$MeO_2C(CH_2)_4$	65
3	$MeOC(CH_2)_2$	$EtOC(CH_2)_2$	55
4	$HO(CH_2)_5$	$MeO_2C(CH_2)_2$	56
5	$CH_3(OCH_2CH_2O)CCH_2$	$MeO_2C(CH_2)_2$	80
6	$CH_3(CH_2)_3$	$EtOC(CH_2)_2$	59

Under basic condition, the formed nitronate allows a complex with DBU, this complex converts into the oxaziridine and then into a hydroxynitroso derivative. The latter would eliminate hyponitrous acid affording ketones as final products.

3.4 Other Methods for the Nef Reaction

Alternative methods for the Nef reaction have been reported treating nitroalkanes with sodium nitrite or chlorotrimethylsilane or SiO_2–TBD.

3.4.1 Method by $NaNO_2$

Sodium nitrite has demonstrated to be an efficient reagent in order to convert nitroalkanes into carbonyl derivatives [25, 26]. In fact, it has been found that treatment of primary nitroalkanes with $NaNO_2$ under acidic conditions (AcOH), in dimethylsulfoxide as solvent and at room temperature for eight hours, converts [25] primary nitro groups into carboxylic acids.

Table 3.18 presents the possible mechanism and the main carboxylic acids prepared (72–96% yields). Concerning the mechanism, under acidic conditions the formation of the *aci*-form takes place and reacts with nitrosating agent to allow nitrolic acid. Successive transformation of nitrolic acid into hydroxamic acid proceeds via the formation of a reactive nitrile oxide intermediate. The latter, by nucleophilic attack of $NaNO_2$ and hydrolysis, yields carboxylic acids.

The same group, few years later, expanded the use of sodium nitrite to the formation of ketones from secondary nitroalkanes [26]. This was possible because secondary nitroalkanes, in marked contrast to the primary ones, do not need any acid catalysis since the reaction can be easily carried out under neutral conditions.

Table 3.18 Nef reaction promoted by $NaNO_2$/AcOH (representative examples).

Entry	R	Yield (%)
1	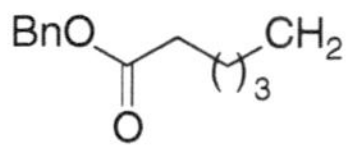	96
2	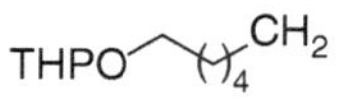	90
3	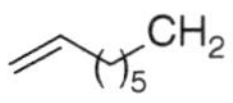	95
4	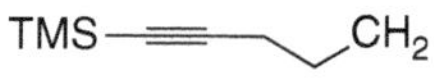	98
5	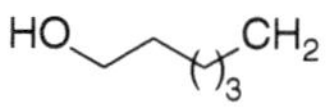	72

The reaction is conducted adding 2 equiv of $NaNO_2$ to a H_2O/DMSO solution of 1 equiv of *sec*-nitroalkane at 65 °C.

3.4.2 Method by Me_3SiCl

The reaction of a variety of *sec*-nitroalkanes (1 equiv) with KH (1 equiv), in dioxane, and with 0.1 equiv of Me_3SiCl, at reflux temperature, give, after 15 hours, ketones in very good yields [27].

Table 3.19 shows the probable mechanism and the main results obtained. Thus, the potassium salt derived from the nitronate structure could decompose upon heating, producing an α-siloxy-α-nitroso intermediate prone to convert into ketones in good yields (62–82%).

The success of this method was then applied, in the same paper, to a new process for the preparation of "poly(1,3-diketones)," an important class of polymers strategic in many industrial fields [28]. As reported in Scheme 3.4, this new way to poly-carbonyls starts with the polymerization of conjugated nitro alkenes in the presence of sodium ethoxide as the initiator. The obtained poly-nitroalkanes are treated with KH/Me_3SiCl, in 1,4-dioxane as solvent, affording poly(1,3-diketones) in very good yields.

Table 3.19 Nef reaction promoted by KH/Me_3SiCl (representative examples).

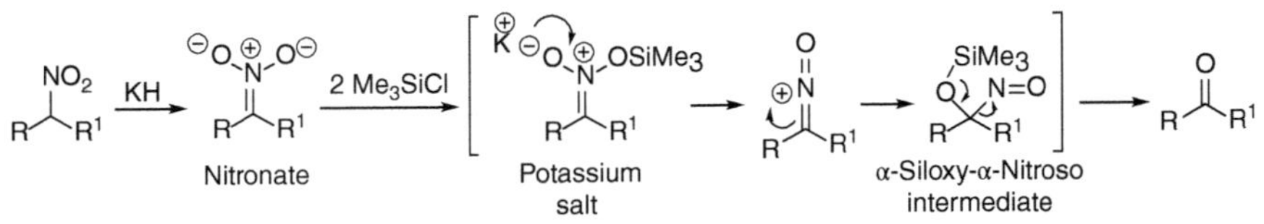

Entry	Nitroalkane	Product	Yield (%)
1	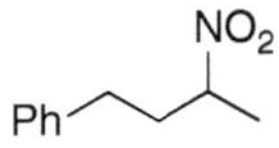	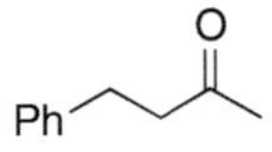	82
2	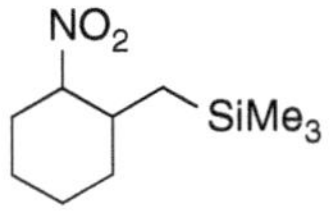	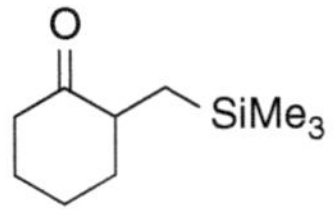	80
3	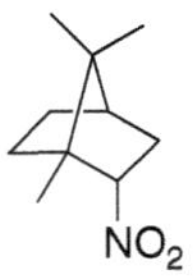	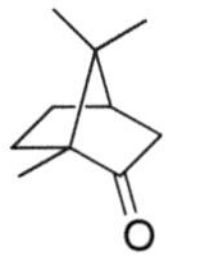	62
4	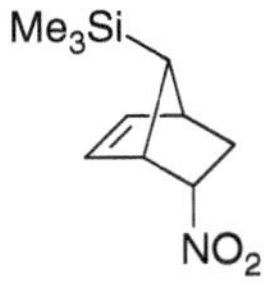	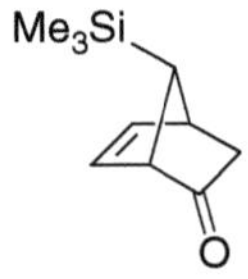	81

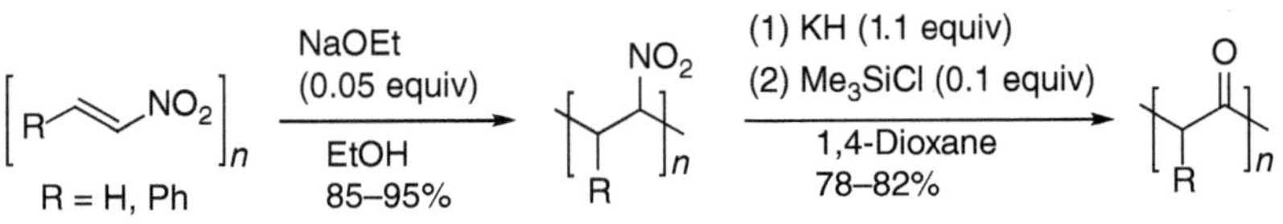

Scheme 3.4 Nef reaction by Me_3SiCl: synthesis of poly(1,3-diketones).

3.4.3 Method by SiO_2/TBD

Secondary nitroalkanes convert into carbonyl with neat silica-supported 1,5,7-triazabicyclo[4.4.0]dec-5-ene (TBD), under neat conditions and at room temperature [29]. The procedure is selective since primary nitro groups are inert to the reaction conditions. As reported in Table 3.20, satisfactory to good yields are obtained in few days.

Table 3.20 Nef reaction promoted by SiO_2–TBD (representative examples).

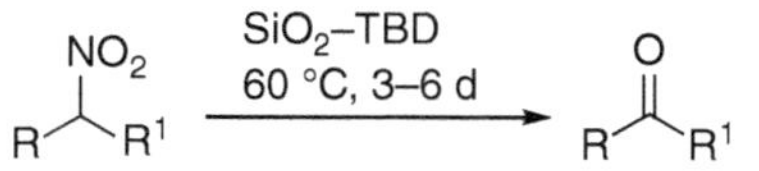

Entry	R	R^1	Yield (%) (reaction time)
1	$Ph(CH_2)_2$	$CH_3(CH_2)_2$	88 (4 d)
2	$CH_3(CH_2)_4$	$MeOC(CH_2)_2$	74 (4 d)
3	$CH_3(CH_2)_{18}$	$CH_3(CH_2)_9$	62 (4 d)
4	$CH_3(CH_2)_2$	$MeO_2C(CH_2)_2$	60 (3 d)
5	$CH_3(CH_2)_8$	CH_3	77 (5 d)

Despite the consistent reaction times, the reaction is very simple and efficient even with compounds containing an additional functional group. However, if primary nitroalkanes are treated, a mixture of unidentified products is obtained.

In addition, the catalyst can be separated and reused, with similar yields, for other two times.

3.5 Synthetic Applications of the Nef Reaction (Representative Examples)

The Nef reaction is one of the most important transformations in organic synthesis and this is why it is often the key step in many synthetic processes.

3.5.1 Solvolytic Methods

The conjugate addition of methylzinc to 1-nitrocyclododecene **1**, in the presence of both $(CuOTf)_2$–C_6H_6 and an appropriate chiral phosphine ligand, under acidic (H_2SO_4) conditions, allows a one-pot process that includes a conjugate addition followed by the Nef reaction with the direct formation of (*R*)-2-methylcyclododecanone **2** (Scheme 3.5) [30].

The hydrolytic Nef procedure has been the key step for the one-pot formation of indoles from nitroalkanes, as reported in Scheme 3.6 [31].

Firstly, the nitroalkanes **4** convert into the corresponding carbonyl derivatives **6** followed by their condensation with the arylhydrazine **3**. The formed hydrazones undergo the Fischer reaction with the direct formation of the indoles **7** in moderate to good yields.

Benzofurans are central intermediates for the preparation of furocoumarins, an important class of molecules due to their importance for the treatment of several skin

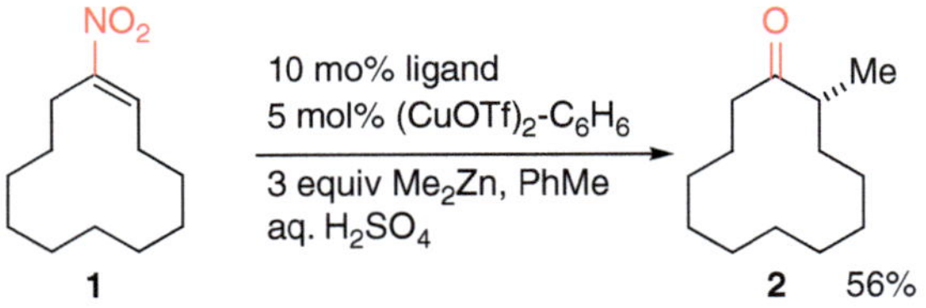

Scheme 3.5 One-pot synthesis of (*R*)-2-methylcyclododecanone.

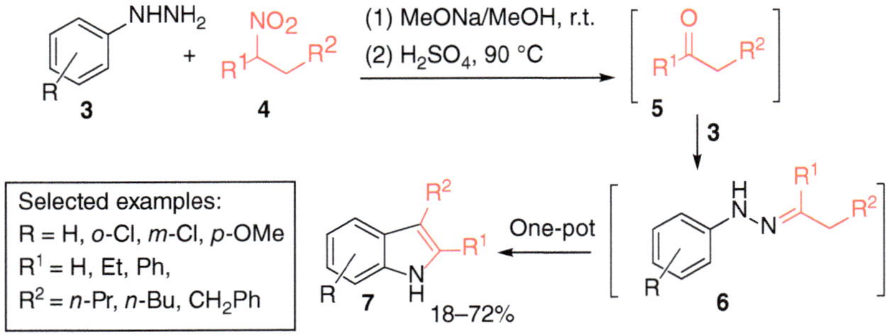

Scheme 3.6 One-pot synthesis of indoles **7**.

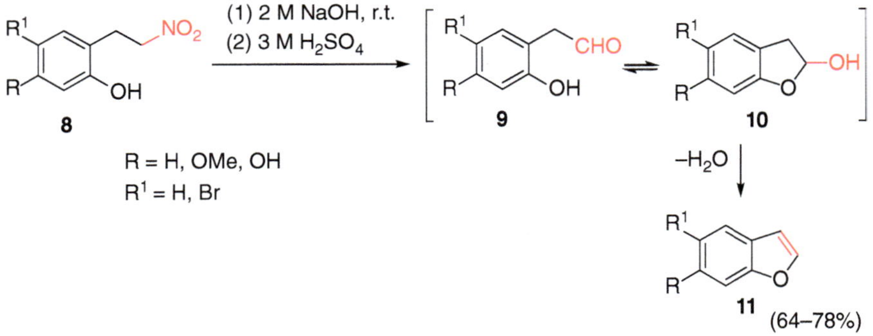

Scheme 3.7 One-pot synthesis of benzofurans **11**.

diseases. An interesting approach to the synthesis of benzofurans has been realized [32] using the acidic Nef reaction, as reported in Scheme 3.7.

The nitro group of the nitrophenols **8**, easily prepared from salicylaldehyde derivatives, is submitted to the Nef reaction allowing the aldehydes **9** that quickly give the formation of the lactols **10** and spontaneous water elimination, yielding (64–78%) the benzofurans **11** in a one-pot process.

3.5.2 Oxidative Methods

Kinamycin antibiotics show benzocarbazoledione structure and a recent synthesis has been developed via the Nef reaction [33]. As reported in Scheme 3.8, the preparation starts with the conjugate addition of aryllithium structure **12** to

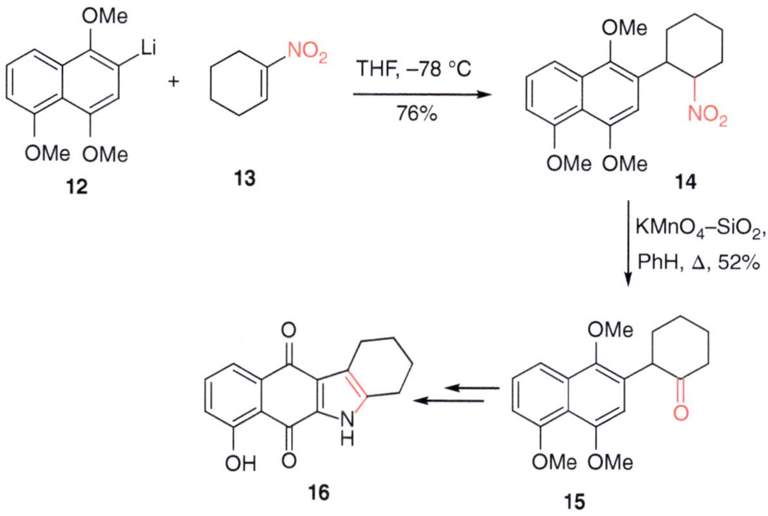

Scheme 3.8 Synthesis of Kinamycin antibiotic analogs **16**.

nitrocyclohexene **13**. The nitro alkane obtained is submitted to the Nef reaction using $KMnO_4$–SiO_2, allowing the cyclohexanone structure **15**. The latter can be transformed, in few synthetic steps, to the target tetracyclic compound **16**.

The oxidative Nef reaction has been used in the synthetic process devoted to the preparation of γ-lactams, as described in Scheme 3.9 [34]. The first step consists in the conjugate addition of ethyl nitroacetates **18** to enone **17**, under Et_3N catalysis. Adducts **19** are submitted to oxidative Nef reaction (H_2O_2/K_2CO_3) yielding the intermediate ketonic compounds **20** that, after in situ hydrolysis, give the α-ketoacids **21**. These acids, under oxidative conditions (H_2O_2/K_2CO_3), give elimination of CO_2, with the consequent formation of the γ-ketoacids **22**.

A successive reaction of **22** with primary amines, under microwave irradiation at 150 W, easily achieves γ-lactams **23** in satisfactory yields.

Sarkomycin shows anticarcinogenic and antibiotic activities, thus the development of synthetic processes for its preparation became of great interest.

An important procedure for the synthesis of Sarkomycin has been realized starting from the conjugate addition of nitromethane to 2-hydroxymethylcyclopentanone **24**, under Triton B catalysis with the formation, in high yields and diastereoselectivity of the adduct **25** (Scheme 3.10) [35].

The latter is submitted to the protection of the sensitive functional groups, and then the obtained compound **26** is treated to Oxone oxidation in aqueous medium leading quantitatively to acid an efficient formation of the corresponding carboxylic acids **27**. Conversion of **27** into Sarkomycin **28** consists in a one-pot process that involves the cleavage of both the protecting groups and final dehydration.

(−)-Cyclophellitol **33** is an efficient glycosidase inhibitor that can be prepared starting from the reaction of Conduriol B **29** with sulphonyl-nitronate **30**, in the presence of $Pd_2(dba)_3$ and an appropriate chiral ligand (Scheme 3.11) [36], affording compound **31** in satisfactory yield and enantiostereomeric excess.

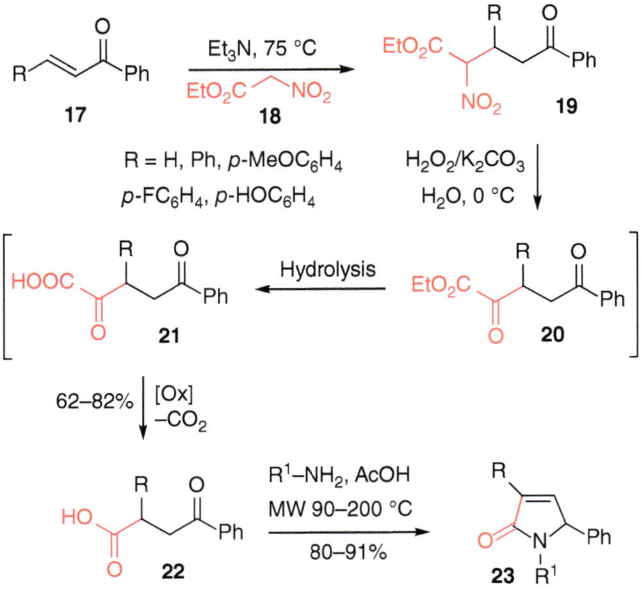

Scheme 3.9 Synthesis of γ-lactams **23**.

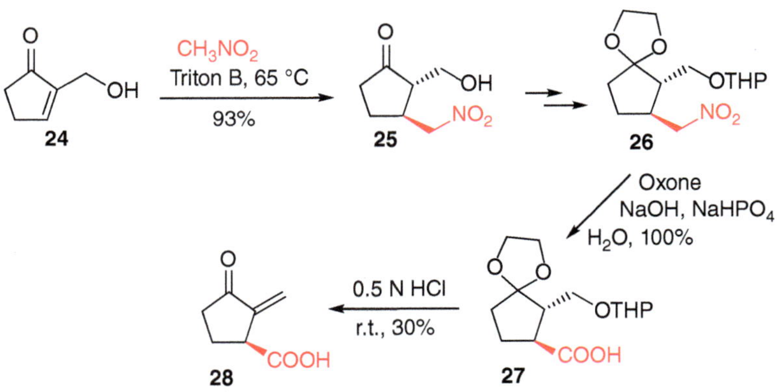

Scheme 3.10 Synthesis of sarkomycin **28**.

Subsequent Nef conversion by dimethyldioxirane under basic conditions (tetramethylguanidine, TMG),allows the acid **32** that is transformed into the target product **33** by further, known synthetic manipulations.

3.5.3 Reductive Methods

Polyfunctionalized molecules bearing labile groups need particular attention when submitted to the nitro to carbonyl conversion. Molecule **35** is an advanced intermediate for the preparation of Manzamine A **35**. Despite the presence of acid-sensitive

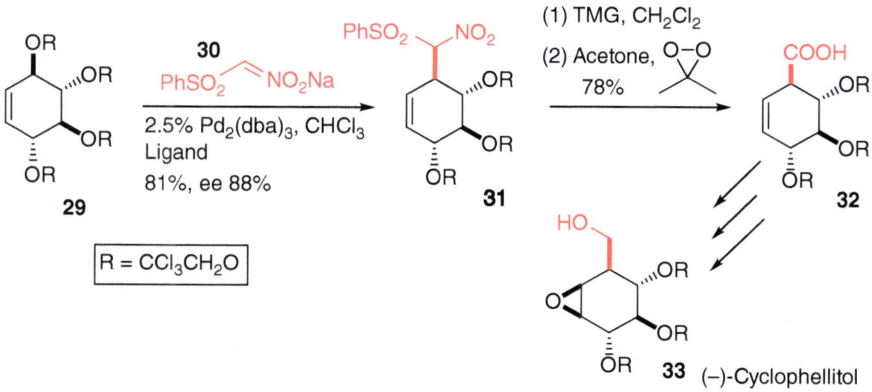

Scheme 3.11 Synthesis of (−)-Cyclophellitol **33**.

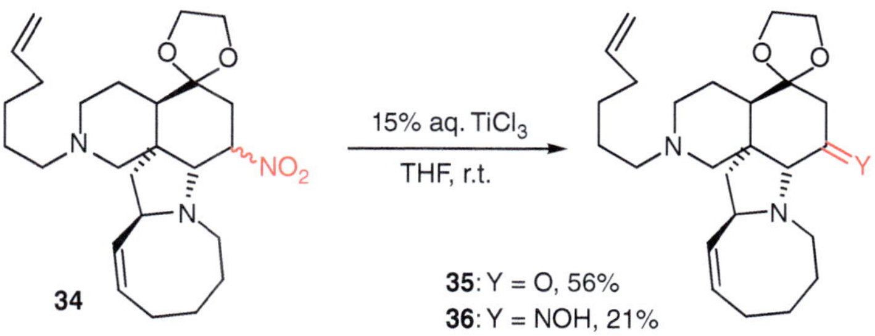

Scheme 3.12 Synthesis of intermediate for Manzamine A **35**.

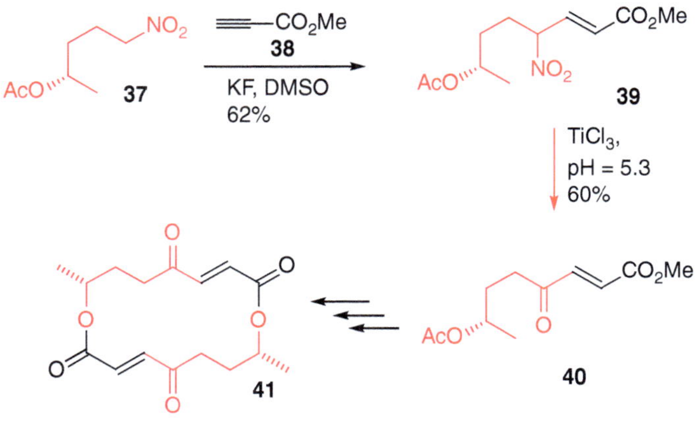

Scheme 3.13 Synthesis of (−)-Pyrenophorin **41**.

groups, the compound **34** undergoes the Nef transformation using $TiCl_3$, at $pH < 1$, gives 35 (56% yield) and traces of **36** (Scheme 3.12) [37].

Pyrenophorin is an antifungal macrolide dilactone that can be obtained from nitroalkane **37** (Scheme 3.13) [38].

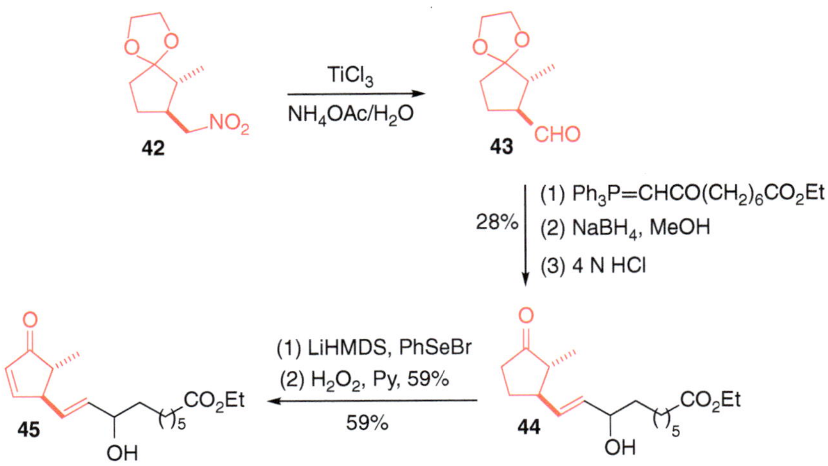

Scheme 3.14 Synthesis of algicidal cyclopentenone **45**.

Conjugate addition of chiral nitroacetate **37** to methylpropiolate **38** allows the adduct **39**. Successive Nef conversion, under reductive conditions ($TiCl_3$), furnishes the enedione **40**. The latter, in few transformations, yields optically active (*R,R*)-(−)-Pyrenophorin **41**.

The polyfunctionalized cyclopentenone **45** is a very important molecule due to its algicidal properties. An efficient synthesis of this target has developed starting from the nitroalkane **42** (Scheme 3.14) [39], easily prepared via conjugate addition of nitromethane to the corresponding cyclopentenone. The Nef conversion of **42**, by $TiCl_3$, allows the aldehyde **43** that is then submitted to the Wittig reaction with the formation of the intermediate **44**. The latter, via elimination of the formed selenyl, under basic conditions, yields (59%) the target structure **45**.

3.5.4 Basic Methods

Lactone **48**, *en route* to the total synthesis of alkaloid (+)-Stemonamide, has been prepared from nitroalkane **47** (Scheme 3.15) [40]. The process starts with the conjugate addition of the nitro ester **47** to enone **46**. The adduct **48** is then submitted to the Nef reaction under basic conditions.

Terpene Isosolanone **53** is an important molecule that has been obtained using reductive Nef reaction as strategic step (Scheme 3.16) [41].

The conjugate addition of nitro-*i*-butane to acrylonitrile is the first step with the formation of the adduct **50**. Successive Nef reaction, using MeONa in dry silica, achieves the cyano-ketone **51**. Wittig olefination and conversion of the cyano group into methyl ketone using MeLi gives **53** as an inseparable diastereomeric mixture ($E/Z = 3:1$).

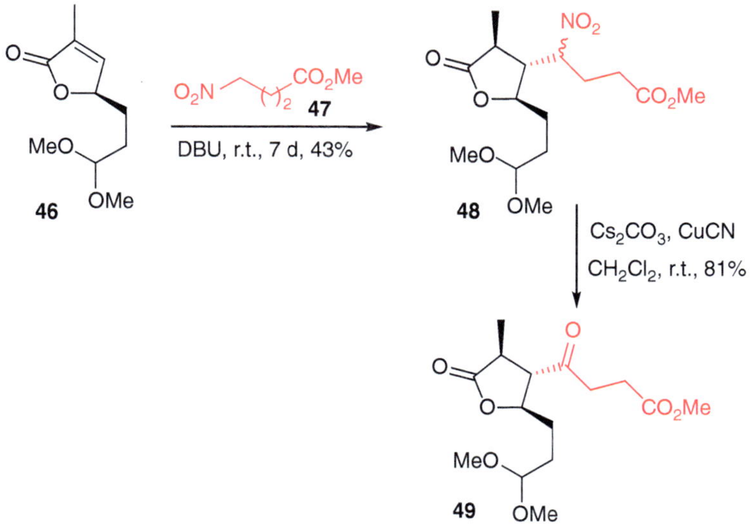

Scheme 3.15 Synthesis of chiral lactone **49**.

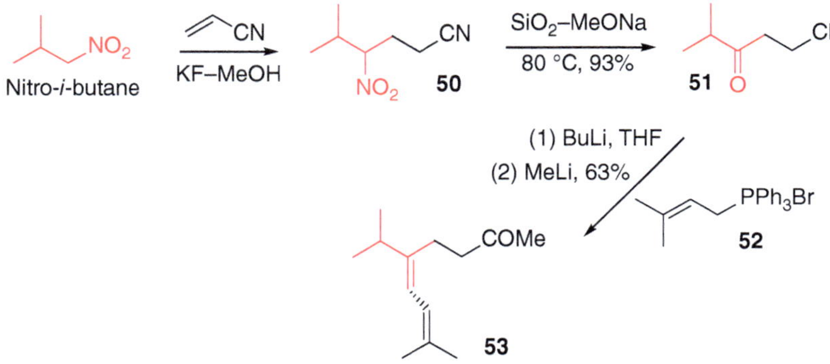

Scheme 3.16 Synthesis of Isosolanone **53**.

3.5.5 $NaNO_2$ Methods

Nitro carbonyl derivative **54** that has been prepared by reaction of nitroalkanes with Baylis–Hilmann adducts undergoes the Nef transformation ($NaNO_2$ in dimethylformamide [DMF]) affording the polycyclic aromatic structure **56** (Scheme 3.17) [42].

It is important to observe that the reaction conditions used for the nitro to carbonyl conversion are even effective for promoting the intramolecular Friedel–Crafts (F–C) process of the intermediate dicarbonyl derivative **55**.

(−)-Botryodiplodin **60** is a natural molecule that exhibits a broad of spectrum of biological activities, including as antibiotic, and this why the development of its synthesis is of great interest. In this context, Alexakis and cowokers reported (Scheme 3.18) [43] an interesting process starting from the conjugate addition of propanal to nitroalkene **57**, in the presence of non-symmetric bis-pyrrolidine

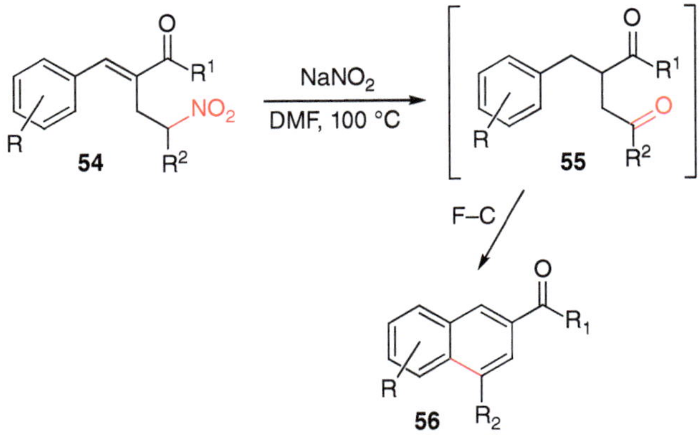

Scheme 3.17 Synthesis of polycyclic aromatic structure **56**.

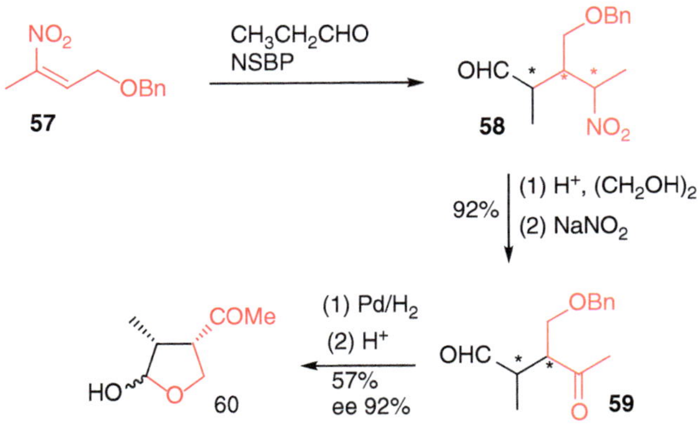

Scheme 3.18 Synthesis of (–)-Botryodiplodin **60**.

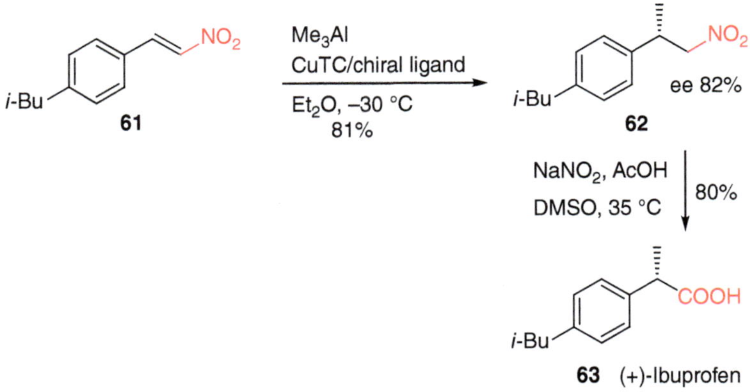

Scheme 3.19 Synthesis of (+)-Ibuprofen **63**.

(NSBP), to give the Michael adduct **58** as a complex mixture of four diastereomers in 92% yield. Nef reaction, promoted by $NaNO_2$, converts the secondary nitro group into ketone **59**, with the consequent presence of just two diastereomers (*syn/anti* = 57 : 43) that can be easily separated with flash chromatography.

Successive hydrogenation and deprotection of the *syn* **59** affords the natural antibiotic **60**. (+)-Ibuprofen **63** is a well-known nonsteroidal *anti*-inflammatory drug (NSAID) that has been conveniently synthesized, few years ago, starting from nitrostyrene **61** (Scheme 3.19) [44].

Conjugate addition of a methyl group, using copper thiophenecarboxylate (CuTC) in the presence of an appropriate chiral ligand, to **61** affords the chiral nitroalkane **62** (ee 82%), prone to be converted into the target compound by Nef reaction under the action of $NaNO_2$/AcOH.

References

1 Nef, J.U. (1894). *Liebigs Ann. Chem.* 280: 264–268.
2 McMurry, J.E., Melton, J., and Padgett, H. (1974). *J. Org. Chem.* 39: 259–260.
3 Olah, G.A. and Gupta, B. (1980). *Synthesis*: 44–45.
4 Kornblum, N., Erickson, A.S., Kelly, W.J., and Henggeler, B. (1982). *J. Org. Chem.* 47: 4534–4538.
5 Steliou, K. and Poupart, M.-A. (1985). *J. Org. Chem.* 50: 4971–4973.
6 Aizpurua, J.M., Oirbide, M., and Palomo, C. (1987). *Tetrahedron Lett.* 28: 5361–5364.
7 Ballini, R. and Petrini, M. (1989). *Tetrahedron Lett.* 30: 5329–5332.
8 Saville-Stones, E. and Lindell, S.D. (1991). *Synlett*: 591–592.
9 Narayana, C., Reddy, N.K., and Kabalka, G.W. (1992). *Synth. Commun.* 22: 2587–2592.
10 Barlogh-Hergovich, Ė., Kaizer, J., and Speier, G. (1996). *Chem. Lett.*: 573–574.
11 Tokunaga, Y., Ihara, M., and Fukumoto, K. (1997). *J. Chem. Soc., Perkin Trans. 1*: 207–209.
12 Ceccherelli, P., Curini, M., Marcotullio, M.C. et al. (1998). *Synth. Commun.* 28: 3057–3064.
13 Adam, W., Mąkosza, M., Saha-Möller, C.R., and Zhao, C.-G. (1998). *Synlett*: 1335–1336.
14 Shahi, S.P. and Vankar, Y.D. (1999). *Synth. Commun.* 29: 4321–4325.
15 Nikalje, M.D., Ali, I.S., Dewkar, G.K., and Sudalai, A. (2000). *Tetrahedron Lett.* 41: 959–961.
16 Pradhan, P.K., Dey, S., Jaisankar, P., and Giri, V.S. (2005). *Synth. Commun.* 35: 913–922.
17 Bortolini, O., De Nino, A., Garofalo, A. et al. (2010). *Synth. Commun.* 40: 2483–2487.
18 Sedelmeier, J., Ley, S.V., Baxendale, I.R., and Baumann, M. (2010). *Org. Lett.* 12: 3618–3621.
19 McMurry, J.E. and Melton, J. (1973). *J. Org. Chem.* 38: 4367–4373.
20 Kirchhoff, R. (1976). *Tetrahedron Lett.* 17: 2533–2534.

21 Burés, J. and Vilarrasa, J. (2008). *Tetrahedron Lett.* 49: 441–444.
22 Durchschein, K., Ferreira-da Silva, B., Wallner, S. et al. (2010). *Green Chem.* 12: 616–619.
23 Keinan, E. and Mazur, Y. (1977). *J. Am. Chem. Soc.* 99: 3861–3862.
24 Ballini, R., Bosica, G., Fiorini, D., and Petrini, M. (2002). *Tetrahedron Lett.* 43: 5233–5235.
25 Matt, C., Wagner, A., and Mioskowski, C. (1997). *J. Org. Chem.* 62: 234–235.
26 Gissot, A., N'Gouela, S., Matt, C. et al. (2004). *J. Org. Chem.* 69: 8997–9001.
27 Hwu, J.R., Josephrajan, T., and Tsay, S.-C. (2006). *Synthesis* 19: 3305–3308.
28 (a) Holdcroft, S. (2001). *Adv. Mater.* 13: 1753–1765. (b) Jager, E.W.H., Smela, E., and Ingañas, O. (2000). *Science* 290: 1540–1545.
29 Ballini, R., Fiorini, D., Maggi, R. et al. (2006). *Synlett* 12: 1849–1850.
30 Luchaco-Cullis, C.A. and Hoveyda, A.H. (2002). *J. Am. Chem. Soc.* 124: 8192–8193.
31 Simoneau, C.A., Strohl, A.M., and Ganem, B. (2007). *Tetrahedron Lett.* 48: 1809–1811.
32 Zhang, B.-L., Wang, F.-D., and Yue, J.-M. (2006). *Synlett*: 567–570.
33 Barcia, J.C., Otero, J.M., Estévez, J.C., and Estévez, R.J. (2007). *Synlett*: 1399–1402.
34 Aginagalde, M., Bello, T., Masdeu, C. et al. (2010). *J. Org. Chem.* 75: 7435–7438.
35 Kar, A. and Argade, N.P. (2005). *Synthesis*: 1234–1236.
36 Trost, B.M., Patterson, D.E., and Hembre, E.J. (2001). *Chem. Eur. J.* 7: 3768–3775.
37 Jakubes, P., Hawkins, A., Felzmann, W., and Dixon, D.J. (2012). *J. Am. Chem. Soc.* 134: 17482–17485.
38 Kalita, D., Khan, A.T., Saikia, A.K. et al. (1998). *Synthesis*: 975–976.
39 Crombie, L. and Heavers, A.D. (1992). *J. Chem. Soc., Perkin Trans. 1*: 2683–2687.
40 Li, Z., Zhang, L., and Qiu, F.G. (2014). *Asian J. Org. Chem.* 3: 52–54.
41 Lee, W.Y., Jang, S.Y., Chae, W.K., and Park, O.S. (1993). *Synth. Commun.* 23: 3037–3046.
42 Basavaiah, D. and Reddy, D.M. (2014). *RCS Adv.* 4: 23966–23970.
43 Andrey, O., Vidonne, A., and Alexakis, A. (2003). *Tetrahedron Lett.* 44: 7901–7904.
44 Polet, D. and Alexakis, A. (2005). *Tetrahedron Lett.* 46: 1529–1532.

4

Nitroaldol (Henry) Reaction

Nitroalkanes bearing α-hydrogen (primary and secondary) are valuable sources of stabilized carbanions since the high electron-withdrawing effect of the nitro group provides an outstanding enhancement of the hydrogen activity at the α-position (CH_3NO_2, $pK_a = 10$). Thus, under basic conditions, a carbanion can be generated with the resultant formation of the corresponding nitronate anion that acts as carbon nucleophile with a variety of electrophiles. Consequently, the utilization of carbanion, stabilized by the electron withdrawing effect of the nitro group, occupies a central position in organic synthesis for the formation of new carbon–carbon bond, including the nitroaldol (Henry) reaction (Scheme 4.1).

The nitroaldol (Henry) reaction is a very old reaction, discovered more than one century ago [1], in which primary or secondary nitroalkanes react, under basic conditions, with carbonyl derivatives (mainly aldehydes) giving access to the formation of nitroalkanols. However, despite the old age of the Henry reaction, it remains one of the most important and strategic opportunity in organic synthesis for the formation of new C—C bonds, so its use grows year by year and a number of related reviews have been published [2a, 3].

The nitroaldol reaction usually allows a diastereomeric mixture of β-nitro alkanols. The lack of stereoselectivity is due to the reversibility of the nitroaldol reaction and the easy epimerization at the nitro-substituted C-atom.

Anyway, the obtained β-nitro alcohols possess two stereogenic centers (except for nitromethane [$R = R^1 = H$] because only one stereogenic center can be generated); however the formed nitroalkanols are mainly employed in three different applications (Figure 4.1): (i) dehydration to conjugate nitroalkenes, (ii) oxidation to α-nitro ketones, and (iii) reduction to the corresponding β-amino alcohols.

So, it is clear that diastereoselective syntheses of nitroalkanols are needed just for specific purposes; in addition, these methods are highly laborious, allow moderate yields, and are not suitable for large scale applications.

Thus, in this chapter we report, separately, the general procedures and the diastereo- or enantioselective ones.

Nitroalkanes: Synthesis, Reactivity, and Applications, First Edition.
Roberto Ballini and Alessandro Palmieri.

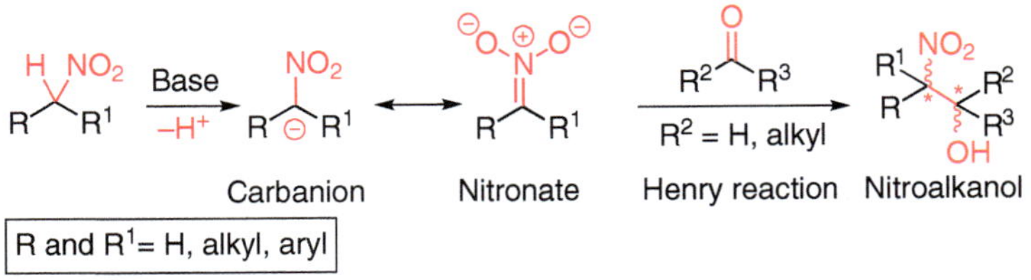

Scheme 4.1 Nitroaldol (Henry) reaction.

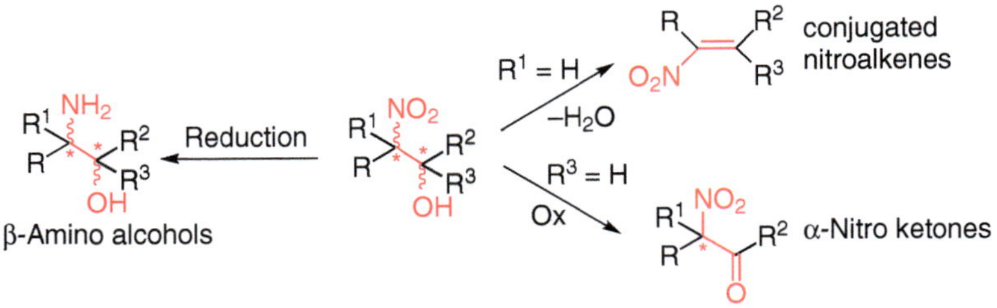

Figure 4.1 Transformation of β-nitro alcohol.

4.1 General Catalysts and Promoters

The classical nitroaldol reaction is routinely performed under homogeneous conditions using bases such as alkali metal hydroxides, carbonates, bicarbonates, alkoxides, calcium and barium hydroxides, and magnesium or aluminum ethoxides. In addition, organic bases such as amines, ammonium acetate, and fluoride have proved to be useful [2]. However, the aforementioned procedures suffer of important limitations such as low yields, poor chemo-selectivity, dehydration process with the formation of nitroalkenes as by-products, retroaldol reaction with the regeneration of the starting materials, or a mixture of products.

4.1.1 Heterogeneous Catalysts and Promoters

Successively, in order to minimize the aforementioned drawbacks and with the scope to develop sustainable procedures, heterogeneous catalysts {basic alumina [4], alumina-supported KF [5], or layered [6] $Zr(KOPO_3)_2$} under neat conditions have been introduced. These new approaches would be particularly beneficial in industrial processes.

As improvement of the heterogeneous procedures Amberlyst A21, a macroreticular resin with $-N(CH_3)_2$ as functional group has been proposed as promoter for nitroaldol reactions under neat conditions [7]. Amberlyst A21 has demonstrated to be far superior as general catalyst because it allows high yields (70–95%), in short reaction times (few hours), with a large variety of substrates including short and long chain nitroalkanes (primary and secondary) and a variety of aldehydes (Table 4.1). Moreover, the mildness of Amberlyst A21 is demonstrated by the stability of the

Table 4.1 Henry reaction promoted by Amberlyst A21 (representative examples).

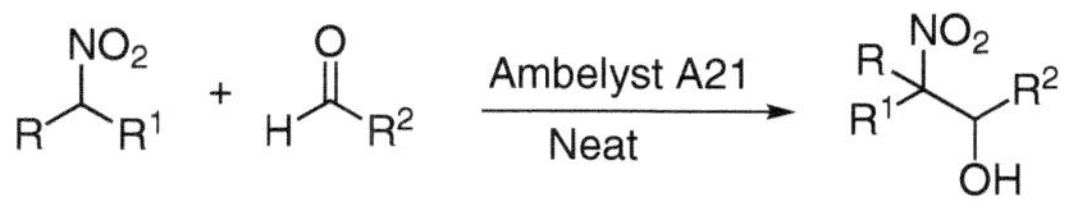

Entry	R	R^1	R^2	Yield (%)
1	$CH_3CH(OH)CH_2$	H	Me	70
2	$HO(CH_2)_6$	H	Me	70
3	CH_3	H	*sec*-Butyl	80
4	CH_3	CH_3	$CH_3(CH_2)_5$	80
5	CH_3	CH_3	(Z)-$CH_3CH_2CH{=}CH(CH_2)_5$	80
6	$CH_3(CH_2)_2$	H	$CH_3(CH_2)_5$	83
7	$CH_3CO(CH_2)_2$	H	$Ph(CH_2)_2$	95
8	$CH_3CO_2(CH_2)_2$	H	2-Furanyl	76
9	$-(CH_2)_5-$		$Ph(CH_2)_2$	95
10	Br	H	$Ph(CH_2)_2$	79

Table 4.2 Henry reaction promoted by SiO_2/MW (representative examples).

Entry	Nitroalkane	Aldehyde	Yield (%) of nitroalcohol
1	$CH_3CH_2NO_2$	PhCHO	75
2	CH_3NO_2	p-$NO_2C_6H_4CHO$	79
3	$CH_3CH_2NO_2$	p-BrC_6H_4CHO	70
4	$CH_3CH_2NO_2$	Furfural	56

bromohydrin (entry 10) to the epoxide formation and the avoidance of dehydration process of nitroalcohols to nitroalkenes even if aromatic aldehydes are employed.

The reaction was also tested in the presence of solvents (Et_2O, CH_2Cl_2, THF); however no substantial changes of the yields are observed.

A quick heterogeneous method for the Henry reaction was then found by treatment of primary nitroalkanes with aromatic aldehydes on surface of SiO_2 (<200 mesh), promoted by microwave (MW) (600 W, three to four minutes) under neat conditions (Table 4.2) [8]. The catalyst can be recovered and reused after activation.

A simple and easy procedure has been prepared through the nitroaldol reaction catalyzed with stoichiometric amount of powered KOH, in dry media [9]. The products are obtained (Table 4.3) in good yields and in few minutes. It is important to note the simple work-up needed at the end of the reaction since the obtained nitroalkanols are simply dissolved in water and then extracted with Et_2O.

Table 4.3 Henry reaction promoted by powered KOH (representative examples).

Entry	Nitroalkane	Aldehyde	Yield (%) of nitroalcohol
1	$(CH_3)_2CHNO_2$	$CH_3(CH_2)_3CHO$	65
2	$BrCH_2NO_2$	PhCHO	99
3	$BrCH_2NO_2$	$CH_3(CH_2)_3CHO$	70
4	$CH_3CH_2NO_2$	$CH_3(CH_2)_2CHO$	75
5	CH_3NO_2	Furfural	64

Table 4.4 Henry reaction promoted by Mg–Al (3 : 1) hydrotalcite (representative examples).

Entry	Nitroalkane	Aldehyde	Yield (%) of nitroalcohol	Threo/erytro
1	$CH_3CH_2NO_2$	*m*-$NO_2C_6H_4CHO$	95	12.5 : 1
2	$CH_3(CH_2)_2NO_2$	*o*-ClC_6H_4CHO	89	1.53 : 1
3	$CH_3CH_2NO_2$	PhCHO	87	3.25 : 1
4	$CH_3CH_2NO_2$	*o*-ClC_6H_4CHO	88	100 : 1
5	$CH_3CH_2NO_2$	*p*-$NO_2C_6H_4CHO$	84	100 : 1

N,N-Diethylpropylamine supported on amorphous silica (KG-60-NEt_2) was found as very efficient and reusable solid catalyst for the Henry reaction under solvent free conditions [10]. Simple nitroalkanols are prepared in very good yields (80–95%).

Mg–Al hydrotalcite, upon thermal calcination at 450 °C, converts into active mixed oxides with strong catalytic activity, high surface area, and a large number of OH groups. Thus, Mg–Al hydrotalcite has been found an excellent catalyst for the Henry reaction [11] assuring the formation of nitroalkanols (Table 4.4) in good yields (usually >80%). Moreover, high chemoselectivity is observed since *threo* form is predominant.

Further results of the Henry reaction of nitromethane with a variety of aldehydes, catalyzed by available Mg–Al 2 : 1 hydrotalcite, were reported few years later [12].

Interesting studies concerning the heterogeneous Henry reaction were reported testing nitromethane with EtCHO promoted with different solid bases [13]. The catalytic activities among alkaline earth oxides were found in the order: MgO > CaO > SrO > BaO. This order is opposite to those of the basic strengths of the alkaline earth oxides.

Commercial ISOLUTE® Si-carbonate (silica trimethylammonium carbonate [Si-TMS$(CO_3^{-2})_{0.5}$]) is a bond equivalent of tetramethylammonium carbonate, which was traditionally used as quaternary anion-exchanger for quenching, neutralizing, and desalting organic mixture and as organic acid scavenging for fast purification of amide and biaryl libraries. During studies into the chemistry

Table 4.5 Henry reaction promoted by ISOLUTE Si carbonate (representative examples).

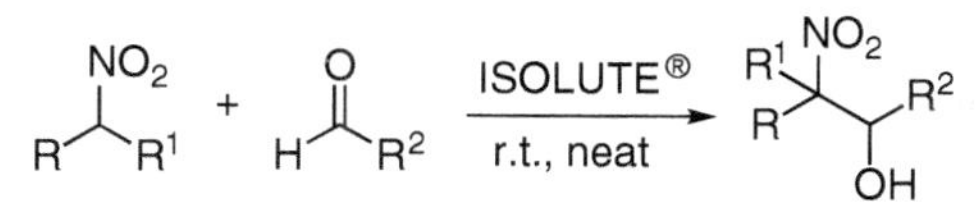

Entry	R	R^1	R^2	Yield (%)	Reaction time (h)
1	H	$Ph(CH_2)_2$	$CH_3(CH_2)_4$	82	5
2	H	$CH_3CO_2(CH_2)_4$	Ph	69	16
3	H	$CH_3(CH_2)_3$	$CH_3(CH_2)_4$	83	8
4	H	Br	$CH_3(CH_2)_3$	71	9
5	H	$CH_3(CH_2)_2$	Ph	78	16
6	CH_3	CH_3	CH_3CH_2	77	16

Table 4.6 Henry reaction promoted by PEI (representative examples).

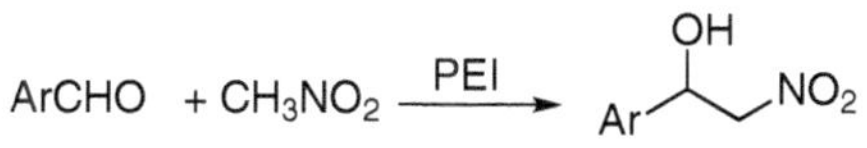

Entry	Ar	Yield (%)	Reaction time (min)
1	*p*-$NO_2C_6H_4$	95	20
2	*p*-BrC_6H_4	78	180
3	*p*-FC_6H_4	83	180
4	*o*-$NO_2C_6H_4$	99	20
5	*p*-C_5H_4N	99	20

of nitroalkanes, it has been found that ISOLUTE Si-carbonate is an efficient base prone to forms nitronates from nitroalkanes under solvent-free conditions, giving access to a new sustainable procedure for the nitroaldol reaction [14].

As reported in Table 4.5, good results have been obtained by the reaction of a variety of primary and secondary nitroalkanes with aromatic and aliphatic aldehydes, in few hours.

Hyperbranched polyethylenimine (PEI) derivatives were successfully employed as catalyst for the Henry reaction [15]. A variety of aromatic aldehydes were reacted with an excess of nitromethane that acts even as solvent. The products are achieved in excellent yields within short reaction times (Table 4.6).

Polymer supported catalysts are considered as borderline class of catalysts that retain the advantages of homogeneous catalysts while securing the ease of recovery by simple filtration and work-up of homogeneous systems. Additionally, such catalysts are less hygroscopic due to the long polymer backbone. Moreover, polymer supported catalysts can be recovered (filtration) and reused several times, reducing

Table 4.7 Henry reaction promoted by PS-DMAP (representative examples).

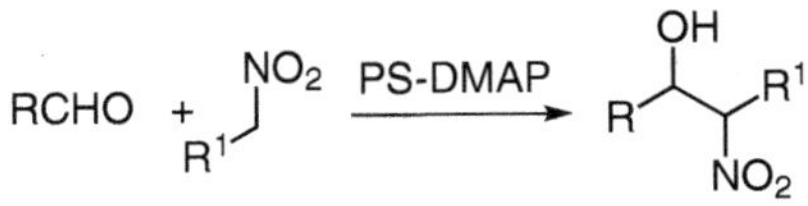

Entry	R	R^1	Yield (%)	Reaction time (min)
1	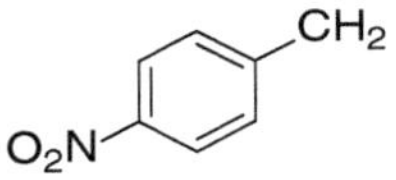	CH_3CH_2	100	25
2	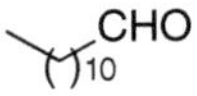	CH_3CH_2	89	45
3	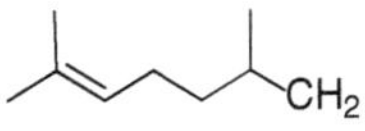	CH_3CH_2	85	45
4	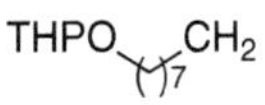	CH_3CH_2	85	40
5	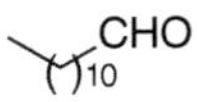	$(CH_3)_2CH$	89	50
6	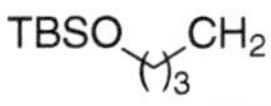	CH_3CH_2	88	45

the operational cost. In this context, polymer supported DMAP (PS-DMAP) has been successfully tested as catalyst of the nitroaldol reaction, under solvent free conditions [16]. As reported in Table 4.7, a variety of products are achieved in near quantitative yields and in very short reaction times.

The catalyst can be recovered (filtration) and reused at least five times without loss of the activity. Moreover, the mild reaction conditions grant high chemoselectivity since a number of functionalities can be preserved.

A simple and cheap method for the Henry reaction has been recently developed (Bosica and Polidano) [17] using 10 mol% of K_2CO_3, at 60 °C, under neat conditions. Satisfactory yields are recovered with both primary and secondary nitroalkanes and a collection of aldehydes (Table 4.8).

Guanidine **I** is a strong base used as catalyst in many aspects of the organic synthesis, recently it has been found that **this base** catalyzes the Henry reaction of $MeNO_2$ with aromatic and aliphatic aldehydes, affording quantitative yields of products (Table 4.9) [18].

The Henry reactions can be accelerated adding silica gel. Thus, following this result found very recently, a considerable array of nitroalcohols has been prepared under Et_3N/SiO_2 catalysis [19]. As reported in Table 4.10, the yields are excellent with short reaction times.

Recently, it has been reported that the synthesis of KI loaded Fe_3O_4 nanoparticles ($KI@Fe_3O_4{\cdot}NPs$) and its applications as catalyst for the Henry reaction [20]. In addition, the activity of the catalyst, varying the amount of KI loading, was also investigated in order to find the best reaction conditions.

Table 4.8 Henry reaction promoted by K_2CO_3 (representative examples).

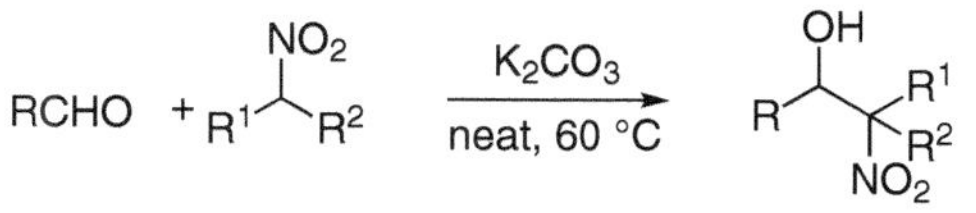

Entry	R	R^1	R^2	Yield (%)	Reaction time (h)
1	CH_3	CH_3	H	75	4.5
2	Ph	CH_3	H	60	4.5
3	$CH_3(CH_2)_3$	CH_3CH_2	H	60	4.5
4	$CH_3(CH_2)_4$	CH_3CH_2	H	72	8
5	*p*-$NO_2C_6H_4$	$(CH_3)_2CH$	H	66	4

Table 4.9 Henry reaction promoted by guanidine (**I**) (representative examples).

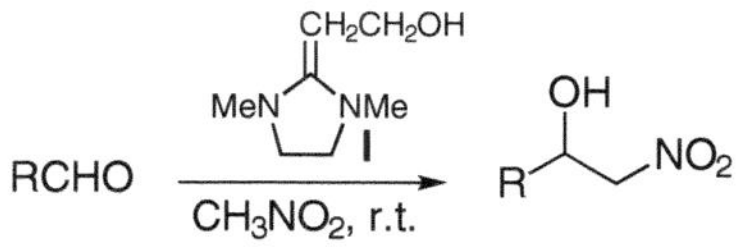

Entry	R	Yield (%)	Reaction time (min)
1	*m*-$NO_2C_6H_4$	98	5
2	*o*-BrC_6H_4	91	2
3	*o*-$MeOC_6H_4$	80	2
4	Ph	83	1
5	Furfural	88	1
6	$CH_3(CH_2)_6$	73	2
7	Cyclohexyl	85	2

As reported in Table 4.11, near quantitative yields are observed using of $MeNO_2$ or $EtNO_2$ with a large number of aldehydes (aromatic and aliphatic).

After completion of the reaction, the catalyst is separated applying a permanent magnet that was successively washed with EtOAc and dried at 95 °C, before being used again in a new reaction for at least five times with the same results.

As reported in Chapter 2, nitroalkanes can be prepared nitrating the corresponding haloderivatives, thus this result has been the key point for the realization of an unprecedented one-pot synthetic process [21] for the Henry reaction, starting from primary halo-compounds and aldehydes, in presence of both IRA-402 nitrite [22] and Amberlyst A21 as heterogeneous catalysts, and dimethyl formamide (DMF) as solvent. In fact, mixing in the same flask 1 equiv of primary haloalkane and aldehyde, the presence of IRA-402 nitrite converts the halo group into nitroalkanes

Table 4.10 Henry reaction promoted by Et_3N/SiO_2 (representative examples).

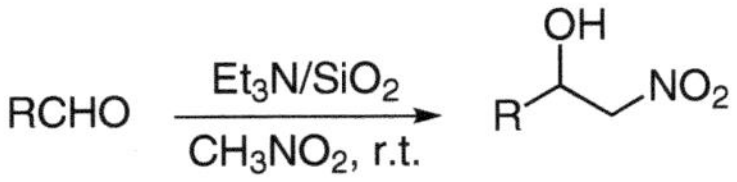

Entry	R	Yield (%)	Reaction time (min)
1	$Ph(CH_2)_2$	99	2
2	$CH_3(CH_2)_5$	95	3
3	Cyclohexyl	96	6
4	$PhCH(CH_3)$	96	5
5	$(CH_3)_3C$	83	96
6	PhCH=CH	83	3
7	Cyclohexyl	85	2

Table 4.11 Henry reaction promoted by $KI@Fe_3O_4$·NPs (representative examples).

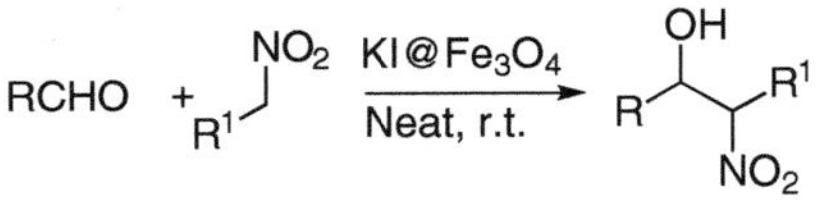

Entry	R	R^1	Yield (%)	Reaction time (min)
1	*p*-BrC_6H_4	H	97	60
2	*o*-$NO_2C_6H_4$	H	96	60
3	*o*-HOC_6H_4	H	92	74
4	$(CH_3)_2CH(CH_2)_6$	H	84	92
5	*p*-$CH_3C_6H_4$	CH_3	77	115
6	*p*-NCC_6H_4	CH_3	94	80
7	$CH_3(CH_2)_{10}$	CH_3	89	87

[22], followed by in situ nitroaldol reaction in the presence of Amberlyst A21 [7] (Scheme 4.2).

The procedure is mild and avoids complicate work-up and complex purification steps. In addition, the amount of waste and energy consumption is strongly reduced with respect to the standard procedures with important advantages from ecological point of view.

4.1.2 Green Solvents

The grown interest for sustainable procedures in organic synthesis has been one of the main goals in the recent years. The need to use less toxic and environmentally

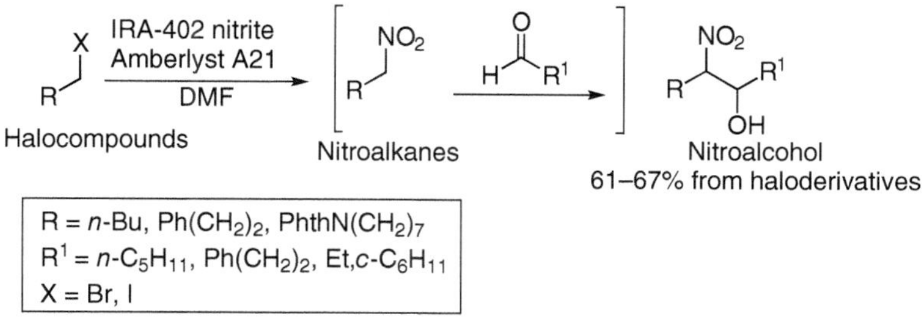

Scheme 4.2 One-pot nitroaldol reaction from haloderivatives.

Table 4.12 Henry reaction promoted by NaOH/CTACl/H_2O (representative examples).

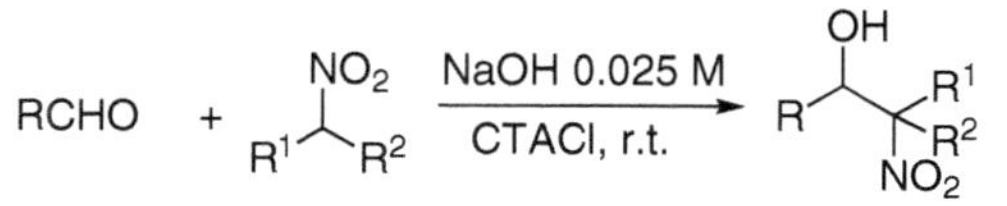

Entry	R	R^1	R^2	Yield (%)	Reaction time (min)
1	H	H	Ph	70	2
2	H	CH_3CH_2	Ph	71	2
3	H	$CH_3(CH_2)_2$	$Ph(CH_2)_2$	85	5
4	H	$THPOCH_2$	$Ph(CH_2)_2$	85	3
5	H	$CH_3(CH_2)_4$	Cyclohexyl	85	2
6	H	$CH_3OC(CH_2)_2$	$CH_3(CH_2)_2CH(CH_3)$	85	2
7	$-(CH_2)_5-$		CH_3CH_2	85	3

compatible materials can be easily performed by the right choice of the solvent. For this purpose, water and ionic liquids (ILs) have been tested in the Henry reaction.

4.1.2.1 Nitroaldol Reaction in Water

The aqueous medium with respect to organic solvent is less expensive, less dangerous, and environment-friendly, while it allows the control of pH and the use of microaggregates such as surfactants. Generally, the low solubility of most reagents in water is not an obstacle to the reactivity while, on the contrary, is reduced with the use of co-solvents.

Two decades ago the first procedure for the nitroaldol reaction in aqueous media [23], using a stoichiometric amount of nitroalkane and aldehyde, in NaOH 0.025 M, and in the presence of cetyltrimethylammonium chloride (CTACl) as cationic surfactant and at room temperature was developed. As reported in Table 4.12, the yields are from good to excellent with aliphatic aldehydes, while slightly decrease with the aromatic ones.

Table 4.13 Henry reaction promoted by CTAOH (representative examples).

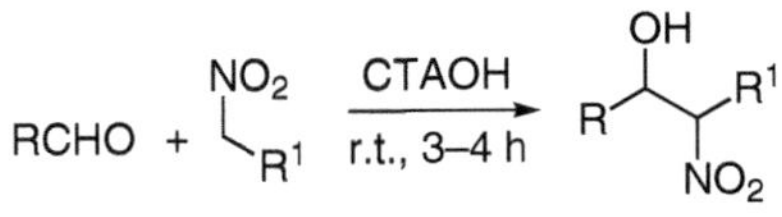

Entry	R	R^1	Yield (%)
1	$Ph(CH_2)_2$	$CH_3(CH_2)_5$	75
2	$CH_3(CH_2)_4$	$CH_3O_2C(CH_2)_4$	77
3	$CH_3(CH_2)_2$	Cyclopentyl	83
4	(*E*)-$CH_3(CH_2)_4CH{=}CHCH_2$	$Ph(CH_2)_2$	85
5	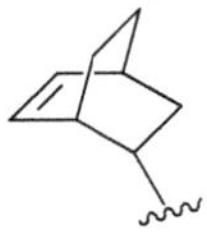	$CH_3(CH_2)_3$	85

Contrary to other methods, the success of this procedure is independent from the ratio catalyst/substrates and does not need of (i) long reaction times, (ii) tedious preparation of the catalyst, (iii) large excess of the nitroalkane, or (iv) severe reaction conditions that are too cumbersome for large-scale preparations. Additionally, the very mild reaction conditions prevent the typical side reactions such as retro-aldol or dehydration. Compared with the conventional methods, the procedure gives better yields in shorter reaction times.

As improvement of the aforementioned approach, more recently it has been reported that the nitroaldol reaction can be performed, simply, using catalytic amount of aqueous cetyltrimethylammoniumhydroxide (CTAOH, 10% water solution) [24]. As reported in Table 4.13, products are obtained in good yields and with short reaction times (three to four hours). Moreover, high chemoselectivity is observed.

Thus, this approach represents an improved, cheap, and ecological process.

In the same years one more report of the Henry reaction in water, catalyzed with Et_3N, has been published [25].

During studies concerning the application of $Ba(OH)_2$ as base catalysis in aqueous medium, it has been found that this base catalyzes the nitroaldol reaction allowing high yields of the adducts, at room temperature and in very short reaction times (Table 4.14) [26].

Although the nitroalkanes tested are just $MeNO_2$ and $EtNO_2$, the procedure seems to be extendable to other substrates. It was fruitful to notice that water, as a green solvent, drastically accelerated the reaction rate.

Bora and Bez reported [27] the results of the Henry reaction performed in aqueous media, at neutral pH, without the need of any catalyst and in the presence of *N,N,N,N*,-cetyltrimethylammonium bromide (CTAB) as phase transfer reagent. Although ketones are unreactive under these conditions, the method works well

Table 4.14 Henry reaction promoted by $Ba(OH)_2$ (representative examples).

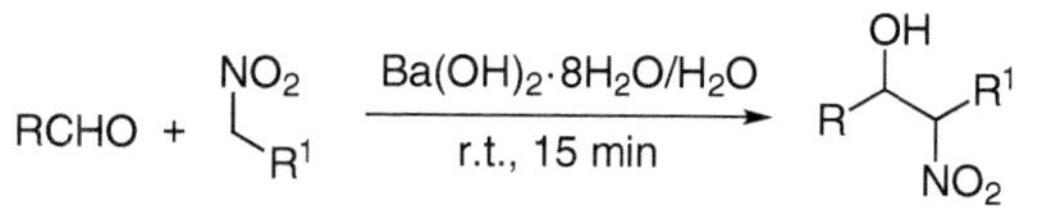

Entry	R	R^1	Yield (%)
1	p-$NO_2C_6H_4$	H	81
2	p-NCC_6H_4	H	83
3	Furfural	H	82
4	2-Pyridyl	H	96
5	$(CH_3)_2CH$	H	82
6	$CH_3(CH_2)_2$	H	87
7	p-$NO_2C_6H_4$	CH_3	83

Table 4.15 Henry reaction promoted by CTAB/phosphate buffer/H_2O, pH = 7 (representative examples).

Entry	R	R^1	Yield (%)	Reaction time (h)	*syn/anti*
1	p-$NO_2C_6H_4$	H	95	14	
2	p-ClC_6H_4	H	87	18	
3	$CH_3(CH_2)_4$	H	90	12	
4	2-Pyridyl	H	82	40	
5	$Ph(CH_2)_2$	H	60	40	
6	$CH_3CH_2O_2C$	H	90	20	
7	p-$CH_3OC_6H_4$	CH_3	94	30	94 : 6
8	Ph	CH_3	86	50	73 : 27
9	m-$NO_2C_6H_4$	CH_3	92	30	78 : 22
10	m-ClC_6H_4	CH_3CH_2	87	50	94: 6
11	p-$CH_3C_6H_4$	CH_3CH_2	70	72	67 : 33

with both aromatic and aliphatic aldehydes. Good yields are obtained using $MeNO_2$, $EtNO_2$, and n-$PrNO_2$. Good diastereoselectivity are obtained with nitroethane and nitropropane (Table 4.15, entries 7–11).

Other methods of nitroaldol reactions in water have been reported testing the reaction using simple nitroalkanes (mainly $MeNO_2$) with a variety of aldehydes, as follow:

Henry reaction using [28] nanocatalysis [$Co_2(bdda)_{1.5}(OAc)_1 \cdot 5H_2O$ (UoB-3)], with antibacterial activity toward Gram-negative bacteria *Escherichia coli* and *Bacillus cereus*, as efficient catalyst in aqueous medium at 70 °C. UoB-3 is highly stable and grants good yields of products. In addition, it could be recycled and reused a number of times without loss of performances. However, the drawback of this procedure is the need of a large excess of the nitroalkane (nitroalkane/aldehyde = 5).

Henry reaction [29] under *Daucus carota* root enzyme catalysis, in water and in the presence of phosphate buffer at pH = 7. The method was tested with the addition of CH_3NO_2 to a variety of aromatic aldehydes, yielding nitroalkanols in 8–10 hours.

Henry reaction [30] catalyzed by I_2/K_2CO_3 in water and at room temperature. The authors tested just nitromethane as nucleophile and a variety of functionalized aromatic aldehydes. Good yields of nitroalcohols are obtained in few hours (mainly one to two hours). No reaction is observed with ketones.

Zinc–Proline complex is an effective catalyst for the nitroaldol reaction of $MeNO_2$ with aromatic aldehydes [31]. The catalyst can be readily prepared in methanol reacting the amino acid with zinc acetate in the presence of Et_3N. The procedure needs of a large excess of nitromethane ($MeNO_2$/aldehyde = 20), probably because the latter acts both as reactant and solvent. However excellent yields of the adducts are granted (near quantitative yields) in few hours. The high activity of the system could be associated to the bifunctional nature of the catalyst because proline assists the generation of carbanion and zinc ion the activation of the carbonyl group.

Benzyltrimethylammonium hydroxide was found to be very efficient catalyst for nitroaldol reaction if THF/H_2O as solvents [32]. The reaction is performed at room temperature involving simple nitroalkanes and aromatic aldehydes, allowing good yields of nitroalkanols in short reaction times (few minutes).

4.1.2.2 Nitroaldol Reaction in Ionic Liquids

Ionic liquids are an emerging class of green solvents and, due to the great importance of the nitroaldol reaction, a number of publications appeared in the last years concerning their use in the Henry reaction.

In 2004, Jiang et al. [33] published the employment of 1,1,3,3-tetramethyl guanidine (TMG) trifluoroacetate [(TMG)(F_3Ac)] and TMG lactate [(TMG)(Lac)], as IL in the nitroaldol reaction using $MeNO_2$ and $EtNO_2$ as nucleophiles.

As depicted in Table 4.16, mild conditions and short reaction times are needed, thus no side products were found, and satisfactory yields are obtained. In addition, the catalyst can be reused repeatedly. However, a large excess of nitroalkane (*ratio* nitroalkane/aldehyde = 20) is required with evident problems in the amount of waste production.

In the same year it has been reported that Ni–Al (Ni:Al = 3 : 1) hydrotalcite in [bmim][BF_4] was efficiently used and recycled (four times) to furnish nitroaldol products derived by the reaction of $MeNO_2$ with aromatic and aliphatic aldehydes [34].

An efficient method was developed [35a] using the same IL [bmim][BF_4] of the above procedure, combined with 1,8-diazabicyclo[5.4.0]undec-7-ene (DBU) as base.

Table 4.16 Henry reaction in (i) TMG·F_3Ac or (ii) TMG Lac (representative examples).

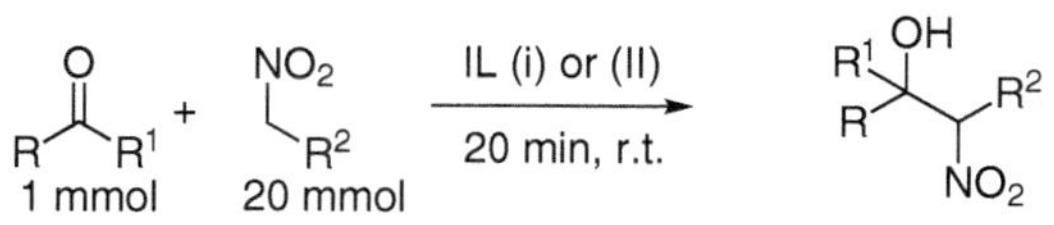

Entry	R	R^1	R^2	Yield (%)	
				IL (i)	IL (ii)
1	CH_3CH_2	H	H	73	72
2	Ph	H	H	47	55
3	$-(CH_2)_5-$		H	20	
4	CH_3CH_2	H	CH_3	57	77
5	$CH_3(CH_2)_2$	CH_3CH_2	H	41	

Table 4.17 Henry reaction promoted by [bmim][BF_4]/DBU (representative examples).

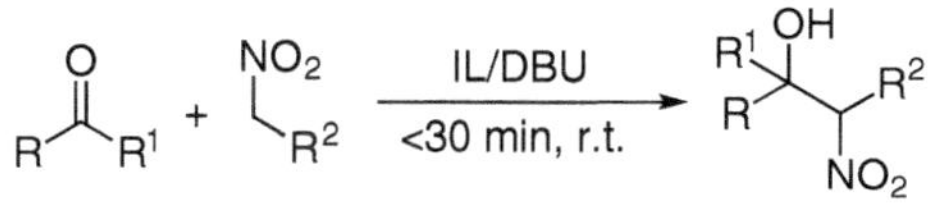

Entry	R	R^1	R^2	Yield (%)
1	$CH_3(CH_2)_3$	H	H	98
2	$CH_3(CH_2)_3$	H	CH_3	95
3	$-(CH_2)_5-$		H	87
4	$-(CH_2)_4-$		H	90
5	*p*-$NO_2C_6H_4$	H	CH_3	80

Following this procedure, the nitroalkanols are obtained, at room temperature, in good yields and in short reaction times (few minutes <30, Table 4.17).

Even ketones afford nitroaldol reaction in good yields (entries 3 and 4) but just with nitromethane, probably due to the steric hindrance of more hindered nitroalkane. The short reaction times and the good yields obtained probably indicate that the Henry reaction in IL could be accelerated in comparison with traditional methods.

One more procedure proposed the same IL and Na_2CO_3 as base [35b]. Satisfactory to good results of nitroalcohols have been obtained with a good number of aldehydes with $MeNO_2$.

Environmental friendly enzyme-promoted procedure of nitroaldol reaction was developed [36] using water in [bmim][PF_6] microemulsion as reaction medium, in the presence of amano acylase (from *Aspergillus oryzae*) as catalyst. A variety of aromatic aldehydes have been reacted with nitromethane furnishing moderate to good yields of nitroalkanols.

Ionic liquid gel [bmim]OH@agar prepared entrapping 1-butyl-3-methylimidazolium hydroxide ([bmim]H) in an aqueous agar gel [37] has successfully employed as heterogeneous catalyst, at room temperature, in the reaction of nitromethane with a variety of aromatic aldehydes. The expected nitroalcohols are obtained in good yields (72–94%) and in short reaction times (0.25–2 hours). The work-up of the reaction is very simple and consists in the filtration of the catalyst to get the crude nitroalkanol, which was simply purified by column chromatography over silica gel.

Other procedures have been reported under electrochemical [38] or reductive ($LiAlH_4$) conditions [39].

4.2 Nitroaldol Condensation

"Nitroaldol condensation" consists in the direct synthesis of conjugated nitroalkenes by in situ dehydration of the formed nitroalkanols during the Henry reaction of primary nitroalkanes with aldehydes or ketones (Figure 4.2). The process is strongly favored if aromatic aldehydes are employed.

An interesting method for the nitroaldol condensation was realized three decades ago [40] reacting furyl aldehydes with nitroalkanes, under basic alumina and in the absence of solvents. Good yields of nitroalkenes are achieved without any polymerization (Table 4.18).

CsNaX (alkaline)zeolite promotes the nitroaldolic condensation reacting aromatic aldehydes with an excess of primary nitroalkanes [$CH_3(CH_2)_nNO_2$, $n = 0$–8], giving the formation of nitroalkenes in satisfactory yields. The reaction is helped by the intrinsic basicity of the zeolite [41].

Hyperbranched polyamides assist, in the presence of $ZnCl_2$, the direct formation of β-nitrostyrenes treating a variety of benzaldehydes with nitromethane at 60 or 140 °C (Scheme 4.3) [15].

Following the previous reports for the Henry reaction in $scCO_2$ [42a, b], an interesting study for the selective nitroaldol condensation vs. nitroaldol reaction in the presence of heterogeneous base (basic Al_2O_3) and supercritical carbon dioxide as solvent was reported a decade ago [42c]. The goal of this report concerns the study of more eco-friendly conditions able to modulate the selectivity between the nitroaldol reaction and the nitroaldol condensation (Scheme 4.4).

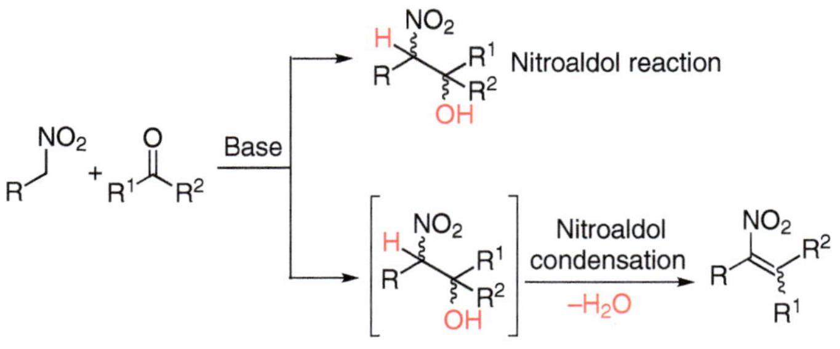

Figure 4.2 Nitroaldol reaction vs. nitroaldol condensation.

Table 4.18 Nitroaldol condensation promoted by basic Al_2O_3 (representative examples).

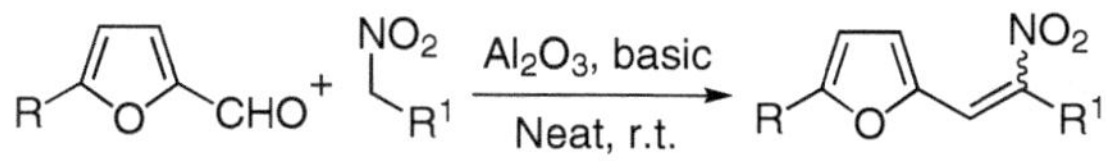

Entry	R	R^1	Yield (%)
1	H	CH_3	70
2	H	CH_3CH_2	78
3	CH_3	$CH_3(CH_2)_3$	85
4	H	$CH_3O_2C(CH_2)_5$	78
5	H	$CH_3O_2CCH_2$	93

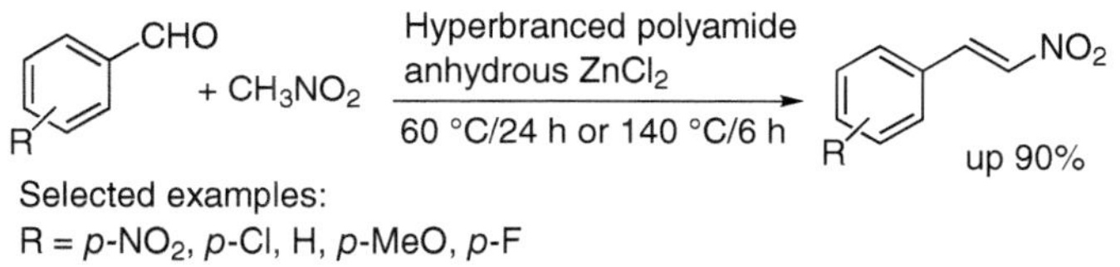

Scheme 4.3 Nitroaldol (Henry) condensation promoted by hyperbranched polyamides.

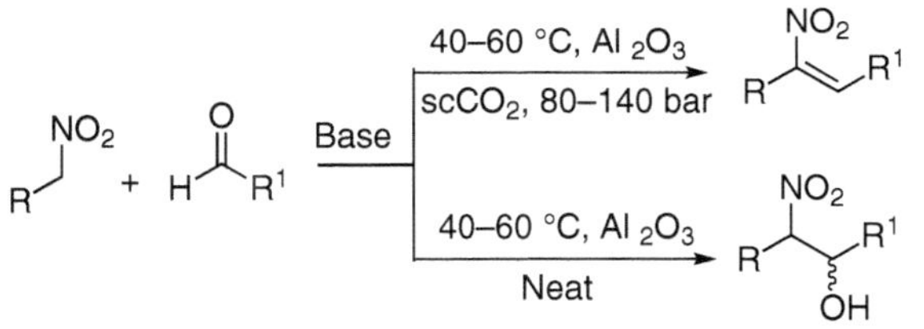

Scheme 4.4 Nitroaldol reaction promoted by Al_2O_3 in $scCO_2$.

In particular, when the pressure and the density of supercritical phase are enhanced, the percentage of nitroalkenes increases. On the other hand, the absence of solvent favors the formation of the Henry reaction.

α-Bromonitroalkenes are a very important class of functionalized nitroalkenes because they are the key starting molecules for the preparation of complex structures and possess antimicrobial activity. Fully heterogeneous diastereoselective, direct synthesis of these nitroalkenes has been recently developed [43a]. The method consists in the reaction of (i) bromonitromethane with aldehydes under neat conditions, at room temperature, promoted by carbonate on silica, and (ii) in situ treatment of the formed nitroalcohol with Amberlyst 15, Ac_2O in EtOAc as solvent (Scheme 4.5).

Thus, the bromonitroalkenes are obtained directly, in situ, in good yields (55–85%) and in moderate reaction times (four to six hours) from the starting molecules.

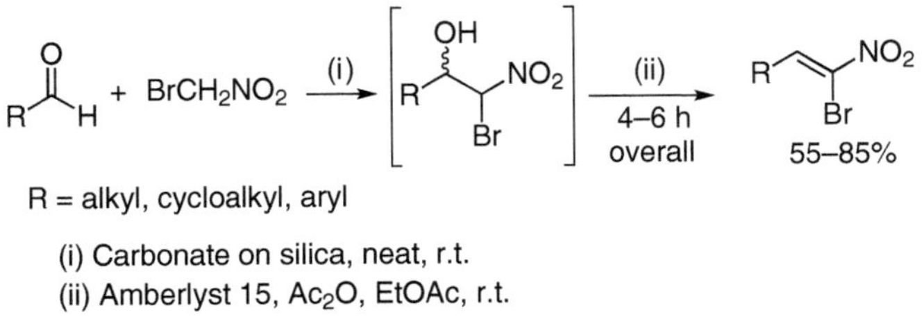

Scheme 4.5 One-pot nitroaldol condensation using bromonitromethane.

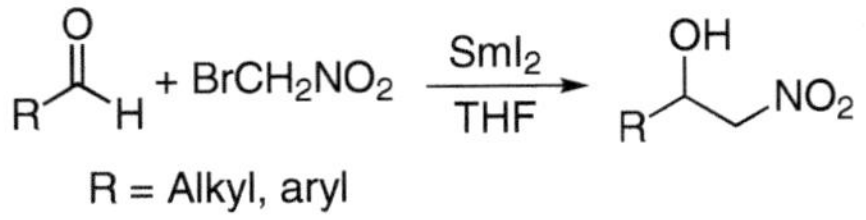

Scheme 4.6 Nitroaldol (Henry) reaction of $BrCH_2NO_2$ under SmI_2 catalysis.

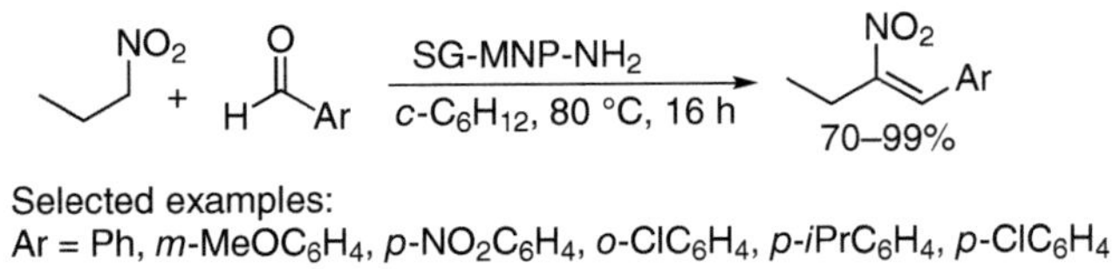

Scheme 4.7 Nitroaldol condensation by SG-MNP-NH_2.

Nitroaldol reaction of bromonitromethane with aldehydes was previously reported using SmI_2 furnishing simple nitroalkanols without the permanence of bromine (Scheme 4.6)[43b].

Silica gel magnetic nanoparticles NH_2-base (SG-MNP-NH_2) is a separable and recyclable catalyst prone to allow nitroaldol condensation reacting primary nitroalkanes with a variety of aromatic aldehydes (Scheme 4.7) [44].

Microwaves are able to assist the nitroaldol condensation of $MeNO_2$ with a variety of aromatic aldehydes, promoted by nickel hydroxyapatite nanocomposite (Ni-HAp) under solvent-free conditions [45]. The yields range from 47% to 96% depending from the nature of other functional groups present in the structure of the aromatic aldehydes. Moreover, the catalyst can be reused at least four times without decreasing yields.

In a previous paper [46], MW assisted nitroaldol condensation, under neat conditions and in the presence of a catalytic amount of ammonium acetate. $MeNO_2$ and $EtNO_2$ have been tested with a variety of benzaldehydes, granting excellent yields of nitroalkenes in few minutes (Table 4.19).

Thus, pulsed MW irradiation is found to be an ideal condition that affords high yields of the conjugated nitroalkenes in a one-pot way. This effect is not purely thermal, in fact using an alternative heating mode (i.e. oil bath) at the same temperature of 90 °C, the reaction could be completed in 18 hours (vs 3–8 minutes).

Ultrasound is an alternative source of energy to perform the nitroaldol condensation by $MeNO_2$ and aromatic aldehydes, in the presence of $NH_4OH/AcOH$ (Table 4.20) [47].

Table 4.19 Nitroaldol condensation promoted under MW, NH_4OAc, and neat reaction conditions (representative examples).

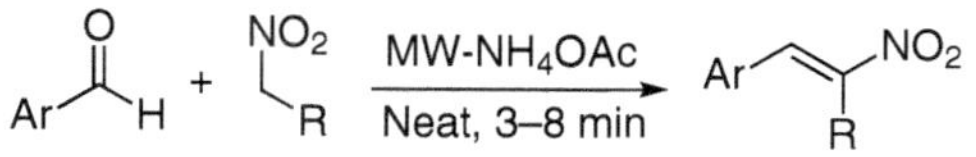

Entry	Aldehyde	R	Yield (%)
1	PhCHO	H	80
2	*p*-HOC_6H_4CHO	H	88
3	3,4-$(MeO)_2C_6H_3CHO$	H	90
4	PhCHO	CH_3	83
5	*p*-$MeOC_6H_4CHO$	CH_3	90
6	1-NapthylCHO	H	80
7	2-NapthylCHO	H	83

Ultrasounds application permits a fast and clean condensation in three hours at room temperature with the direct synthesis of (*E*)-nitroalkenes.

A novel synthesis of (*E*)-nitroalkenes has been developed using a new type of mesoporous phenolic resins that immobilizes the base (1,2-ethylendiamine) catalysts named FDU-ED (*F*u*d*an *U*niversity-*e*thylene*d*iamine) [48]. In fact, the condensation of nitroethane with several aromatic and aliphatic aldehydes (Table 4.21) proceeds with good to high yields, under mild conditions, in six hours.

The catalyst possesses several advantages such as high surface area, large pore volume, and tunable mesostructure. Moreover, the catalyst can be reused more than seven times without significant loss of activity.

An efficient and simple strategy, applying the concept of dual catalysis, has been recently reported. In fact, the condensation of nitroalkanes with aldehydes has been performed using a combination of iron ($FeCl_3$) and secondary amines (piperidine, Table 4.22) [49]. This dual catalysis simultaneously activates both electrophile and nucleophile.

The method shows several advantages such as the use of cheap catalyst, simple work-up, and tolerance of many sensitive groups.

4.2.1 Application of General Henry Reaction

General nitroaldol reaction (or condensation) is often the key step in organic synthesis. An interesting application concerns the synthesis of the A-ring of Taxane diterpenes [50]. As reported in Scheme 4.8, the nitroaldehyde **2**, obtained by diisobutylaluminium hydride (DIBAL) reduction of **1**, is prone to give an intramolecular Henry reaction under tetramethylguanidine with the formation of the α,β-nitroalcohol **3**.

Successive manipulation, including the Nef reaction, converts into the desired compound **4**.

Table 4.20 Nitroaldol condensation promoted under ultrasound, NH_4OAc, and neat reaction conditions (representative examples).

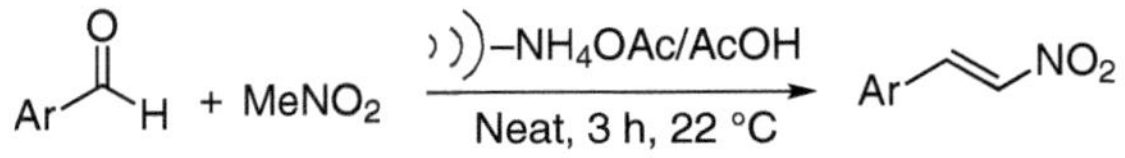

Entry	Aldehyde	Yield (%)
1	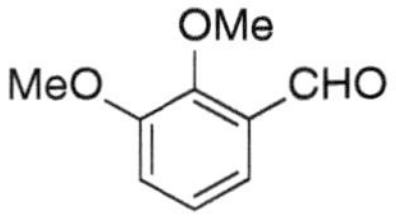	99
2	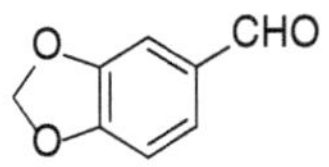	99
3	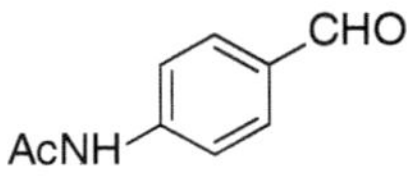	95
4	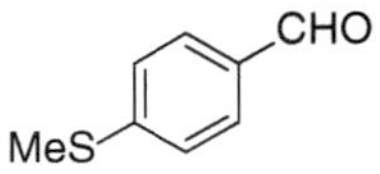	96
5	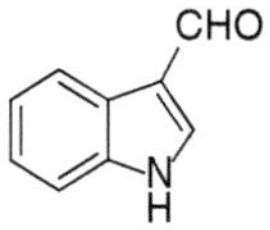	93

Table 4.21 Nitroaldol condensation under FDU-ED catalysis (representative examples).

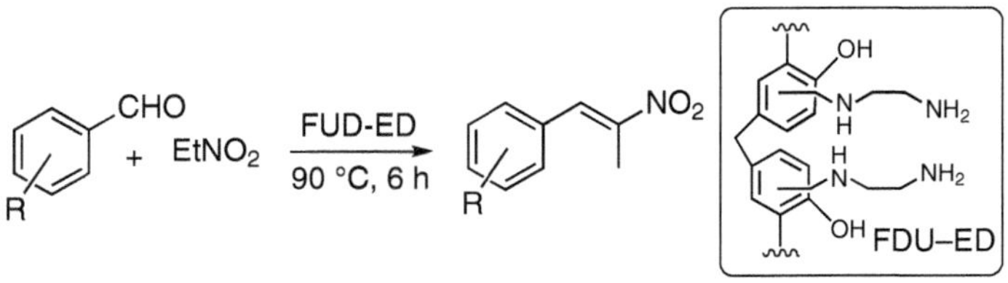

Entry	R	Yield (%)
1	*o*-Cl	77
2	*o*-Br	89
3	*m*-NO_2	70
4	*p*-F	80
5	*m*-MeO	85
6	*p*-OH	93
7	*m*-$(CH_3)_2N$	96
8	2,3,5-MeO	81

Table 4.22 Nitroaldol condensation promoted by $FeCl_3$-piperidine (representative examples).

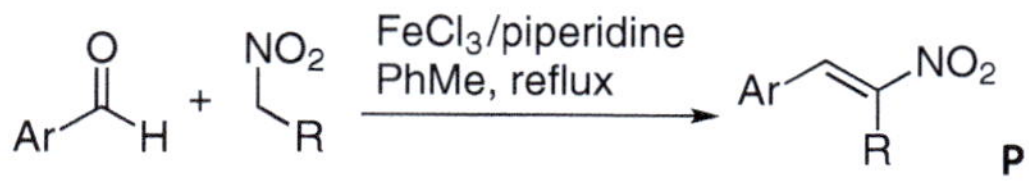

Entry	Aldehyde	R	Yield (%)
1	PhCHO	H	95
2	*p*-$MeOC_6H_4CHO$	H	94
3	3,4,5-$(MeO)_3C_6H_2CHO$	H	82
4	*o*-HOC_6H_4CHO	H	71
5	*o*-$BnOC_6H_4CHO$	H	80
6	*o*-$AllylOC_6H_4CHO$	H	85
7	*p*-BrC_6H_4CHO	CH_3	85
8	*p*-NCC_6H_4CHO	CH_3	84
9	*p*-FC_6H_4CHO	CH_3	95

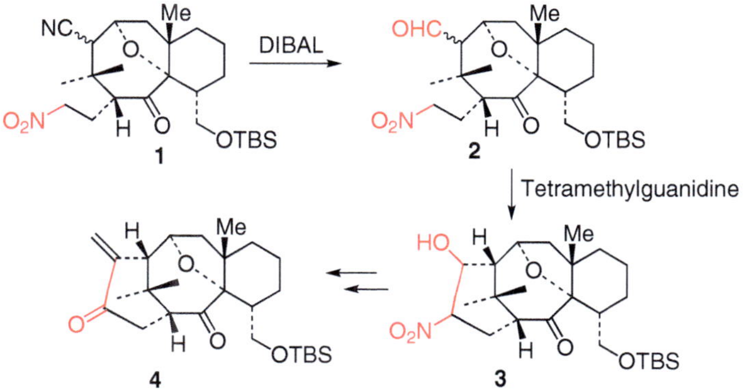

Scheme 4.8 Synthesis of A-ring of Taxane diterpene, via the Henry reaction.

Conjugate reduction of keto-nitroalkenes **5** using Stryker's reagent $(Ph_3PCuH)_6$ provides an intramolecular Henry reaction with the formation cyclic structures **6** [51]. The procedure is executed under dry conditions, in toluene at −40 °C. After few hours good results of cyclic structures are obtained as reported in Scheme 4.9.

D-Glucosamine monosaccharides **11** can be easily prepared by a diastereoselective domino Michael/Henry reaction [52]. The process (Scheme 4.10) starts with the Michael addition, under lithiumhexamethyldisiloxane (LiHMDS), of β-hydroxyaldehyde **7** to nitroalkene **8**, with the formation of polyfunctionalized nitroalkane **9** prone to give an intramolecular Henry reaction with the formation of the target compound **10**.

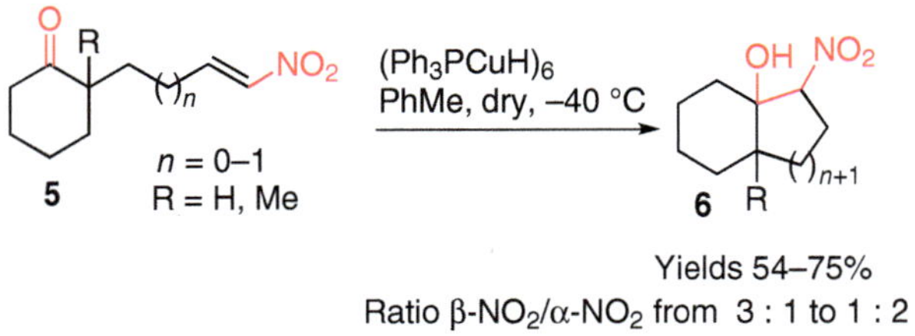

Scheme 4.9 Intramolecular Henry reaction with Stryker's reagent.

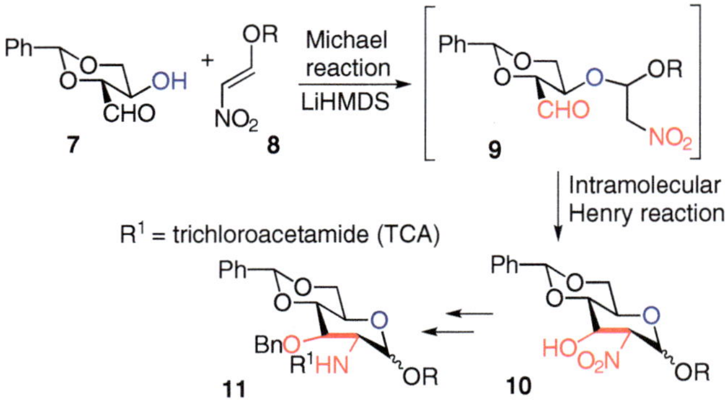

Scheme 4.10 Synthesis of D-Glucosamine monosaccharides.

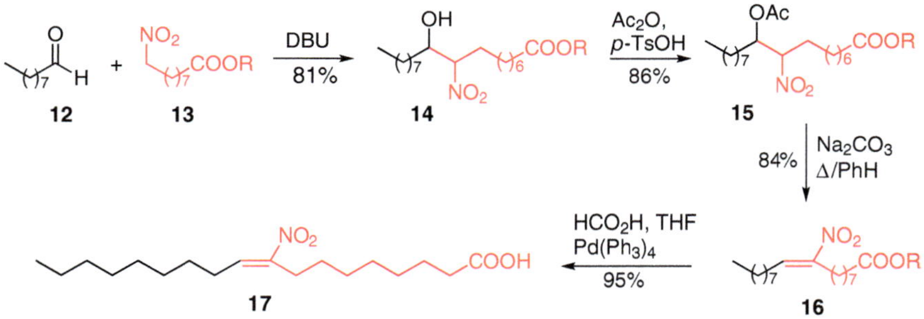

Scheme 4.11 Synthesis of (*E*)-9-nitrooleic acid **17**.

Successive manipulation, including the reduction–protection of the nitro to the amino group, allows the target compounds **11**.

Unsaturated fatty acids are nitrated endogenously to produce nitrated lipids that have high chemical reactivity and profound biological implications. Thus, their availability became of great interest in order to test the properties of these molecules. In this contest, a decade ago an efficient synthesis of two important nitro lipids derived from oleic acid, such as **17** and **20,** with the Henry reaction as key step (Schemes 4.11 and 4.12) has been elaborated [53].

The method reported for the nitro lipid **17** starts with the nitroaldol reaction of the aldehyde **12** with the nitroalkane **13** under DBU catalysis (nitroalkanes can be easily

Scheme 4.12 Synthesis of (*E*)-9-nitrooleic acid **20**.

prepared following the procedures reported in Chapter 1). The obtained nitroalkanol **14** is dehydrated to **16** derivating the hydroxyl group and treatment with Na_2CO_3 as base.

The ester functionality of the obtained nitroalkene **16** is finally hydrolyzed with formic acid, in the presence of palladium, allowing good overall yields of the target 9-nitrooleic acid **17**.

According with the aforementioned procedure, 10-nitrooleic acid **20** can be easily obtained from nitroalkane **18** and aldehyde **19**.

Nitro sugars are valuable compounds and strategic intermediates in organic synthesis. As application of the Henry reaction of $BrCH_2NO_2$ with aldehydes under SmI_2 catalysis [43b], it has been reported the synthesis of sugar-derived 2-nitroalkanols (Table 4.23) [54].

The obtained stereoselectivity can be explained through the Felkin–Anh model.

The reaction was then tested using indium instead of SmI_2, under sonication. Comparable yields and stereoselectivity were obtained.

α-Cedrene **26** is a cedar-wood oil and vetiver oil. This interesting sesquiterpene possesses a relatively rare tricycle[5.3.1.$0^{1.5}$]undecane skeleton. This target has been prepared starting from a nitroaldol reaction, under Amberlyst A21 catalysis, of nitrocyclohexane derivative **21** with the aldehyde **22** (Scheme 4.13), yielding nitroalkanol **23** [55]. The successive tandem radical denitration–cyclization of **23** was effected treating tributyltin hydride in the presence of 20 mol% of azoisobutyronitrile (AIBN) in benzene at 80 °C yielding **24** [56]. Successive oxidation of **24** (Jone's reagent) at 0–5 °C, and treatment with catalytic amount of HI in benzene at 30 °C, gives 87% yield of norcedrenone **25**. The latter can be converted into α-Cedrene **26** under Buchi's procedure.

The same authors reported the synthesis of (+)-Acoradiene **32a**, a spirocyclic sesquiterpene isolated from vetiver oil, starting from nitroaldol reaction of readily accessible cyclic nitroalkane **27** and aldehyde **22** (Scheme 4.14) under Amberlyst A21 catalysis [57].

Radical denitration–cyclization (**28** and **29**) is carried out with Bu_3SnH/AIBN affording a separable mixture of spirocyclic diols **29a–d**. Oxidation of the latter with CrO_3 produces two isomers (**a** and **b**) of the hydroxyl ketones **30** that are then submitted to dehydration with the formation of enones **31a,b**. Finally, treatment with MeLi and $CeCl_3$ and dehydration achieves the formation of a mixture of **32a,b** (68% yields, *ratio* = 68 : 32) that can be easily separated (chromatography).

Lycoricidine **36** is an important alkaloid analog that has been prepared starting from the Henry reaction (Scheme 4.15) [58]. In fact, basic alumina (activity I) promoted 6-*exo-trig* intramolecular nitroaldol reaction cyclization of the nitroaldehyde **34** (easily prepared from the nitroalkene **33**) in a highly diastereoselective manner.

Table 4.23 Henry reaction of sugars with $BrCH_2NO_2$ under SmI catalysis (selected examples).

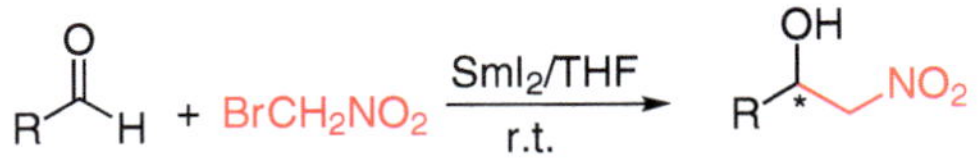

Entry	Aldehyde	Yield (%)	*anti/syn*
1	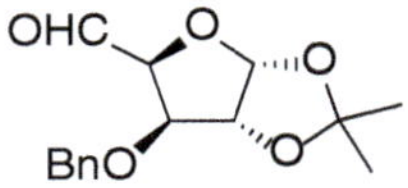	75	100 : 0
2	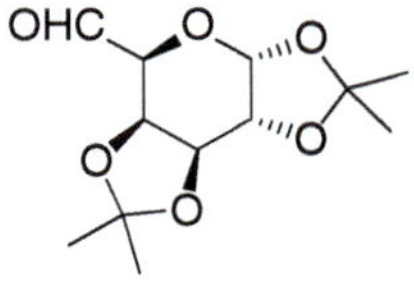 P	83	85 : 15
3	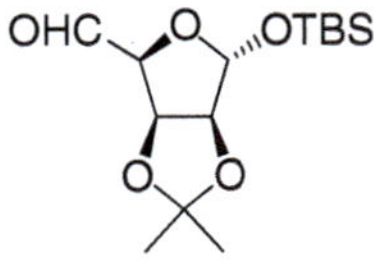	91	91 : 9
4	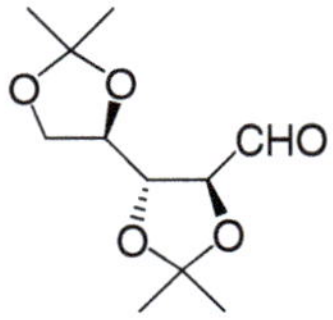	86	61 : 39
6	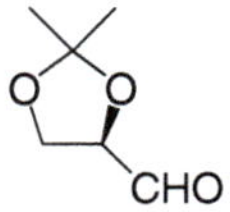	89	81 : 19

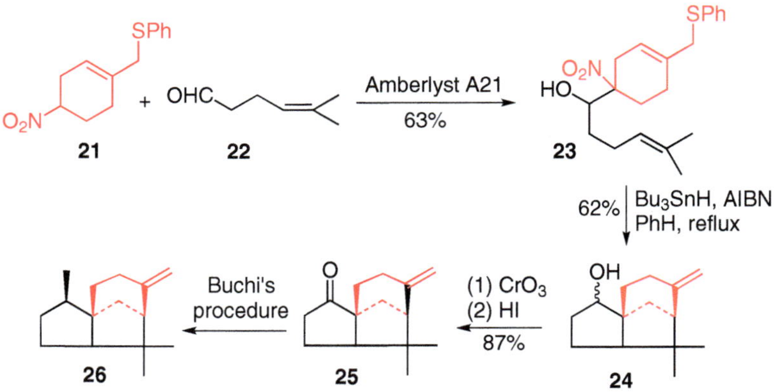

Scheme 4.13 Synthesis of α-Cedrene **26**.

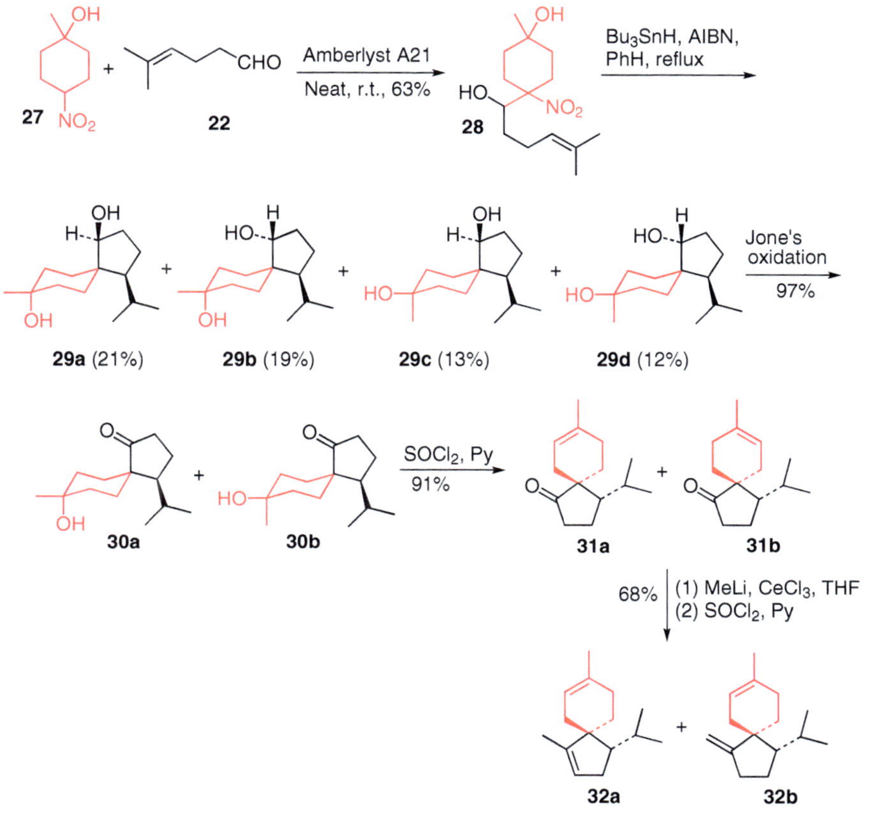

Scheme 4.14 Synthesis of (±)-Acoradiene **32a**.

The obtained nitroalkanols **35** is an important alkaloid analog that can be prepared from 6-*exo-trig* intramolecular nitroaldol reaction cyclization, under basic alumina, of nitroaldehyde **34** (easily obtained from nitroalkene **33**) in a highly diastereoselective manner. The compound **35a** can be easily converted into the target compound **36** in only two further steps.

4-Hydroxyheptadecan-7-one **40** and 14-hydroxyoctadecan-8-one **43** are the main components of the roots of *Chiococca alba* (*Anguifuga*, *Brachiata*, *Racemosa*, and *Trisperma*) of the Rubiaceae family. Their synthesis has been realized using the nitroaldol condensation catalyzed under basic alumina [59]. In fact (Scheme 4.16), nitroaldol condensation of the nitroalcohol **37** and aldehyde **38** affords the hydroxynitroalkene **39** that, via Nef reaction, converts into the target hydroxyketone **40**.

In the similar way, hydroxynitroalkane **41** and heptanal **42** yield nitroalkene **43** precursor of the target 14-hydroxyoctadecan-8-one **44** (Scheme 4.17).

The starting nitroalkanes **37** and **41** are easily prepared following the procedures reported in Chapter 1.

C_{11}–C_{23} fragment **48** of marine macrolide Swinholide A has been prepared though a double nitroaldol reaction of 1 mol of nitromethane with 2 mol of aldehyde, under high pressure and in the presence of Et_3N as base (Scheme 4.18) [60]. The symmetric

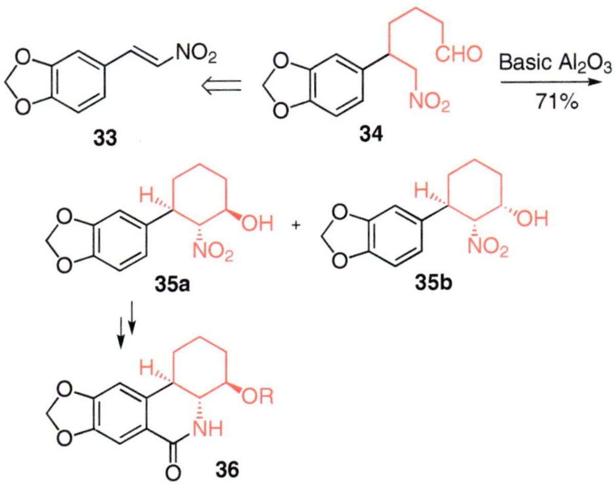

Scheme 4.15 Synthesis of Lycoricidine **36**.

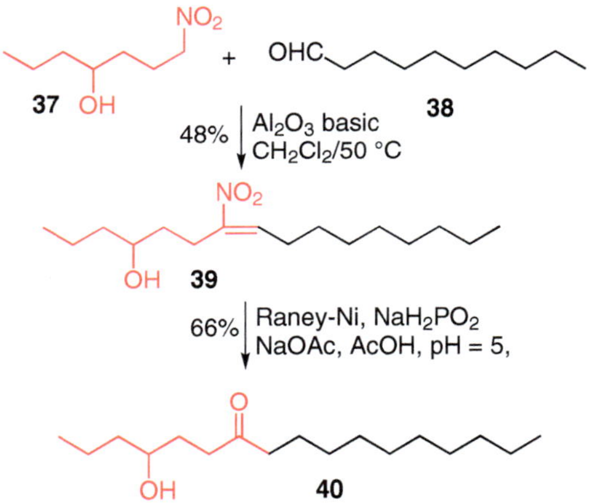

Scheme 4.16 Synthesis of 4-hydroxyheptadecan-7-one **40**.

nitrodiol **46**, obtained in 90% yield, is converted into the keto diol **47** (C_{11}–C_{21} fragment) by Nef reaction. Further manipulation allows the target structure **48** in 54% overall yield.

4.3 Asymmetric Henry Reaction

Contrary to standard nitroaldol reaction developed long time ago by Henry in 1894 [1], the first report of catalytic asymmetric nitroaldol reaction was published one century lather [61] using (*S*)-(−)Binaphthol in conjunction with lanthanum alkoxide as catalyst. However, in the following years there was a real explosion of new catalytic procedures for the development of asymmetric nitroaldol reactions. In fact,

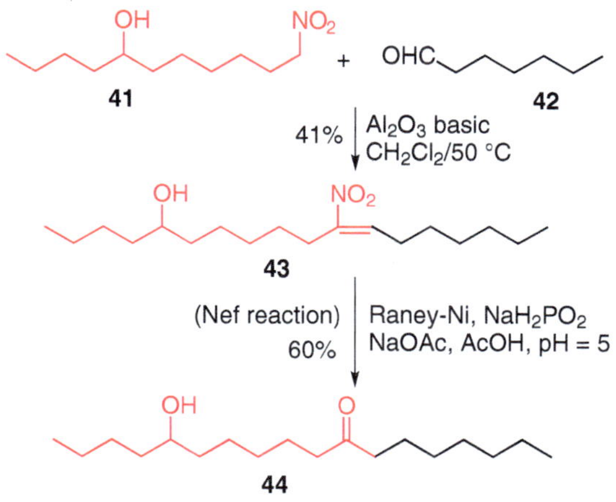

Scheme 4.17 Synthesis of 14-hydroxyoctadecan-8-one **44**.

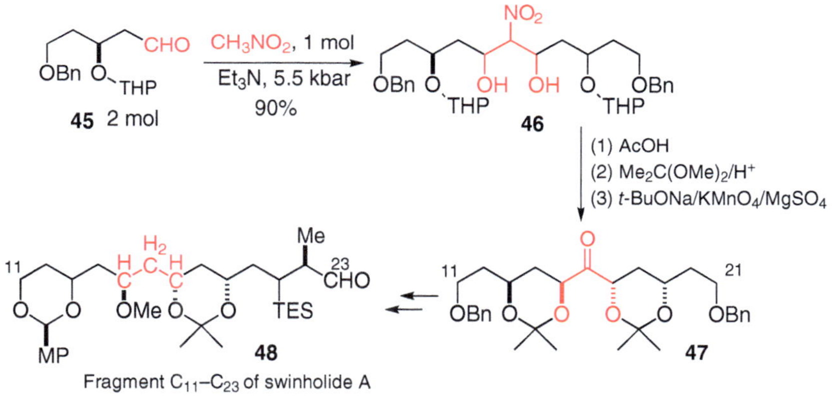

Scheme 4.18 Synthesis of segment **48** of Swinholide A.

in the new century three reviews have been published [3b–d] completely dedicated to the asymmetric nitroaldol reaction. Therefore, our choice will be to report some of the most recent and most significant applications of asymmetric Henry reactions:

A recent catalytic asymmetric Henry reaction has been developed using C_1-symmetric chiral aminopyridine ligand **A** derived from camphor and picolidine (Scheme 4.19) [62]. A variety of aromatic and aliphatic aldehydes react with nitroalkanes in the presence of diisopropylethylamine (DIPEA, 1 equiv), $Cu(OAc)_2 \cdot H_2O$ (5 mol%), and ligand **A** (5 mol%).

The products are obtained in high yields (>90%) and moderate to good diastereoselectivities (>82 : 18, R = alkyl) and excellent enantioselectivities (>98%, R = H).

The method was applied to the enantioselective synthesis of (*S*)-Miconazole (**53**), a potent antifungal agent (Scheme 4.20).

(*S*)-Nitroalcohol **50**, prepared by Henry reaction of $MeNO_2$ with aldehyde **49** (under ligand **A**/$Cu(OAc)_2$/DIPEA), is reduced to hydroxylamine **51** with

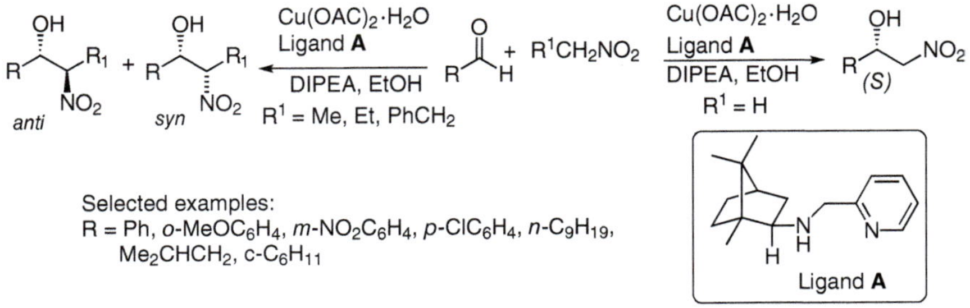

Scheme 4.19 Asymmetric Henry reaction by ligand **A**.

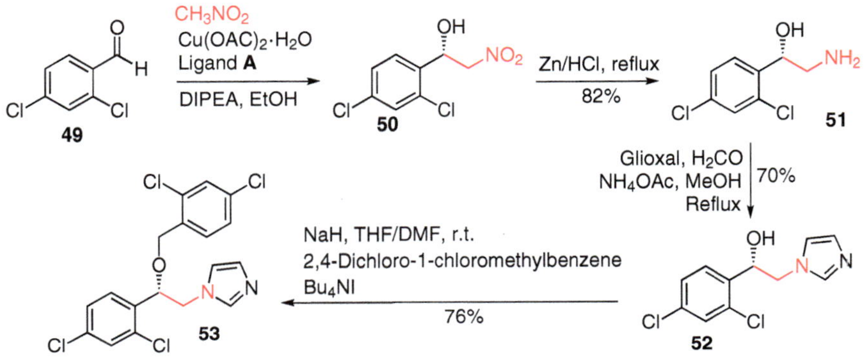

Scheme 4.20 Synthesis of (*S*)-Miconazole **53**.

Zn/HCl. The imidazole ring in **52** was then synthesized from **51** with glyoxal, H_2CO, and ammonium nitrate. Finally, alkylation of the hydroxyl with 2,4-dichloro-1-chloromethyl benzene, in presence of NaH and Bu_4NI, allows the target **53** in 76% yield and with an ee of 98%, without loss of optical purity.

Cu(II)-macrocyclic [H_4]salen complex, obtained by ligand **B** and $Cu(OAc)_2 \cdot H_2O$, catalyzes asymmetric nitroaldol reaction aliphatic and aromatic aldehydes with CH_3NO_2 [63]. Excellent yields (80–96%) and high enantioselectivity (ee ~ 99%) are achieved. It is important to point out the recyclability of the catalyst for up to eight cycles.

This method has been applied to the synthesis of (*R*)-Phenylephrine **57**, an α_1-adrenergic receptor agonist receptor, from nitroaldol reaction of $MeNO_2$ with *m*-methoxybenzaldehyde **54** (Scheme 4.21).

(*R*)-Nitroalcohol **55** is than reduced to the hydroxylamine **56** precursor, in only two steps, of the target molecule **57**.

Highly enantioselective Henry reaction in water was reported [64] under catalyzed copper-tertiary amine complex, obtained by a mixture of ligand **C**, $CuBr_2$, Cs_2CO_3, *p*-*t*-butylphenol, and Bu_4NBr. The reaction was tested reacting both aromatic and aliphatic aldehydes with nitromethane yielding (>80%) (*R*)-nitroalkanols (ee > 91%).

As application of the aforementioned method, *p*-hydroxyaldehyde **58** was directly employed as starting molecule for the synthesis of (*S*)-*N*-*trans*-Feruloyl octopamine

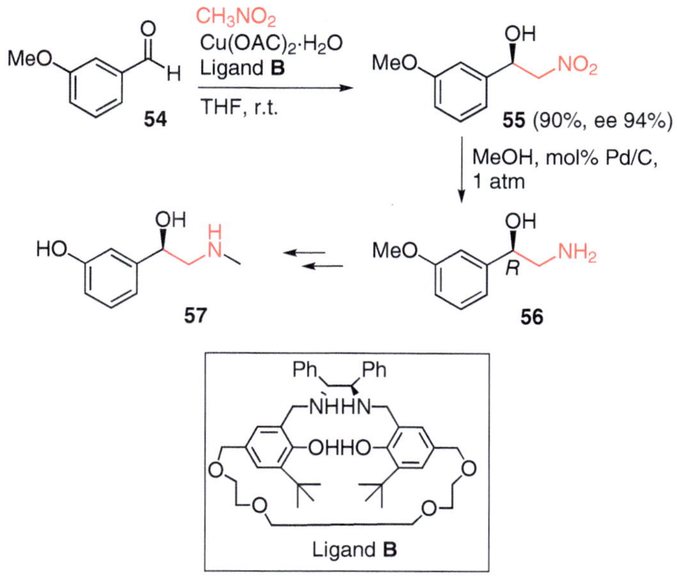

Scheme 4.21 Synthesis of (*R*)-Phenylephrine **57**.

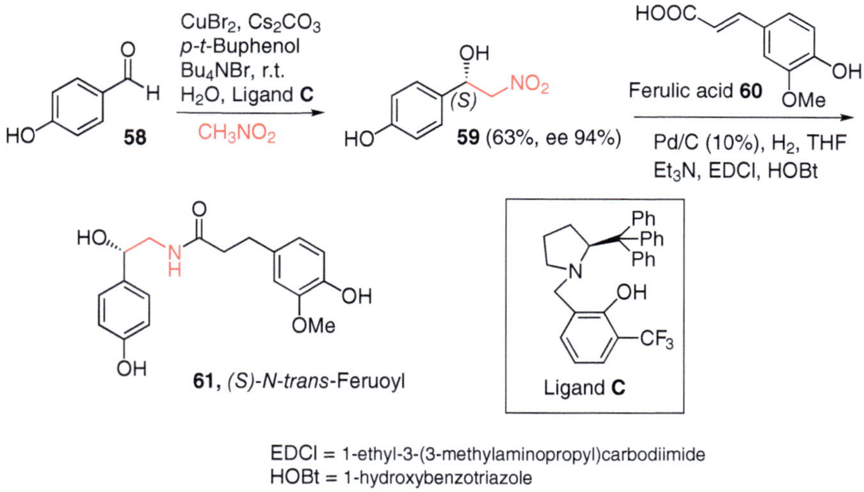

Scheme 4.22 Synthesis of (*S*)-*N*-*trans*-feruloyl **61**.

(**61**, Scheme 4.22), an important natural product and a potent inducing death of cancerous white blood cells.

Thus, enantioselective nitroaldol reaction of **58** with $MeNO_2$ allows (*S*)-nitroalkanol **59**, successive reduction, and amidation with ferulic acid **60**, and affords **61** in overall 64% yield and 90% ee.

A series of bis(sulfonamide)-diamine (BSDA) ligands have been tested for asymmetric Henry reaction and the best results have been obtained using pyridine as base,

Table 4.24 Henry reaction promoted by CuBr, Py, MeOH, and ligand **D** (selected examples).

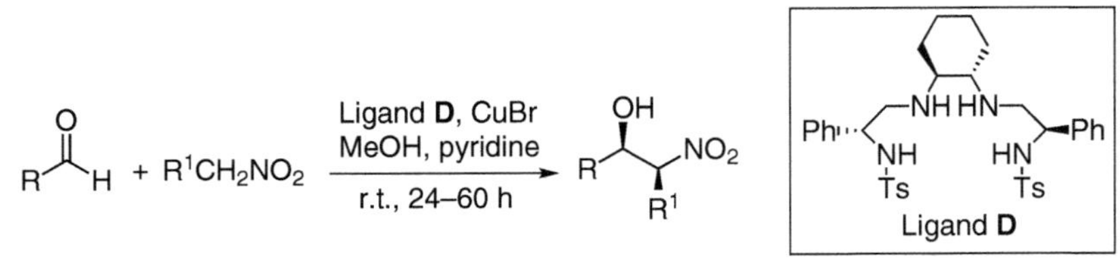

Entry	R	R^1	Yield (%)	ee (%)	*anti/syn*	ee of *syn* (%)
1	*p*-$NO_2C_6H_4$	H	98	95 (*R*)		
2	Ph	H	93	96 (*R*)		
3	*o*-$CH_3C_6H_4$	H	92	97 (*R*)		
4	1-Naphthyl	H	85	93 (*R*)		
5	Bn	H	99	97 (*R*)		
6	Cyclohexyl	H	99	98 (*R*)		
7	$(CH_3)_2CHCH_2$	H	99	97 (*R*)		
8	Ph	CH_3	95		2.2 : 1	87
9	*p*-$CH_3OC_6H_4$	CH_3	99		3.9 : 1	77
10	*p*-ClC_6H_4	CH_3	98		2.2 : 1	72
11	Cyclohexyl	CH_3	71		8.1 : 1	94
12	Cyclohexyl	CH_3CH_2	72		32.3 : 1	97
13	Cyclohexyl	Ph	90		5.7 : 1	77

ligand **D** and CuBr in methanol, at room temperature [65]. Excellent enantioselectivity and diastereoselectivity were obtained (Table 4.24).

An unprecedented diastereoselective nitroaldol reaction was successfully tested in the synthesis core fragment of HIV protease inhibitor Ritonavir and Lopinavir **66** (Scheme 4.23) [66]. In fact, nitroaldol reaction of β-amino aldehyde **63** (easily prepared from (*S*)-phenyl alanine **62**), and 3-phenylnitroethane **64**, under isopropyl bisoxazoline ligand/$Cu(OAc)_2$ catalysis in *i*-PrOH, allows nitroalcohol **65** in 76% yield (dr 62 : 29 : 4 : 5). Reduction of **65** affords the pure diastereomer 1,4-diamino-3-hydroxy derivative **66** as a core structure of both Ritonavir and Lopinavir. This approach was the first example of nitroaldol reaction involving an enantiopure protected β-amino aldehyde **63**.

Anti-diastereoselective Henry reaction is generally difficult to achieve and this limit could be attributed to the fact that metal-chelation transition state would favor a *syn* diastereomer, thus an open antiparallel transition state would be strategic for *anti*-selective nitroaldol reaction. In this context, recently it has been found that [(bisurea-salen)Co]·OBz^F ($BUSCoOBz^F$ were OBz^F = 3,5-bistrifluoromethyl benzoate) could be the right catalyst for the anti-selectivity [67]. Moreover, the best results are observed at −70 °C, with methyl *t*-butyl ether (MTBE) as solvent and *N*-ethylpiperidine (EtPip) as base (Table 4.25). It is possible to observe that

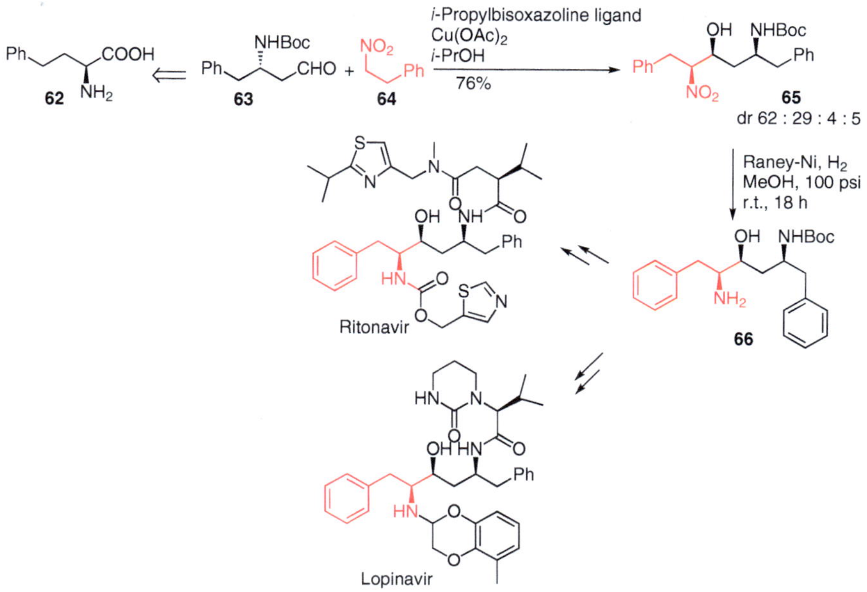

Scheme 4.23 Synthesis of **66**.

Table 4.25 Anti-selective Henry reaction by [(bisurea-salen)Co]·OBz^F (selected examples).

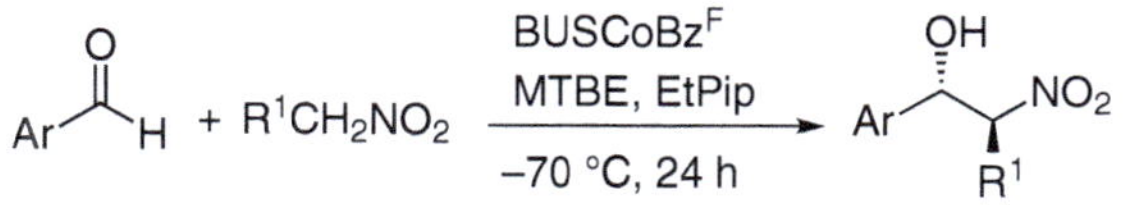

Entry	Ar	R^1	Yield (%)	dr	ee (%)
1	*o*-$MeOC_6H_4$	CH_3	84	48 : 1	96/nd
2	*o*-$BnOC_6H_4$	CH_3	85	11 : 1	96/76
3	*m*-$MeOC_6H_4$	CH_3	95	2 : 1	99/87
4	*o*-FOC_6H_4	CH_3	90	8 :1	94/82
5	*m*-FOC_6H_4	CH_3	82	1.8:1	95/85
6	*o*-MeO-4-C_6H_3	CH_3	94	15:1	95/96
7	*o*-$MeOC_6H_4$	CH_3CH_2	67	6:1	85/88
8	*o*-$MeOC_6H_4$	$TBSOCH_2$	88	50:1	85/88

anti-selectivities were obtained with *ortho* substituted benzaldehydes (Table 4.25, entries 1, 2, 4, 6–8), while significantly decreased with benzaldehydes that lack *ortho* substitution.

The synthetic utility of the above procedure has been demonstrated by a stereoselective preparation of (1*R*,2*S*)-methoxamine hydrochloride **69**, an important α_1-adrenergic agonist (Scheme 4.24).

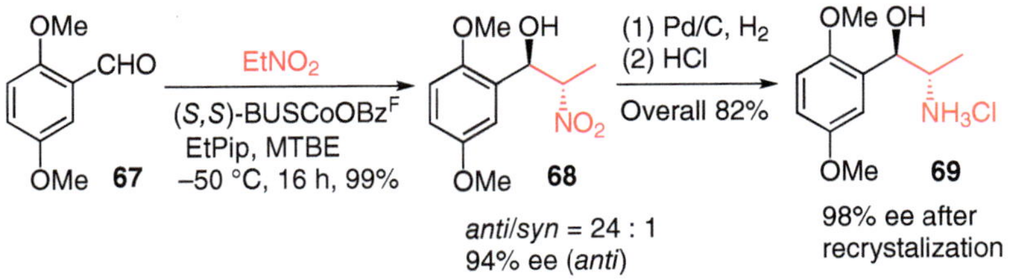

Scheme 4.24 Asymmetric synthesis of (1*R*, 2*S*)-methoxamine·HCl.

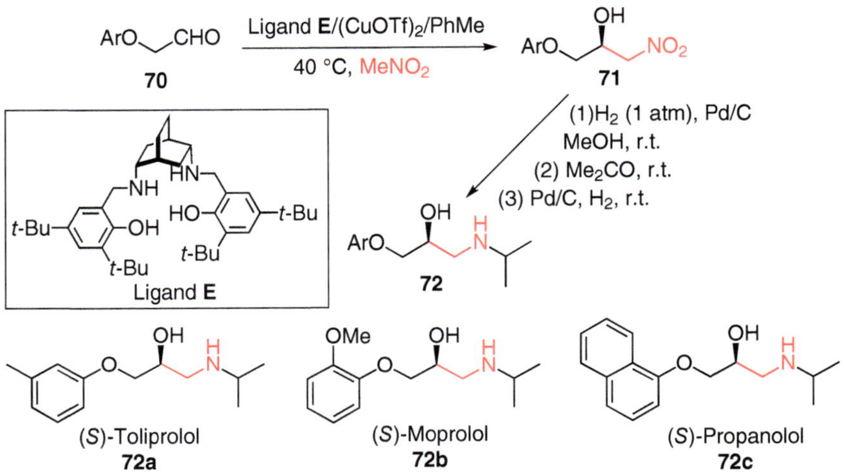

Scheme 4.25 Asymmetric synthesis of (*S*)-Toliprolol **72a**, (*S*)-Moprolol **72b**, and (*S*)-Propranolol **72c**.

Excellent stereocontrol (*anti/syn* = 24 : 1, 94% ee) is obtained using 5 mol% of the antipode catalyst (*S*,*S*)-BUSCoOBz^F.

A new chiral tetrahydrosalen ligand (ligand **E**) was used for the preparation of a complex with $(CuOTf)_2$·PhMe. The latter is able to catalyze the asymmetric nitroaldol reaction of nitromethane, at 40 °C in MeOH and in the presence of molecular sieves (Å MS), with a variety of aromatic and aliphatic aldehydes [68]. Nitroalcohols are obtained in very good yields (81–99%) and excellent ee (>91%). This new enantioselective procedure was then applied to the synthesis of (*S*)-Toliprolol **72a**, (*S*)-Moprolol **72b**, and (*S*)-Propranolol **72c** (Scheme 4.25), important blocking agents of β-adrenergic receptors.

Chiral dinuclear macrocyclic $[H_4]$salen ligand with diphenyldiamine collar (ligand **F**) in combination with $Cu(OAc)_2·H_2O$ works well for asymmetric Henry reaction. This protocol has demonstrated to be highly efficient for the reaction of $MeNO_2$, $EtNO_2$, *i*-$PrNO_2$ with aliphatic, aromatic and heteroaromatic aldehydes [69], to furnish the corresponding nitroalcohols in high yields and high enantioselectivities for *syn* diastereomers. Moreover, the ligand **F**–Cu(II) complex is recyclable up to eight cycles with no significant loss of performances.

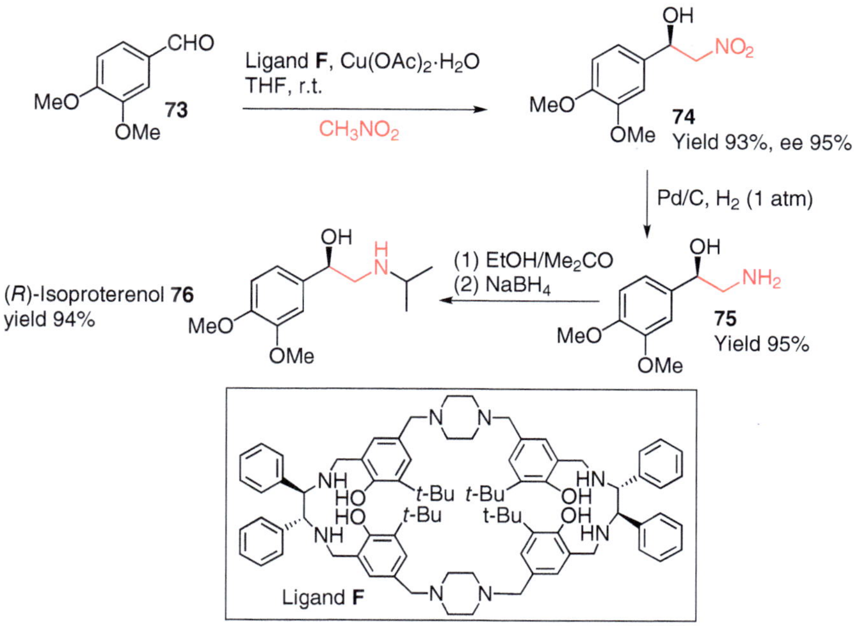

Scheme 4.26 Asymmetric synthesis of (*R*)-Isoproterenol **76**.

Figure 4.3 Ligand **G**.

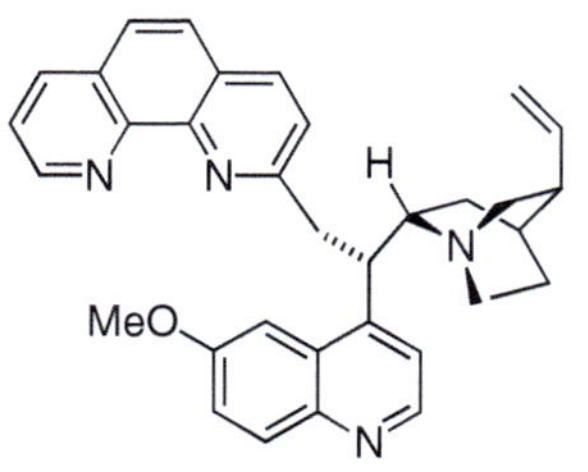

As application of this method, the synthesis of (*R*)-Isoproterenol **76**, β-adrenergic agonist, has been developed (Scheme 4.26).

New ligands, derived from phenanthroline and quinine, have been tested for asymmetric Henry reaction of aromatic, heteroaromatic, and aliphatic aldehydes with $MeNO_2$. During these studies, it has been found that ligand **G** (Figure 4.3) gives selective production of either enantiomeric product without the need for a resolution process [70]. In fact, 50–86% yields and 86–90% ee were obtained.

Simple chiral N^1-Alkyl, N^2-arylmethyl diamine ligand **H** (Figure 4.4), complexed with $Cu(OAc)_2 \cdot H_2O$, has demonstrated to be an efficient and recyclable (at least three times) catalyst for asymmetric Henry reaction of aldehydes with nitromethane, in the presence of *i*-Pr_2NEt as base and *i*-PrOH as solvent [71]. (*S*)-Nitroalkanols are prepared in high yields and enantioselectivity (75–94%).

(2*S*,5*R*)-2-Methylaminomethyl-1-methyl-5-phenylpyrrolidine was found to be highly efficient chiral ligand (ligand **I**, Figure 4.5) for Cu(II)-catalyzed asymmetric Henry reaction, in presence of Et_3N, of aryl and alkyl aldehydes with CH_3NO_2 [72].

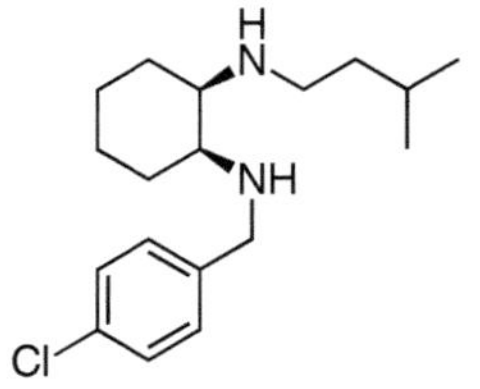

Figure 4.4 Ligand **H**.

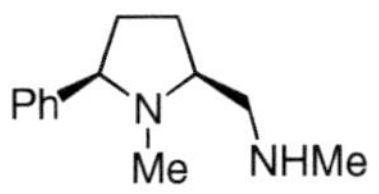

Figure 4.5 Ligand **I**.

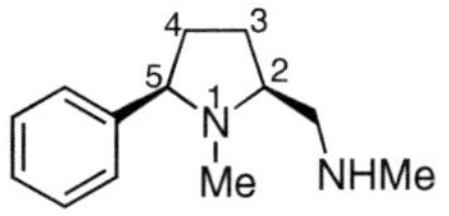

Figure 4.6 Ligand **J**.

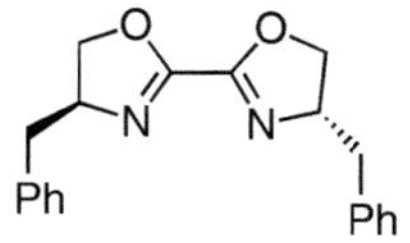

Figure 4.7 Ligand **K**.

(*S*)-Nitroalkanols are obtained in good yields (>91%) and very high ee (>99%) from aromatic, heteroaromatic, and aliphatic aldehydes.

During the studies of new catalytic systems for asymmetric Henry reaction, a series of 5-cis-substituted prolinamines, as ligands, have been tested with some Cu(II) salts, under basic conditions (Et_3N). Ligand **J** (Figure 4.6) demonstrated to be the most efficient in combination with $Cu(OAc)_2$ [73].

The reactions revealed that an aromatic or sterically demanding aliphatic substituents in 5-*cis* position is crucial for high levels of stereocontrol, while bulkier substituents, at the nitrogen atoms, diminish both enantioselectivities and reaction rates.

A number of aldehydes and nitroalkanes RCH_2NO_2 (R = H, Me, Et, CH_2OTBS) have been tested and the procedure grants high dr and ee.

Chiral bisoxazoline ligand (ligand **K**, Figure 4.7) complexed with $Cu(Ac)_2$ was found to work well in Henry reaction of CH_3NO_2 with brominated and fluorinated benzaldehydes [74].

Thus, unbridged ligand **K**-Cu(II) complex results as an excellent catalyst granting high enantioselectivity, in fact nitroalkanols are prepared in excellent yields (up to 99%) and with enantioselectivities up to >99%.

Very recently it has been prepared a new series of 2-(pyridine-2-yl)imidazolidine-4-thione derivatives (ligand **La,b**, Figure 4.8) [75].

Their copper (II) complexes were found to be highly enantioselective catalysts for asymmetric nitroaldol reaction (up to 98% ee). Successively, these complexes have been immobilized by anchoring onto the Merrifield resin with respect to their use as recyclable catalysts.

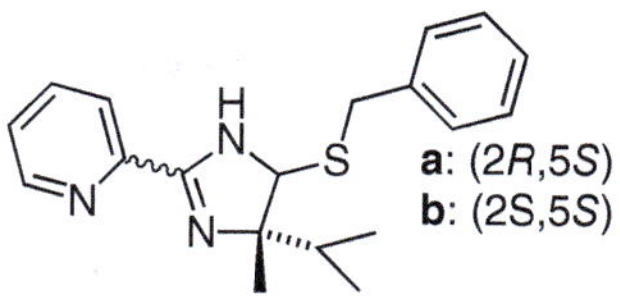

Figure 4.8 Ligand **La,b**.

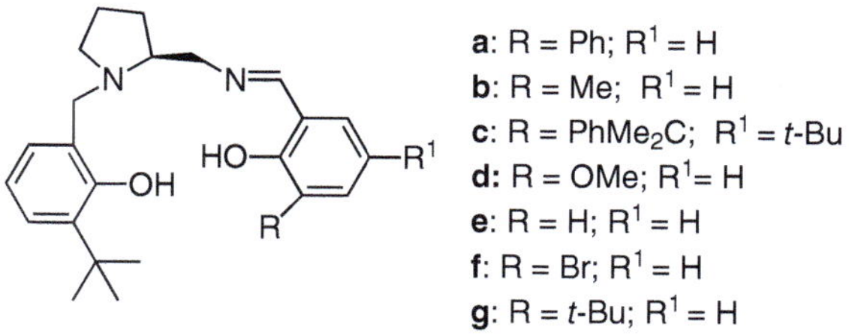

Figure 4.9 Ligand **Ma–g**.

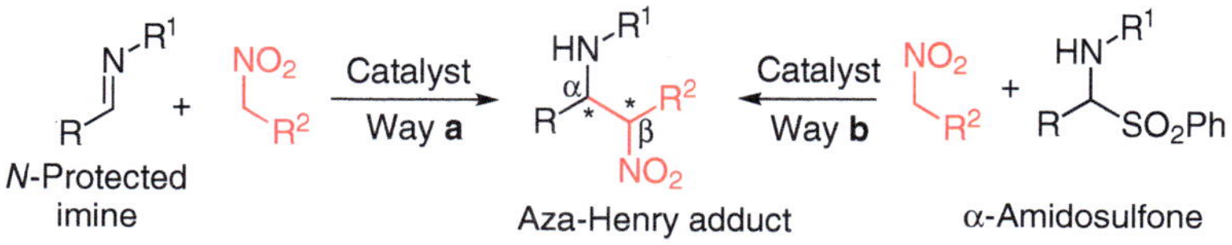

Figure 4.10 Aza-Henry (or nitro-Mannich) reaction.

Chiral salalen ligands (ligand **M**), derived from (*S*)-Proline and derivatives of salicylaldehydes, have been prepared and complexed with Cu(II) (Figure 4.9) [76]. After screening of the ligands, it has been found that ligand **Mb** is the better catalysis for nitroaldol reaction ($MeNO_2$ or $EtNO_2$ with aldehydes) and the best yields are obtained if the ligand is complexed with $Cu(OAc)_2 \cdot H_2O$ in the presence of *i*-$PrNH_2$ as base. Satisfactory to good yields and good ee were obtained. Henry reaction of benzaldehyde derivatives with prochiral $EtNO_2$ was even performed and anti-selective products were obtained especially with aldehydes possessing electron withdrawing groups.

4.4 Aza-Henry Reaction

Aza-Henry reaction, or nitro-Mannich (Figure 4.10) represents one of the most versatile and attractive ways for accessing β-nitroamines, prone to be converted into important building blocks such as biological active vicinal diamines and amino carbonyl derivatives. The aza-Henry reaction can be performed starting from *N*-protected (because of their inherent instability) imines (Figure 4.11, way a) or from α-amidosulfones (way b).

Depending on the nature of nitroalkane, the reaction produces one (R^1 = H) or two (R^1 = alkyl, aryl) stereogenic centers, thus asymmetric aza-Henry would be very strategic. In this contest, Shibasaki and coworkers [77] reported, two decades ago, the first catalytic asymmetric aza-Henry reaction of nitromethane with *N*-protected imines. Few years later two important asymmetric procedures have been published [78, 79].

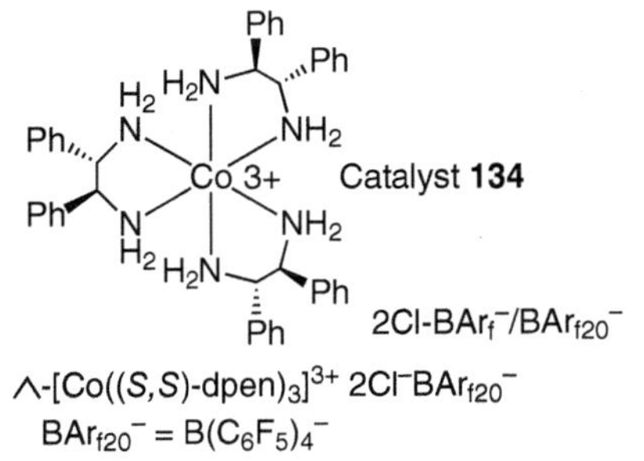

Figure 4.11 Catalyst **134**.

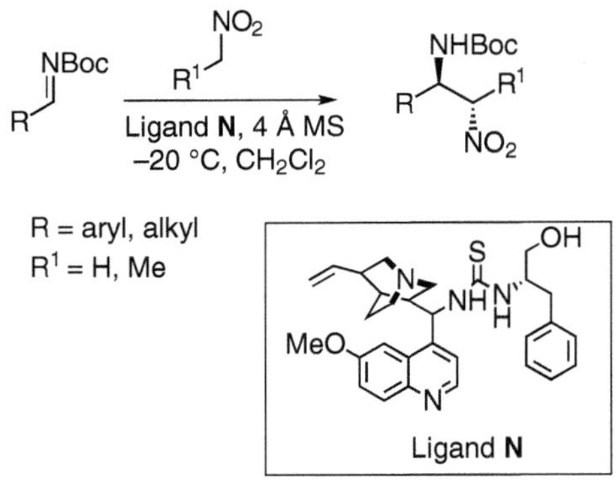

Scheme 4.27 Asymmetric aza-Henry reaction with ligand **N**.

Successively, impressive progress has been made on the development of more selective catalytic systems.

Thus, we report some of the more representative procedures published in the last decade.

4.4.1 Aza-Henry Reaction via *N*-Protected Imines

A variety of modular bifunctional chiral thiourea organocatalysis (derived from *Cinchona* alkaloids and amino acids) have been tested for asymmetric aza-Henry reaction of nitroalkanes with NBoc imines. Ligand **N** demonstrated (Scheme 4.27) to be the more efficient in terms of yields (up to 85%), enantioselectivity (up to 99% ee), and diastereoselectivities (up to 17 : 1) [80].

Indium-mediated reaction of *gem*-bromo-nitroalkanes [$BrCH_2NO_2$, $EtCHBrNO_2$, $(CH_3)_2CBrNO_2$] and imines, under sonication, with remarkably high *anti*-selectivity (Scheme 4.28) [81]. As representative examples we report the reaction of 5-bromo-2,2-dimethyl-5-nitro-1,3-dioxane **78** with aldimine **77**, yielding the adduct **79**.

The latter can be easily reduced to chiral diamine **80** as useful intermediate and precursor of ligands, very important for asymmetric synthesis.

Enantioselective aza-Henry reaction of simple nitroalkanes ($MeNO_2$ and $EtNO_2$) with trifluoromethyl ketimines **81** has been realized using a bifunctional phase-transfer chiral ammonium catalyst, derived from quinine (Scheme 4.29) [82].

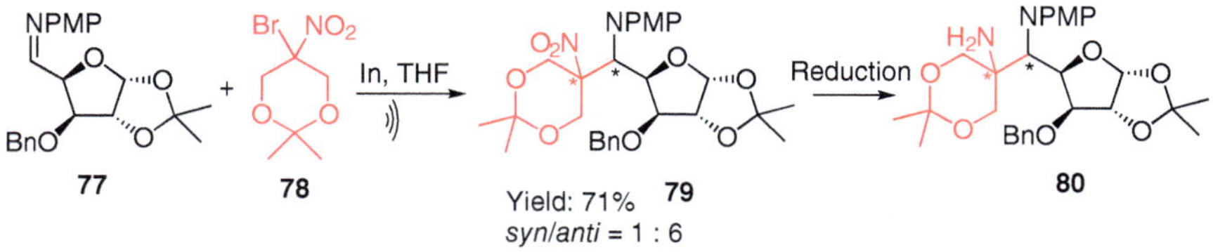

Scheme 4.28 Asymmetric aza-Henry reaction with indium. Selected example.

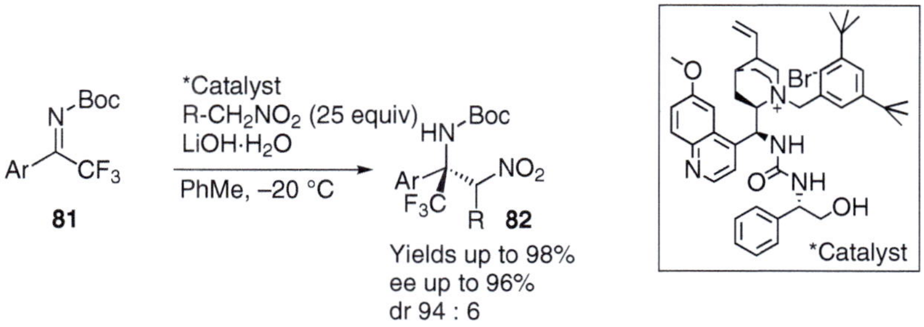

Ar = Ph, *o*-ClC$_6$H$_4$, *m*-FC$_6$H$_4$, *m*-MeOC$_6$H$_4$, *p*-BrC$_6$H$_4$, 1-naphtyl
R = H, Me

Scheme 4.29 Aza-Henry reaction of trifluoromethyl ketimines. Selected examples.

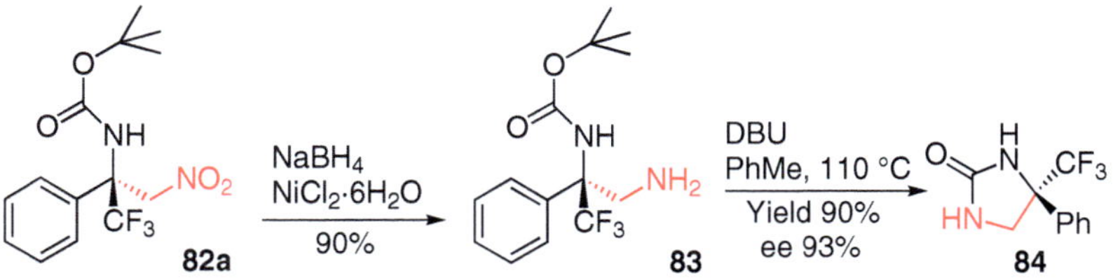

Scheme 4.30 Cyclization of **82a** into imidazoline **84**.

It is important to point out that –CF_3 group containing compounds are extremely important in pharmaceuticals, agrochemical products, and materials. Thus, the compounds **82** are of great importance, mainly as sources of diamino derivatives. As example, **82a** (Ar = Ph) can be reduced (Scheme 4.30) to the diamine **83** in 90% yield.

The simultaneous presence of amino and alkoxycarbonyl groups in **83** creates favorable conditions for intramolecular cyclization to the corresponding imidazoline **84**, in one step and without loss of enantioselectivity.

A variety of chiral Cu(II) complexes of aminoalcohol have been prepared for asymmetric aza-Henry reaction of nitroalkanes with substituted *N*-tosylimines After several tests, ligand **O** was found to be the more efficient [83]. As reported in Scheme 4.31, the method works well with a variety of examples.

The applicability of the procedure was demonstrated by the synthesis of chiral drug (*S*)-Levamisole **89** (Scheme 4.32), an anthelminthic agent used to treat parasitic worm infections in humans.

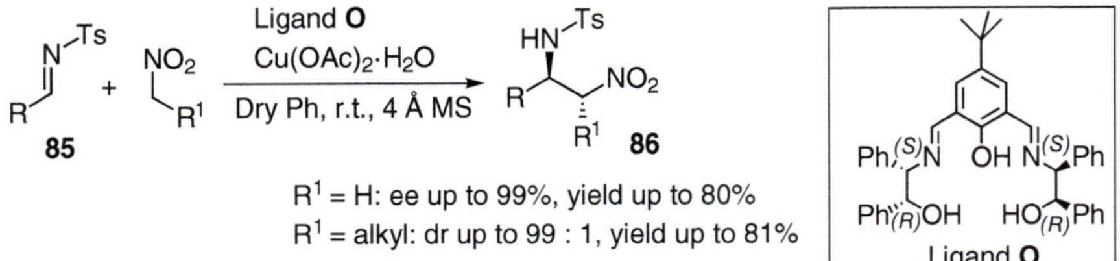

Selected examples:
R = o-MeC$_6$H$_4$, o-FC$_6$H$_4$, p-MeOC$_6$H$_4$, 1-naphtyl, PhCH=CH(Me)
R^1 = H, Me, Et, Bu

Scheme 4.31 Asymmetric aza-Henry reaction with ligand **O**.

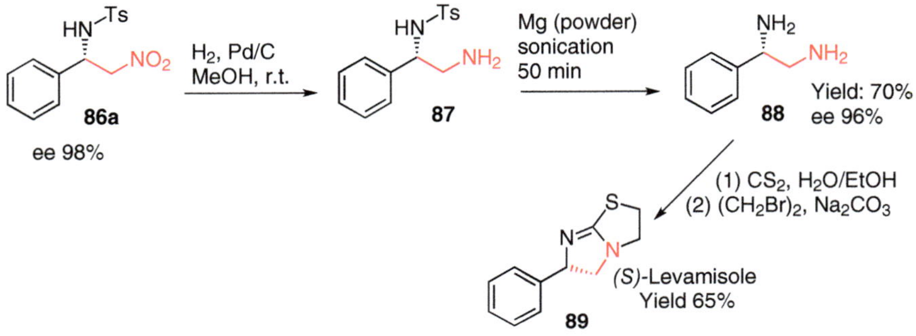

Scheme 4.32 Synthesis of (S)-Levamisole **89**.

Yields up to 74%
R^1 = alkyl: dr 2 : 1

Selected examples:
Ar = Ph, m-MeC$_6$H$_4$, p-MeOC$_6$H$_4$, p-BrC$_6$H$_4$, p-PhC$_6$H$_4$
R = Me, Et, i-Pr, t-Bu, Bn
R^1 = H, Me, Et

Scheme 4.33 Dehydrogenative aza-Henry reaction.

The β-nitro-N-tosylaldamine was hydrogenated with Pd/C and the resulting **87** was deprotected with Mg/sonication to allow the diamine **88** in good overall yields and ee. Finally, treatment of **88** with CS_2 and $(CH_2Br)_2$ produces the target compound **89** in 65% yield.

An innovative, copper-catalyzed dehydrogenative cross-coupling aza-Henry reaction of nitroalkanes with α-amino acid esters was recently reported (Scheme 4.33) [84].

The mechanism probably involves the formation of radicals and the method is, of course, not asymmetric. However, for nitroalkanes different from CH_3NO_2 (R^1 = alkyl), a dr 2 : 1 was observed.

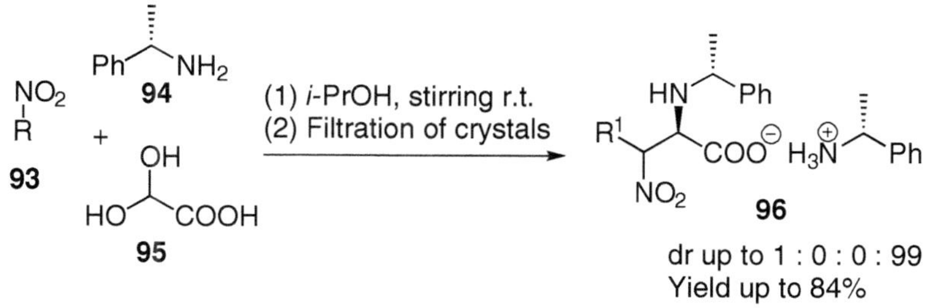

Selected examples:
Nitroalkanes **93**: $MeNO_2$, Me_2CHNO_2, Et_2CHNO_2, c-$C_5H_9NO_2$, c-$C_6H_{11}NO_2$

Scheme 4.34 Three-component aza-Henry reaction via crystallization.

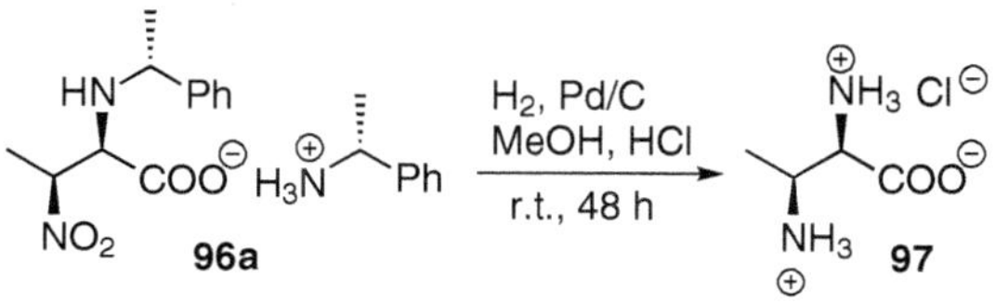

Scheme 4.35 Reduction of amino acid. Synthesis of vicinal diamines.

Recently appeared an article titled: *Crystallization does it all: an alternative strategy for stereoselective aza-Henry reaction* [85]. As reported in Scheme 4.34, the method consists in a three component procedure followed by crystallization that plays a pivotal role in the overall direction of the reaction course, including the crystallization-induced diastereoselective transformations (CIDTs).

The authors took advantages of the Henry-reaction reversibility and, under a simple reaction setup, would yield crystalline α-amino-β-nitro acids in excellent stereoisomeric purities. Moreover, selective crystallization, using *i*-PrOH as solvent, followed by filtration greatly simplifies the reaction work up and represents an attractive separation option.

Thus, the highly stereoselective crystallization-driven reaction is applicable to a broad range of substrates.

Moreover, only 1 equiv of nitroalkane is necessary, while in the previous reported procedures several fold in excess are required, thus hampering applications to more complex and expensive nitroalkanes. The adducts **96** can easily converted into vicinal diamines **97** and a representative example is reported in Scheme 4.35.

Asymmetric aza-Henry reaction of *N*-Boc imines **98** was developed in the presence of 10 mol% dinuclear Zinc–AzePhenol catalyst (ligand **P**), in THF and at room temperature (Scheme 4.36) [86a].

A variety of nitroamines **99** can be prepared in good yields (up to 97%) with excellent enantioselectivity (up to 99% ee) and high diastereoselectivity (up to 14 : 1 dr).

Previously, Perdrosa and coworkers [86b] reported a solvent-free, enantioselective aza-Henry reaction of *N*-Boc imines, catalyzed by supported thioureas.

Epiquinamide **105** is an important alkaloid isolated, a decade ago, from Ecuadorian poison frog, *Epipedobates tricolor*. Due to its importance a synthetic

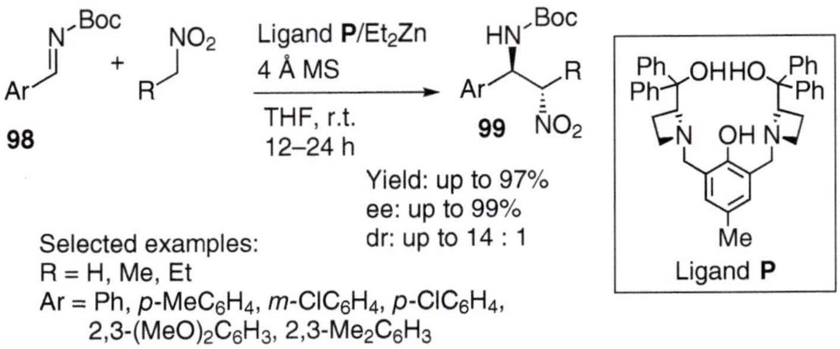

Scheme 4.36 Aza-Henry reaction under ligand **P**-Et_2Zn catalysis.

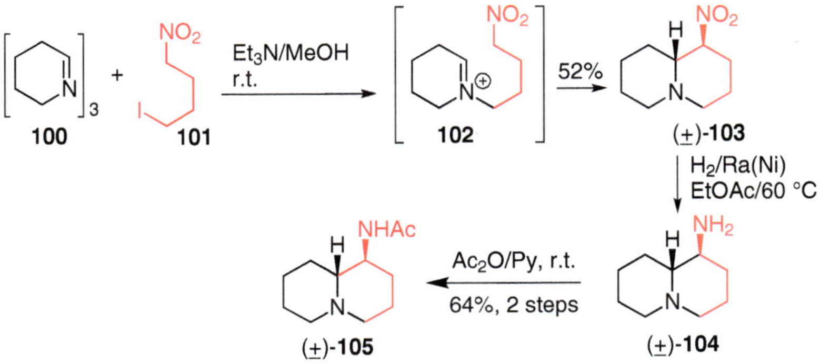

Scheme 4.37 Synthesis of Epiquinamide **105**.

procedure has been reported using an intramolecular aza-Henry reaction (Scheme 4.37) [87]. The process starts with the addition of iodo-nitroalkane **101** to $\Delta^{1,2}$-piperideine **100**, as trimer, according with the literature. The cyclization of the adduct **102** is than performed under basic conditions (Et_3N) to afford racemic 1-nitroquinolizidine **103** as single diastereomer. Subsequent reduction and acetylation afford (1*R**,9*S**)-*cis*-diastereomer **105** (*epi*-Epiquinamide) in 64% yield.

4.4.2 Aza-Henry Reaction via α-Amidosulfones

Rosin-derived thiourea catalyst has been applied to the doubly stereocontrolled asymmetric aza-Henry reaction with the in situ generation of *N*-carbamate imines from the corresponding amidosulfones (Scheme 4.38) [88].

The results show that all reactions with nitromethane (R^1 = H) proceed affording the desired products, with (*S*) or (*R*) configuration, with excellent enantioselectivities ((1*S*)-adducts: 81–98% ee, yields 80–92%; (1*R*) adducts 81–97% ee, yields 81–90). Moreover, the reaction with nitroethane (R^1 = Me) affords *anti*/*syn* up to 98 : 2.

High enantio- and diastereoselective aza-Henry reaction of α-amidosulfones and nitroalkanes has been realized using phase-transfer (PhMe/H_2O) catalysis (catalyst **113**) derived from *Cinchona* alkaloids and *N*-benzotriazole (Scheme 4.39) [89].

The products **112** are obtained in good yields (up to 94%), with excellent enantioselectivities (R^1 = H, up to 99% ee) and diastereoselectivities (up to 99 : 1).

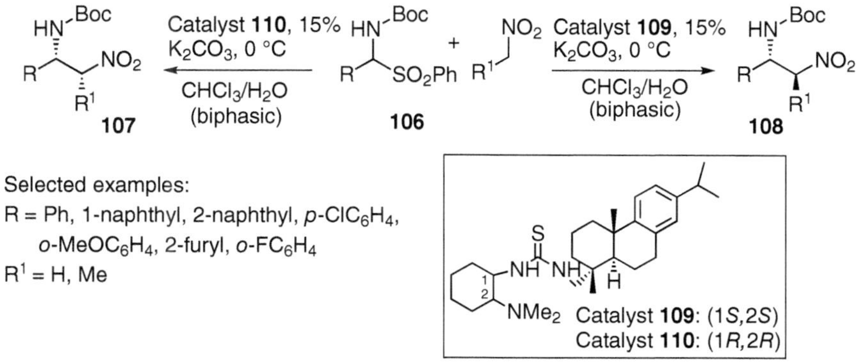

Scheme 4.38 AHR under rosin-derived amine thiourea catalyst **109** and **110**.

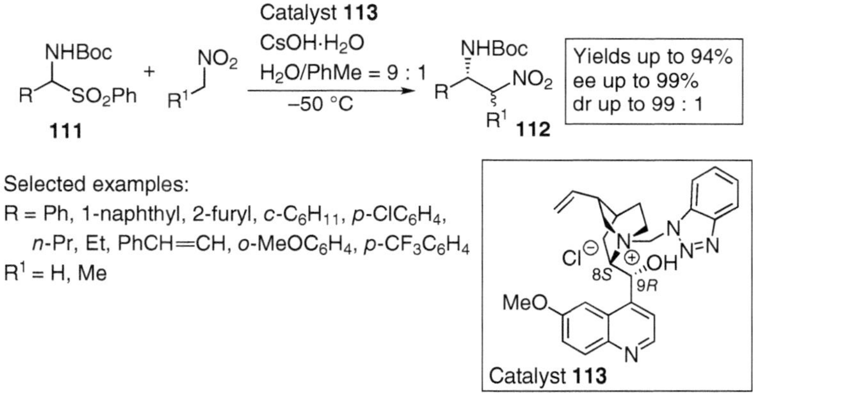

Scheme 4.39 AHR under phase-transfer catalysis **113** derived from *cinchona* alkaloids and *N*-benzotriazole.

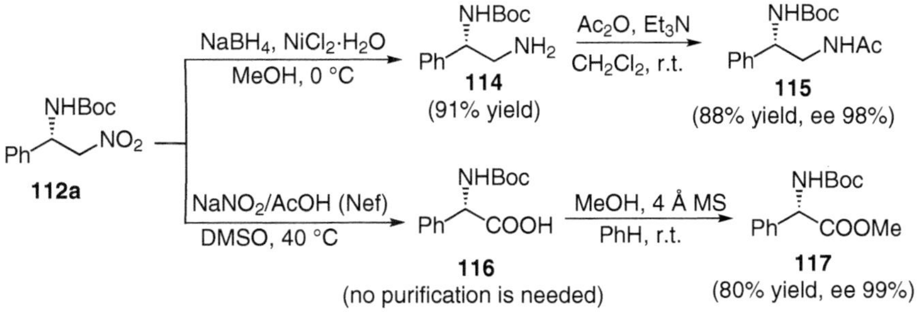

Scheme 4.40 Synthesis of 1,2-diamines **115** and α-aminoester **117**.

As application of this procedure, 1,2-diamines **115** and β-aminoacids **116** have been prepared (Scheme 4.40).

In fact, $NaBH_4/NiCl_2$ reduction of **112a** gives diamine **114**, successively converted into diprotected diamine **115**. On the other hand, Nef transformation (by $NaNO_2$) of

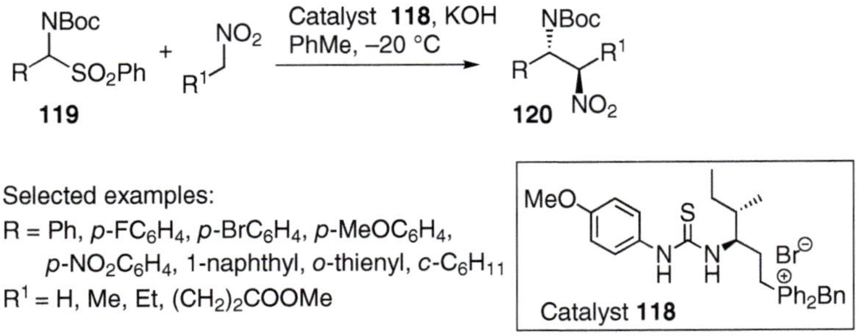

Scheme 4.41 Asymmetric AHR catalyzed by thiourea-phosphonium salt (catalyst **118**).

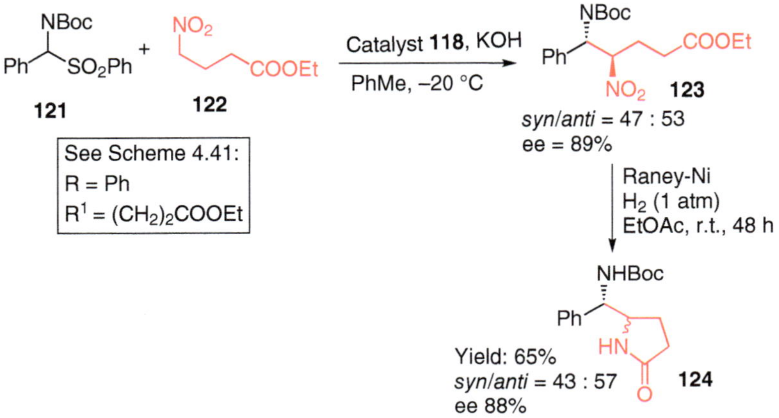

Scheme 4.42 Synthesis of chiral amino-substituted γ-lactam **124**.

112a allows the α-amino acid **116**, than easily converted into the more stable methyl ester **117**.

An interesting chiral bifunctional thiourea-phosphonium salts, obtained from available and cheap chiral amino acids, was proposed as highly efficient phase-transfer catalysts for aza-Henry reaction [90]. As described in Scheme 4.41, the optimum reaction conditions was determined as follow: 5 equiv of KOH and 5 mol% of the catalyst **118** in PhMe, at −20 °C.

Under these conditions the products **120** have been prepared in 77–98% yields and with 88–98 ee. As representative application of this method, chiral amino-substituted γ-lactam **124** has been prepared (Scheme 4.42) from α-amidosulfone **121a** and 4-nitroethylbutyrate.

The obtained nitro adduct **123** is submitted to reduction with consequent cyclization to chiral lactam **124** in 65% yield.

Few years ago Zhao and coworkers reported an interesting study devoted to development of new and efficient phase-transfer catalysis for asymmetric aza-Henry reaction [91]. In this context, a novel class of chiral bifunctional thiourea-ammonium phase-transfer catalysts has been prepared from commercial

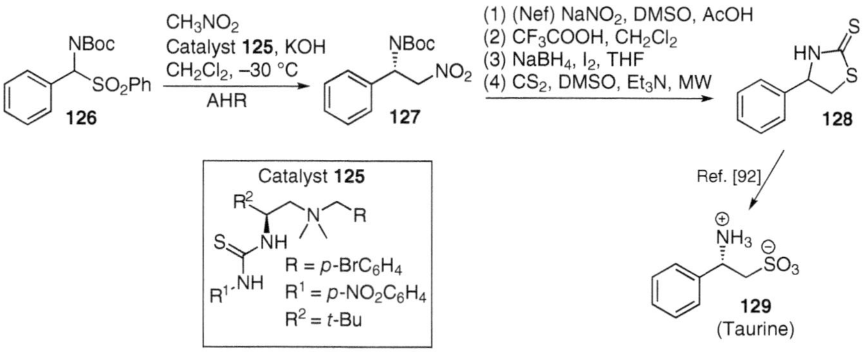

Scheme 4.43 Synthesis of Taurine **129**.

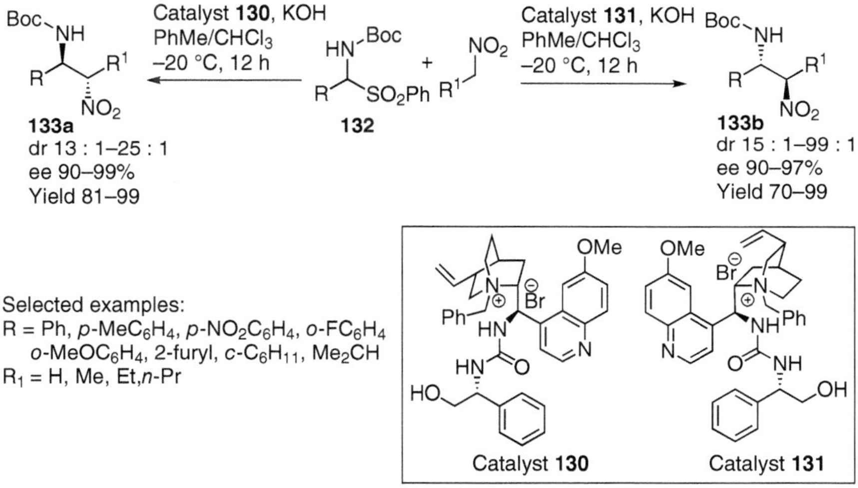

Scheme 4.44 AHR under catalysts **130** and **131**.

α-amino acids and tested. After several trials concerning the reaction of amidosulfones with nitroalkanes ($MeNO_2$, $EtNO_2$, n-$PrNO_2$), the best results (yields up to 99%, ee 99.5%, dr 25 : 1) are obtained under the following reaction conditions: catalyst **125** (5 mol%), KOH, CH_2Cl_2, −30 °C.

As application of the procedure, the amino acid **129** (Taurine) was prepared from **126** (Scheme 4.43; [92]).

Quaternary ammonium type of bifunctional asymmetric phase-transfer catalysts, derived from *Cinchona* alkaloids, has been prepared and tested for aza-Henry reaction of nitroalkanes with amidosulfones [93]. Optimization of the reaction demonstrated that the more efficient catalysts is **130** or **131**, while the best reaction conditions are: 5 mol% of catalyst, KOH as base (5 equiv in water), PhMe/$CHCl_3$ as solvents, at −20 °C (Scheme 4.44).

Depending from the catalyst (**130** or **131**), both enantiomers **133a,b** can be prepared and the products are achieved in high yields (75–99%), high enantioselectivities (90–99% ee) and diastereoselectivity (13 : 1 to 99 : 1 dr).

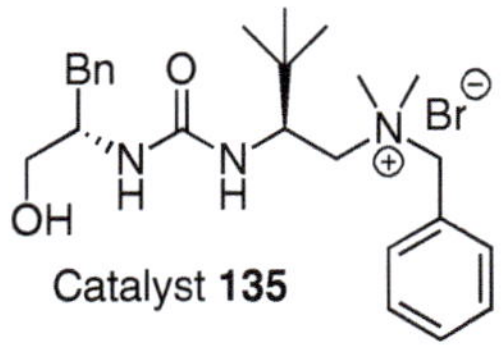

Figure 4.12 Catalyst **135**.

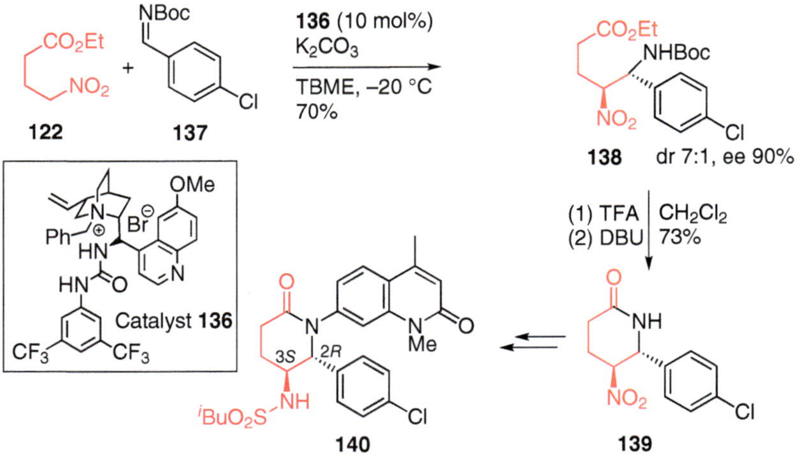

Scheme 4.45 Asymmetric synthesis of **140** under catalyst **136**.

Enantioselective addition of nitroalkanes ($MeNO_2$, $EtNO_2$) to imines generated from α-amidosulfones, lipophilic salts of "chiral tris(1,2-diphenylethylenediamine cobalt(III)) trication" (as hydrogen bond donor catalysts, catalyst **134**, Figure 4.11), allows good yields (89–93%), good ee (70–91%), and high diastereoselectivity (95 : 5) of the products [94].

New quaternary ammonium types of bifunctional asymmetric phase transfer catalyst (Catalyst **135**, Figure 4.12) bearing hydrogen bonding donors, derived from amino acids, have been easily prepared and found to be highly efficient in the asymmetric aza-Henry reaction [95].

A variety of substrates were submitted, and the products were achieved in high enantio- and diastereoselectivities (ee 90–99.9%, dr 90 : 10 to 92 : 2).

Organo catalyst **136**, bifunctional *Cinchona* alkaloid derived phase transfer catalyst, was reported for the asymmetric synthesis of quinolone-fused lactam **140** a potent and selective inhibitor of the BRD7 and BRD9 bromodomains (Scheme 4.45) [96].

The key reaction is the catalytic nitro-Mannich reaction of **122** with **137** that furnish, under catalyst **136** and in *t*-BuOMe (TBME), the adduct **138** as 7 : 1 mixture of diastereomers, both in 90% ee. *N*-Boc deprotection of the latter and simultaneous cyclization with trifluoroacetic acid (TFA), followed by epimerization with DBU, produces the lactam **139** as single diastereomer. The preparation of **140** was completed using conventional reduction, cross-coupling, and sulfonylation reaction.

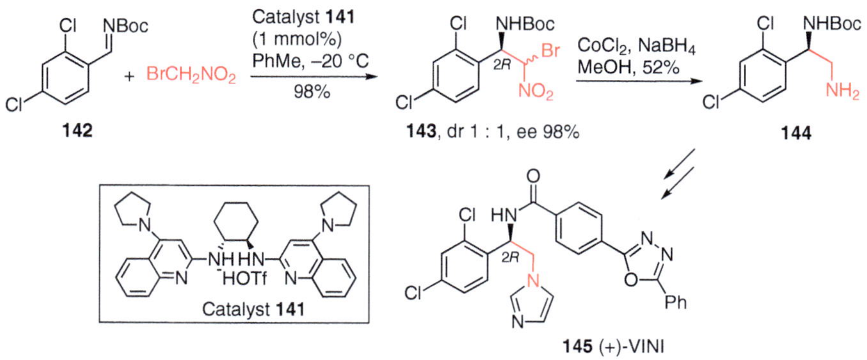

Scheme 4.46 Asymmetric synthesis of **145** under catalyst **141**.

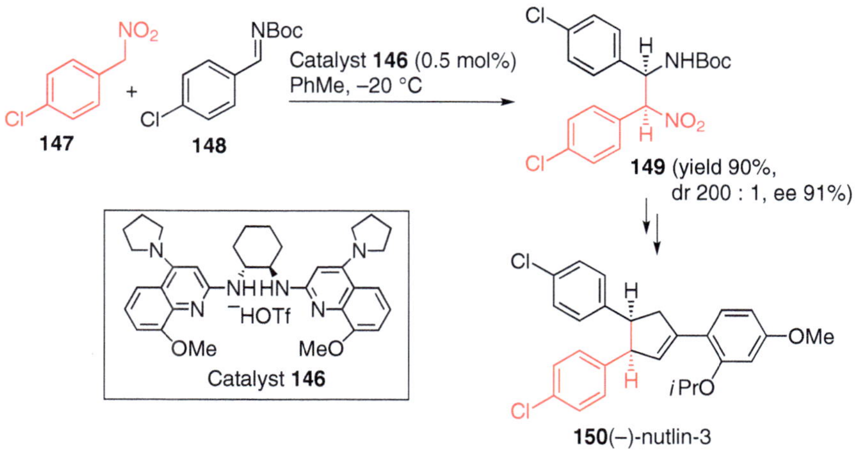

Scheme 4.47 Asymmetric synthesis of **150** under catalyst **146**.

Bisamide catalyst **141** granted the asymmetric synthesis of (+)-*VINI* **145**, a potent inhibitor of CYP51 that is considered as an experimental drug against Chagas' disease caused by I. Cruzi parasites (Scheme 4.46). Aza-Henry reaction of CH_2BrNO_2 (its choice vs $MeNO_2$ is due to the need to suppress the possible double nitro-Mannich reaction) with **142**, is the first step, under catalyst **141** (1 mol%), in PhMe and at −20 °C [97]. The β-nitroamine **143**, obtained with the *R*-configuration of the stereocenter C-2 with excellent enantioselectivity for each diastereomer, is easily reduced to diamine **144**. Successive attachment of the imidazolium fragment, followed by Boc-deprotection and acylation, converts **145** into (+)-VINI.

Catalyst **146** was found highly active for the stereoselective addition of **147** and **148**, the key step for the synthesis of (−)-Nutlin **150** (Scheme 4.47), a promising anti-cancer drug developed by pharmaceutical industry [98].

In fact, the adduct **149** was successively converted into the most reactive levorotatory enantiomer of nutlin-3 by a series of reduction, deprotection, and acylation. The method has been successful used even for a scalable process.

References

1 Henry, L.C.R. (1985). *Acad. Sci.* 120: 1265–1268.
2 (a) Rosini, G. (1991). *Comprehensive Organic Synthesis*, vol. 2 (eds. B.M. Trost and I. Fleming), 321–340. Oxford: Pergamon Press. (b) Rosini, G. and Ballini, R. (1988). *Synthesis*: 833–847.
3 (a) Luzzio, F.A. (2001). *Tetrahedron* 57: 915–945. (b) Boruwa, J., Gogoi, N., Sikia, P.P., and Barua, N.C. (2006). *Tetrahedron: Asymmetry* 17: 3315–3326. (c) Palomo, C., Oiarbide, M., and Laso, A. (2007). *Eur. J. Org. Chem.*: 2561–2574. (d) Alvarez-Casao, Y., Marques-Lopez, E., and Herrera, R.P. (2011). *Symmetry* 3: 220–245.
4 Rosini, G., Ballini, R., and Sorrenti, P. (1983). *Synthesis*: 1014–1016.
5 Mélot, J.-M., Texier-Boullet, F., and Foucaud, A. (1986). *Tetrahedron Lett.* 27: 493–496.
6 Costantino, U., Curini, M., Marmottini, F. et al. (1994). *Chem. Lett.*: 2215–2218.
7 Ballini, R., Bosica, G., and Forconi, P. (1996). *Tetrahedron* 52: 1677–1684.
8 Sampath Kumar, H.M., Subba Reddy, B.V., and Yadav, J.S. (1998). *Chem. Lett.*: 637–638.
9 Ballini, R., Bosica, G., and Parrini, M. (1999). *Chem. Lett.*: 1105–1106.
10 Ballini, R., Bosica, G., Livi, D. et al. (2003). *Tetrahedron Lett.* 44: 2271–2273.
11 Bulbule, V.J., Deshpande, V.H., Velu, S. et al. (1999). *Tetrahedron* 55: 9325–9332.
12 Cwik, A., Fuchs, A., Hell, Z., and Clacenc, J.-M. (2005). *Tetrahedron* 61: 4015–4021.
13 Akutu, K., Kabashima, H., Seki, T., and Hattori, H. (2003). *Appl. Catal., A: General* 247: 65–74.
14 Ballini, R., Bosica, G., Palmieri, A. et al. (2008). *Green Chem.* 10: 541–544.
15 Ganesan, S.S., Ganesan, A., and Kothandapani, I. (2014). *Synlett* 25: 1847–1850.
16 Das, D., Pathak, G., and Rokhum, L. (2016). *RSC Adv.* 6: 104154–104163.
17 Bosica, G. and Polidano, K. (2017). *J. Chem.*: 6267036.
18 Han, J., Yingpeng, X., Su, Y. et al. (2008). *Catal. Commun.* 9: 2077–2079.
19 Tanemura, K. and Suzuki, T. (2018). *Tetrahedron Lett.* 59: 392–396.
20 Rokhum, L. and Pathak, G. (2019). *Catal. Lett.* 149: 2887–2898.
21 Ballini, R., Barboni, L., and Palmieri, A. (2007). *Synlett* 19: 3019–3021.
22 Gelbard, G. and Colonna, S. (1977). *Synthesis*: 113–116.
23 Ballini, R. and Bosica, G. (1997). *J. Org. Chem.* 62: 425–427.
24 Ballini, R., Fiorini, D., Gil, M.V., and Palmieri, A. (2004). *Tetrahedron* 60: 2799–2804.
25 Zhou, C.I., Zhou, Y.Q., and Wang, Z.Y. (2003). *Chin. Chem. Lett.* 14: 355–358.
26 Pandi, M., Chanani, P.K., and Govindasamy, S. (2012). *Appl. Catal., A: General* 441–442: 119–123.
27 Bora, P.P. and Bez, G. (2013). *Eur. J. Org. Chem.*: 2922–2929.
28 Aryanejad, S., Bagherzad, G., and Moudi, M. (2019). *Appl. Organomet. Chem.* 33: 1–11.
29 Archarya, C., Achari, A., and Jaisankar, P. (2018). *Tetrahedron Lett.* 59: 663–665.
30 Ren, Y. and Cai, C. (2007). *Catal. Lett.* 118: 134–138.

31 Rajander Reddy, K., Rajasekhar, C.V., and Gopi Krishna, G. (2007). *Synth. Commun.* 37: 1971–1976.

32 Bulbule, V.J., Jnaneshwara, G.K., Deshmukh, R.R. et al. (2001). *Synth. Commun.* 31: 3623–3626.

33 Jiang, T., Gao, H., Han, B. et al. (2004). *Tetrahedron Lett.* 45: 2699–2701.

34 Khan, F.A., Dash, J., Satapathy, R., and Upadhyay, S.K. (2004). *Tetrahedron Lett.* 45: 3055–3058.

35 (a) Qian, W., Ju, F., Bao, W., and Zhang, Y. (2004). *J. Chem. Res.*: 154–155. (b) Khan, F.A., Sudheer, C., and Sahu, N. (2005). *Synth. Commun.* 35: 201–207.

36 Xia, W.-J., Xie, G.-B., Jang, G.-F., and Le, Z.-G. (2013). *Molecules* 18: 13910–13919.

37 Jagadale, M., Naikwade, A., Salunkhe, R. et al. (2018). *New J. Chem.* 42: 10993–11005.

38 (a) Caruso, T., Feroci, M., Inesi, A. et al. (2006). *Adv. Synth. Catal.* 348: 1942–1947. (b) Elison, M.N., Ilovaisky, A.I., Merkulova, V.M. et al. (2008). *Tetrahedron* 64: 5915–5919.

39 Youn, S.W. and Kim, Y.H. (2000). *Synlett* 6: 880–882.

40 Rosini, G., Ballini, R., Petrini, M., and Sorrenti, P. (1984). *Synthesis*: 515–517.

41 Ballini, R., Bigi, F., Gogni, E. et al. (2000). *J. Catal.* 191: 348–353.

42 (a) Parratt, A.J., Adams, D.J., Clifford, A.A., and Rayner, C.M. (2004). *J. Chem. Soc., Chem. Commun.*: 2720–2721. (b) Rayner, C.M. (2007). *Org. Process Res. Dev.* 11: 121–132. (c) Ballini, R., Noè, M., Perosa, A., and Selva, M. (2008). *J. Org. Chem.* 73: 8520–8528.

43 (a) Palmieri, A., Gabrielli, S., and Ballini, R. (2013). *Synlett* 24: 114–116. (b) Concell6n, J.M., Rodriguez-Solla, H., and Concell6n, C. (2006). *J. Org. Chem.* 71: 7919–7922.

44 Abu-Reziq, R. and Alper, H. (2012). *Appl. Sci.* 2: 260–276.

45 Neelakandeswari, N., Sangami, G., Emayavaramban, P. et al. (2012). *Tetrahedron Lett.* 53: 2980–2984.

46 Varma, R.S., Dahiya, R., and Kumar, S. (1997). *Tetrahedron Lett.* 38: 5131–5134.

47 McNulty, J., Steere, J.A., and Wolf, S. (1998). *Tetrahedron Lett.* 39: 8013–8016.

48 Yan, S., Gao, Y., Xing, R. et al. (2008). *Tetrahedron* 64: 6294–6299.

49 Jalal, S., Sarkar, S., Bera, K. et al. (2013). *Eur. J. Org. Chem.*: 4823–4828.

50 Magnus, P. and Pye, P. (1995). *J. Chem. Soc., Chem. Commun.*: 1933–1934.

51 Chung, W.K. and Chiu, P. (2005). *Synlett* 1: 55–58.

52 Adibekian, A., Timmer, M.S.M., Stallforth, P. et al. (2008). *Chem. Commun.*: 3549–3551.

53 Woodcock, S.R., Marwitz, A.J.V., Bruno, P., and Branchaud, B.P. (2006). *Org. Lett.* 8: 3931–3934.

54 Rodriguez-Solla, H., Alvaredo, N., and Soengas, R.G. (2012). *Synlett* 23: 2083–2086.

55 Chen, Y.-Y. and Lin, W.-Y. (1993). *Tetrahedron Lett.* 34: 2961–2962.

56 Ono, N., Miyake, H., Tamura, R., and Kaji, A. (1981). *Tetrahedron Lett.* 22: 1705–1708.

57 Chen, Y.-J. and Lin, W.Y. (1993). *Tetrahedron* 49: 10263–10270.

58 McNulty, J. and Mo, R. (1998). *Chem. Commun.*: 933–934.

59 Ballini, R., Bosica, G., and Rafaiani, G. (1995). *Helv. Chim. Acta* 78: 879–882.

60 Nakata, T., Komatsu, T., Nagasawa, K. et al. (1994). *Tetrahedron Lett.* 35: 8228–8255.

61 Sasai, H., Suzuki, T., Arai, S., and Shibasaki, M. (1992). *J. Am. Chem. Soc.* 114: 4418–4420.

62 Blay, G., Domingo, L.R., Hernåndez-Olmos, V., and Pedro, J.R. (2008). *Chem. Eur. J.* 14: 4725–4730.

63 Kureshy, R.I., Das, A., Khan, N.H. et al. (2011). *ACS Catal.* 1: 1529–1535.

64 Lai, G., Guo, F., Zheng, Y. et al. (2011). *Chem. Eur. J.* 17: 1114–1117.

65 Jin, W., Li, X., and Wan, B. (2011). *J. Org. Chem.* 76: 484–491.

66 Roy, A., Reddy, A., Dwivedi, N. et al. (2011). *Tetrahedron Lett.* 52: 6968–6970.

67 Lang, K., Park, J., and Hong, S. (2012). *Angew. Chem. Int. Ed.* 51: 1620–1624.

68 White, J.D. and Shaw, S. (2012). *Org. Lett.* 14: 6270–6273.

69 Das, A., Kureshy, R.I., Prathap, K.J. et al. (2013). *Appl. Catal., A: General* 459: 97–105.

70 Zhang, L., Wu, H., Yang, Z. et al. (2013). *Tetrahedron* 69: 10644–10652.

71 Liu, F., Gou, S., Li, L. et al. (2013). *J. Mol. Catal. A: Chem.* 379: 163–168.

72 Scharnagel, D., Prause, F., Kaldun, J. et al. (2014). *Chem. Commun.* 50: 6623–6625.

73 Kaldun, J., Prause, F., Scharnagel, D. et al. (2006). *ChemCatChem* 8: 1846–1856.

74 Cruz, H., Aguirre, G., Madrigal, D. et al. (2016). *Tetrahedron: Asymmetry* 27: 1217–1221.

75 Nováková, G., Drabina, P., Svoboda, J., and Sedlák, M. (2017). *Tetrahedron: Asymmetry* 28: 791–796.

76 Dixit, A., Kumar, P., and Singh, S. (2018). *Chirality* 30: 1257–1268.

77 (a) Yamada, K.-I., Harwood, S.J., Gröger, H., and Shibasaki, M. (1999). *Angew. Chem.* 111: 3713–3715. (b) Yamada, K.-I., Harwood, S.J., Gröger, H., and Shibasaki, M. (1999). *Angew. Chem. Int. Ed.* 38: 3504–3506.

78 Knudsen, K.R., Risgaard, T., Nishiwaki, N. et al. (2001). *J. Am. Chem. Soc.* 123: 5843–5844.

79 Palomo, C., Oirbide, M., Halder, R. et al. (2006). *Angew. Chem. Int. Ed.* 45: 117–120.

80 Li, H., Zhang, X., Shi, X. et al. (2012). *Adv. Synth. Catal.* 354: 2264–2274.

81 Soengas, R.G., Silva, S., Estévez, A.M. et al. (2012). *Eur. J. Org. Chem.*: 4339–4346.

82 Wang, X., Gao, Y., Wei, Z. et al. (2019). *Org. Chem. Front.* 6: 3269–3273.

83 Choudhary, M.K., Tak, R., Kureshy, R.I. et al. (2015). *J. Mol. Catal. A: Chem.* 409: 85–93.

84 Zhu, Z.-Q., Xie, Z.-B., and Le, Z.-G. (2017). *Synlett* 28: 485–488.

85 Marčeková, M., Gerža, P., Šoral, M. et al. (2019). *Org. Lett.* 21: 4580–4584.

86 (a) Liu, S., Gao, W.-C., Miao, Y.-H., and Wang, M.C. (2019). *J. Org. Chem.* 84: 2652–2659. (b) Andrés, J.M., de La Cruz, N., Valle, M., and Perdrosa, R. (2016). *ChemPlusChem* 81: 86–92.

87 Fitch, R.W., Sturgeon, G.D., Patel, S.R. et al. (2019). *J. Nat. Prod.* 72: 243–247.

88 Jiang, X., Zhang, Y., Wu, L. et al. (2019). *Adv. Synth. Catal.* 351: 2096–2100.

89 Wei, Y., He, W., Liu, Y. et al. (2012). *Org. Lett.* 14: 704–707.

90 Cao, D., Chai, Z., Zhang, J. et al. (2013). *Chem. Commun.* 49: 5972–5974.

91 Wang, H.-Y., Chai, Z., and Zhao, G. (2013). *Tetrahedron* 69: 5104–5111.

92 Chen, N., Jia, W., and Xu, J. (2009). *Eur. J. Org. Chem.*: 5841–5846.

93 Wang, B., Liu, Y., Sun, C. et al. (2014). *Org. Lett.* 16: 6432–6435.

94 Joshi, H., Ghosh, S.K., and Gladysz, J.A. (2017). *Synthesis* 49: 3905–3915.

95 Liu, Y., Wei, Z., Liu, Y. et al. (2017). *Org. Biomol. Chem.* 15: 9234–9242.

96 Clark, P.G.K., Vieira, L.C.C., Tallant, C. et al. (2015). *Angew. Chem. Int. Ed.* 54: 6217–6224.

97 Clayton, J. (2010). *Nature* 465: 4–5.

98 Davis, T.A., Vilgeim, A.E., Richmond, A., and Johnston, J.N. (2013). *J. Org. Chem.* 78: 10605–10616.

5

Conjugate Addition of Nitroalkanes to Electron-Poor Alkenes (Michael Reaction)

Conjugate addition of nitroalkanes to electron-poor alkenes (Michael reaction) is one of the most popular reactions (Figure 5.1, way a); however, the nature of nitroalkane (primary, secondary, or CH_3NO_2) and the ratio nitroalkane/electrophile are very important in order to avoid or favor multiple Michael reactions (Figure 5.1, way b,c) [1]. As reported in Chapter 4, nitroalkanes are important source of stabilized carbanions since the electron withdrawing power of the nitro group provides an outstanding enhancement of the geminal hydrogen acidity. Thus, they are valuable precursors of stabilized carbanions under basic conditions, prone to generate a nitronate anion that acts as strategic nucleophile in the conjugate additions (or Michael reaction) to electron-poor alkenes. The anion-solvating predilection of the solvent and the electrophilicity of the alkene are fundamental for the amounts of adducts obtained. In addition, it is important to point out that strong bases are needed when poorly electrophilic alkenes are used, while weak bases are more appropriate to promote the conjugate addition of activated nitroalkanes (i.e. α-nitro ketones) or when reactive electron-deficient alkenes are employed. Moreover, highly electrophilic acceptors are so reactive that basic catalysis is not needed. Thus, a right choice of base and solvent appears strategic for the success of the reaction.

The following acronyms are used in this chapter

MG	Michael general
MH	Michael heterogeneous
MW	Michael in water
MIL	Michael in ionic liquid
MA	Michael asymmetric

5.1 General Homogeneous Procedures

A number of bases in homogeneous solutions, such as trialkylamines [2], triphenylphosphine [3], fluoride salts [4], ammonium hydroxides [5], alkaline metal hydroxides, and alkoxides [6], 1,8-diazabicyclo[5.4.0]undec-7-ene

Nitroalkanes: Synthesis, Reactivity, and Applications, First Edition.
Roberto Ballini and Alessandro Palmieri.

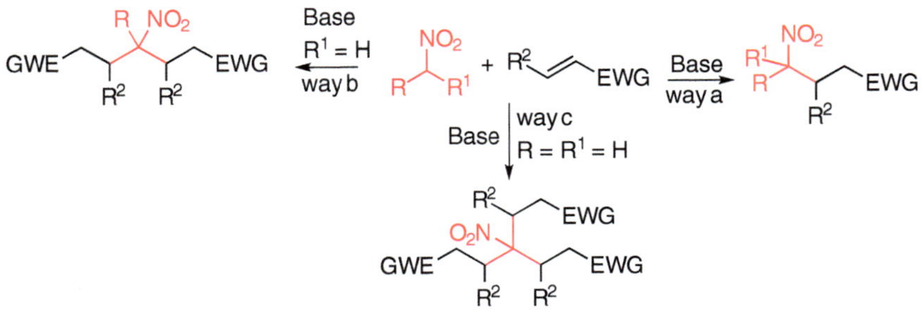

Figure 5.1 Conjugate addition (Michael reaction).

(DBU) or TMG [7], have been originally proposed.

MG1: The in situ Henry–Michael reactions of nitroalkanes **1** with α,β-unsaturated aldehydes **2** under DBU catalysis were even proposed as key step for the preparation multifunctional diamino alcohols **4** (Scheme 5.1) [8].

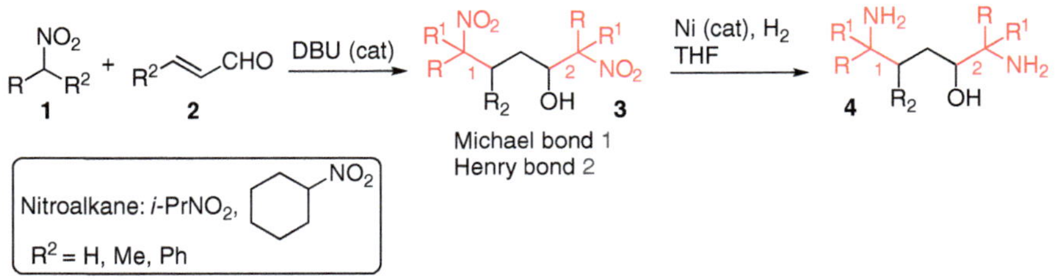

Scheme 5.1 Synthesis of diamino alcohols **4** from in situ Henry–Michael reactions.

In fact, dropwise addition of aldehyde **2** to an excess of nitroalkane **1**, at room temperature and in presence of a catalytic amount of DBU as base, provides the formation of the adducts **3**. The latter, submitted at hydrogenation, allows the target diamino alcohols **4**.

MG2: *exo*-Methylene butyrolactones are an important class of lactones present in a large number of natural products and biologically active compounds. A versatile route for their preparation has reported via a three-step procedure, starting from the Michael reaction of primary nitroalkanes with bromoacrylates **5** (Scheme 5.2) [9]. In fact, the adduct **6** is obtained by a simultaneous formation of C—C bond and HBr elimination.

The obtained α,β-unsaturated esters **6** are of special interest because they can be simultaneously employed as electrophiles (β-position) and/or nucleophiles at carbon bearing the nitro group.

Nef reaction and successive reduction of the formed carbonyl **7** allows, during the acidic work-up, the lactonic systems **8** in good overall yields.

MG3: α,β-Unsaturated nitriles are of great interest in organic synthesis since the nitrile group represents one of the classical functionality in organic chemistry. Thus, the Michael addition of nitroalkanes to compound **9** (Scheme 5.3) [10] is

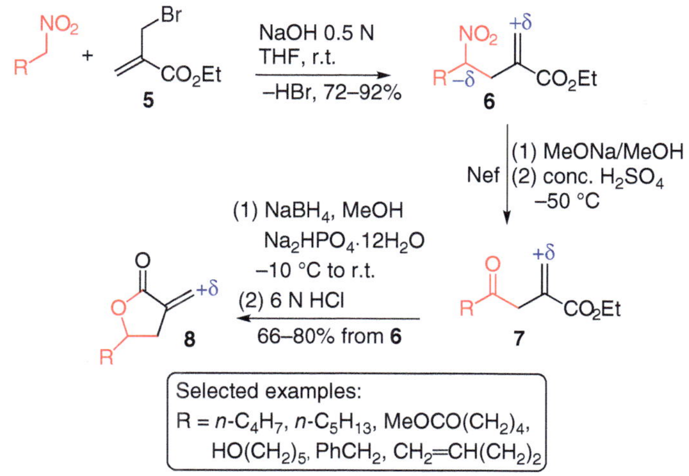

Scheme 5.2 Synthesis of methylene butyrolactones **8**.

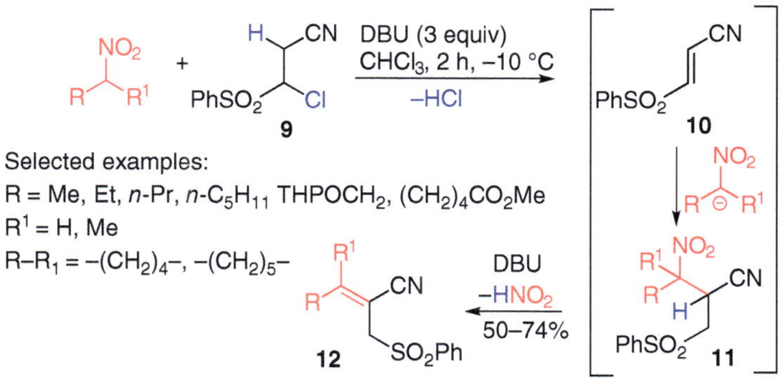

Scheme 5.3 Synthesis of α,β-unsaturated nitriles **12**.

the key step for the synthesis of functionalized α,β-unsaturated nitriles **12**, via elimination of nitrous acid.

Thus, **12** can be prepared in a one-pot process, in short reaction times (two hours) and in good overall yields (50–74%).

5.2 Heterogeneous Procedures

A variety of heterogeneous procedures, mainly under solvent-free conditions (SFCs), have been developed with evident advantages from sustainable point of view. In the first report, basic alumina was used as solid base, in the absence of any solvents, for an efficient conjugate addition of nitroalkanes to a large variety of conjugated enones [11]. Successively, basic alumina doped with KF were reported [12]. Amberlyst A21, a macroreticular ion-exchange resin, supports the Michael addition of simple or functionalized nitroalkanes to methyl acrylates [13]. An easy work-up is needed since a rapid recovery of the crude products can be obtained by filtration.

MH1: A general procedure was achieved with Amberlyst A27 (purchased from Carlo Erba) under neat conditions [14]. In fact, as reported in Scheme 5.4, a large variety of nitroalkanes are prone to afford the conjugate addition to a huge number of electron-poor alkenes **13**, with satisfactory to excellent yields of products **14**.

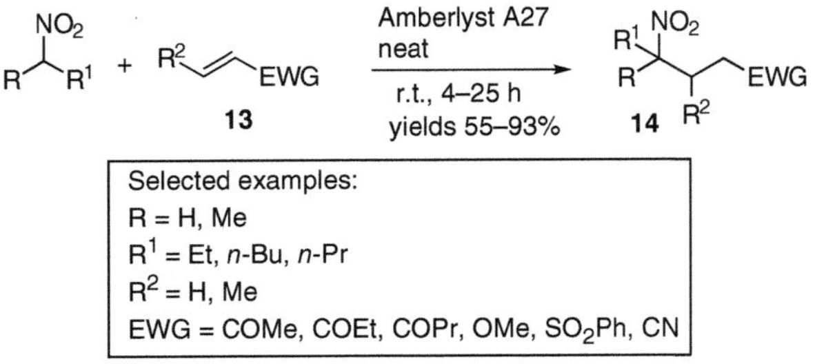

Scheme 5.4 Michael reaction using Amberlyst A27 as promoter.

MH2: Nonionic bases such as $P(RNCH_2CH_2)_3N$ [15] and Merrifield Resin-$C_6H_4CH_2N_3P(MeNCH_2CH_2)_3N$ (Verkade's bases) are effective with a number of examples, including secondary and cyclic nitroalkanes and hindered acceptors. Other catalysts such as $Yb(OTf)_3$ (a mild Lewis catalyst) in H_2O [16] or silica gel 60 under neat conditions [17] are very efficient for 1,4-addition of activated α-nitro carbonyl derivatives (α-nitro ketones and α-nitro esters).

MH3: Isolute® Si-carbonate [silica trimethylammonium carbonate; Si-TMS $(CO_3^{-2})_{0.5}$] catalyzes, under SFCs and at room temperature, nitroalkanes addition to a variety of electron-poor alkenes (Scheme 5.5) [18]. Good yields of products **15** (60–91%) are accomplished in 7–27 hours.

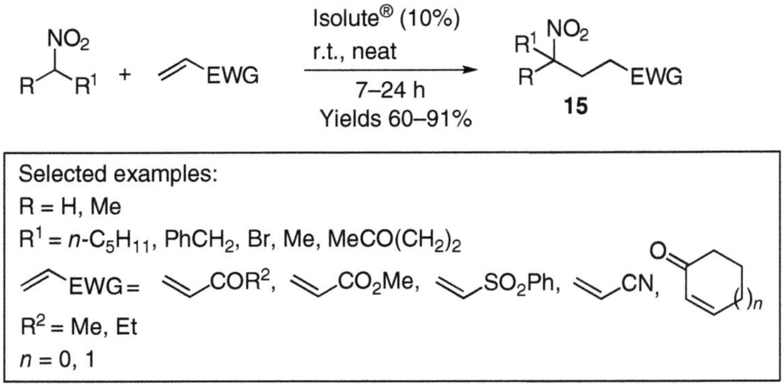

Scheme 5.5 Michael reaction using Isolute® as promoter.

It is important to note that the catalyst is commercially available and can be reused at least six times with modest decreasing of the efficiency. Moreover, because Isolute Si-carbonate works well even in the Henry reaction (see Chapter 4), the use

Figure 5.2 Structure of PS-BEMP.

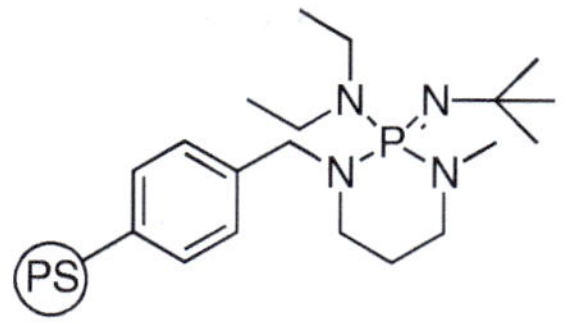

of the same catalyst for both Henry and Michael reactions offers an important opportunity, as reported in the same paper, for a one-pot Henry–Michael solventless process (Scheme 5.6) from **16**, with the formation of the double adduct **17** and traces of the heterocyclic structure **18** (ratio **17**/**18** = 5 : 1).

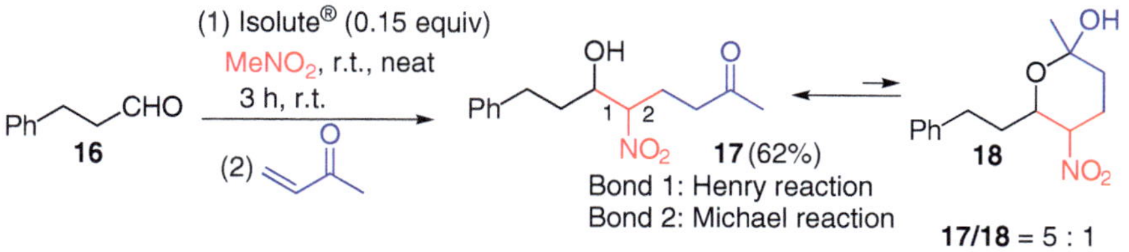

Scheme 5.6 One-pot Henry–Michael reaction promoted by Isolute®.

MH4: 2-*tert*-Butylimino-2-diethylamino-1,3-dimethylperhydro-1,3,2-diazaphosphorine supported on polystyrene (PS-BEMP, Figure 5.2) was found to be a very efficient catalyst for the conjugate addition of nitroalkanes to a variety of α,β-unsaturated carbonyl compounds without the need of solvents [19].

A number of enones are successful tested giving excellent results of Michael adducts (Table 5.1).

The catalyst can be recycled just three times because the magnetic stirring bar causes crushing of the catalyst beads, hampering its complete recovery. However, the reuse of the catalyst in order to improve the environmental efficacy can be solved since the first solvent-free cyclic continuous-flow reactor has been set up.

MH5: α-Nitro ketones possess the simultaneous presence of two electron-withdrawing groups (nitro and carbonyl), linked to the same carbon, that strongly favor the enolic form. Thus, conjugate addition of linear or cyclic α-nitro ketones to α,β-unsaturated carbonyls can be performed under acidic (SiO_2) and neat conditions (Table 5.2) [17].

MH6: Bifunctional heterogeneous catalyst "silica-alumina-supported tertiary amines" with controlled acid–base interactions was reported for the Michael addition of primary and secondary nitroalkanes to enones [20]. The procedure assures good yields (up to 96%) avoiding the formation of undesirable side reactions.

MH7: Poly(propyleneimine) (PPI) dendrimers, functionalized with C_{16} alkyl chain, were found as novel and efficient catalyst for the intramolecular Michael reaction (Scheme 5.7) [21].

Thus, the alkylated PPI dendrimers C_3–C_{16} acts as an organocatalyst and showed a novel dendritic effect in the intramolecular Michael reaction based on substrate orientation within its internal nanocavity.

Table 5.1 Michael reaction promoted by PS-BEMP (selected examples).

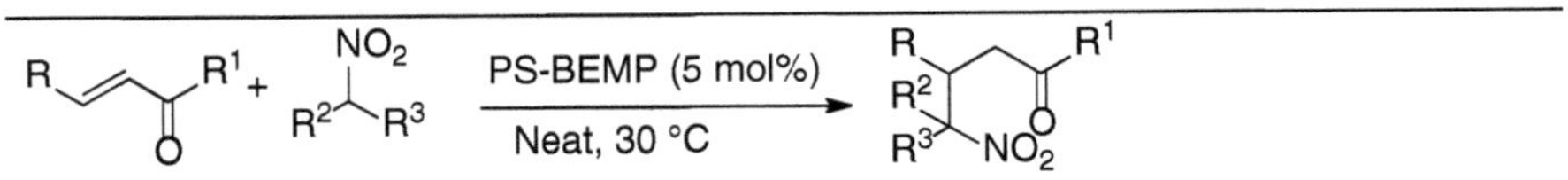

Entry	Enone	Nitroalkane	Product	Yields (%)	Time h
1	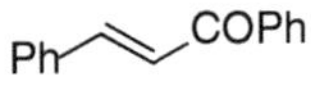	CH_3NO_2	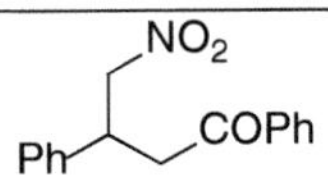	88	18
2	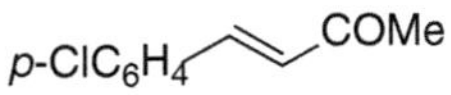	CH_3NO_2	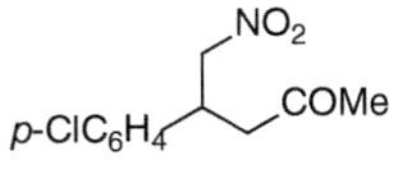	85	4
3	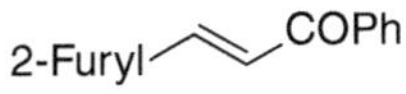	CH_3NO_2	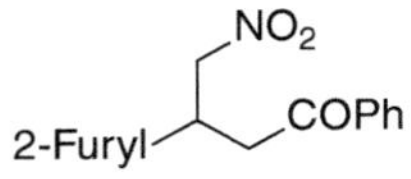	76	15
4	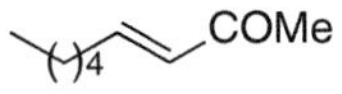	CH_3NO_2	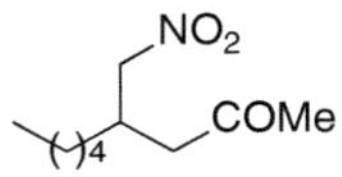	94	4
5	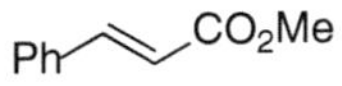	CH_3NO_2	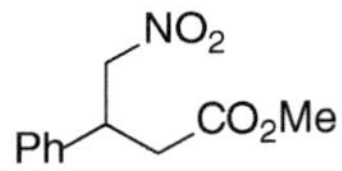	85	18
6	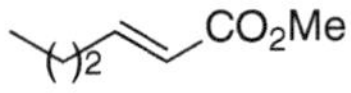	CH_3NO_2	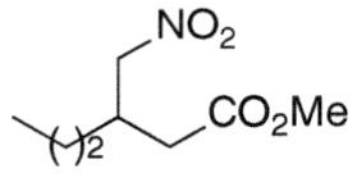	95	22
7	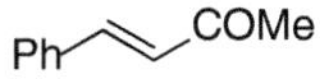	$EtNO_2$	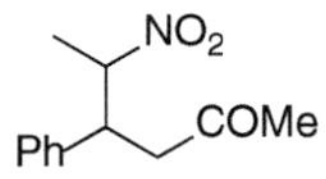	98	4
8	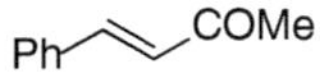	*i*-$PrNO_2$	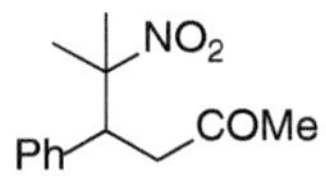	98	20
9	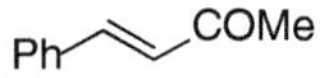	*c*-$C_5H_9NO_2$	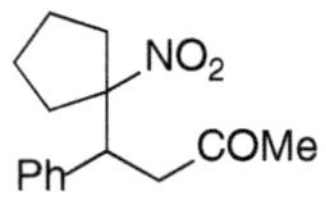	85	4

MH8: A one-pot procedure of Michael reaction from electron-poor alkenes and haloderivatives **21** was developed under heterogeneous catalysis (Scheme 5.8) [22].

Thus, in situ halo-nitro substitution promoted by IRA-402 nitrite produces a nitroalkane that, under Amberlyst A21 catalysis, proceeds with the conjugate addition to the enone and formation of adducts **22** in good overall yields (53–65%).

Table 5.2 Michael reaction of α-nitro ketones under SiO_2 catalysis (selected examples).

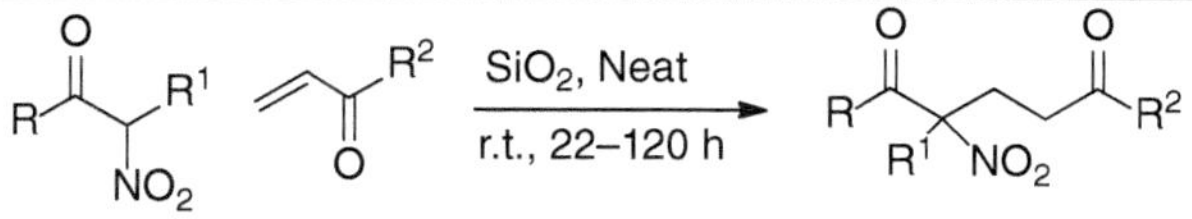

Entry	α-Nitro Ketone	Enone	Yield (%)	Time (h)
1	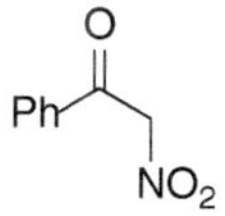	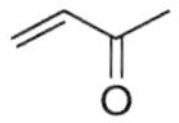	87	22
2	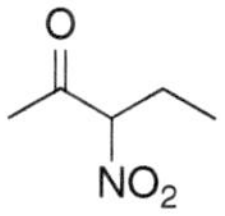	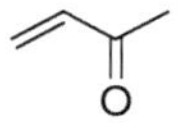	74	100
3	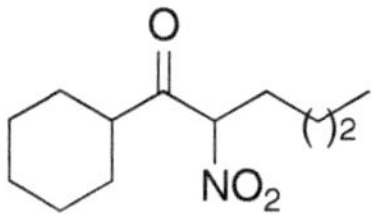	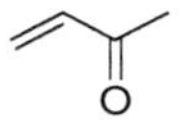	86	100
4	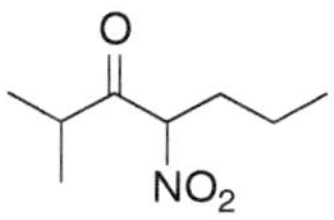	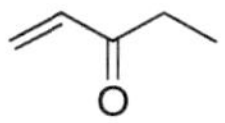	72	120
5	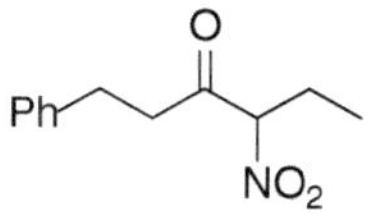	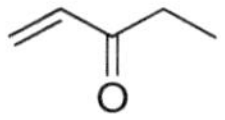	76	120
6	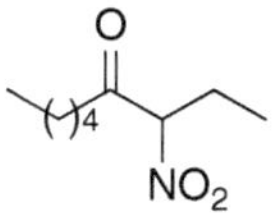	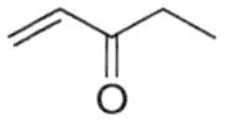	77	50
7	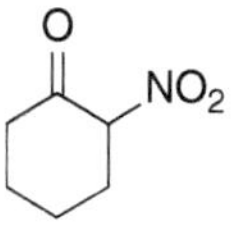	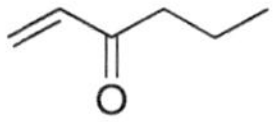	75	44
8	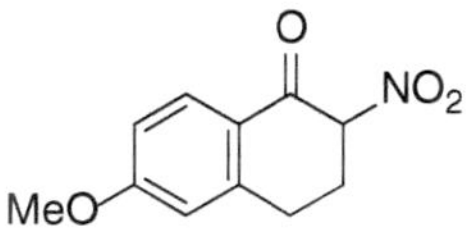	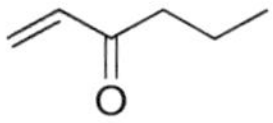	88	70

MH9: Activating techniques, such as microwave and ultrasound irradiation, have been proposed for a more efficiency of the catalyst, mainly those working under heterogeneous and neat conditions [23].

MH10: The conjugate addition of nitroalkanes to dienoic esters **23**, under Amberlyst A27, produces 1,6-adducts as a regiomeric mixture of α,β-**24** (major) and β,γ-unsaturated **25** (minor) adducts (Scheme 5.9) [24].

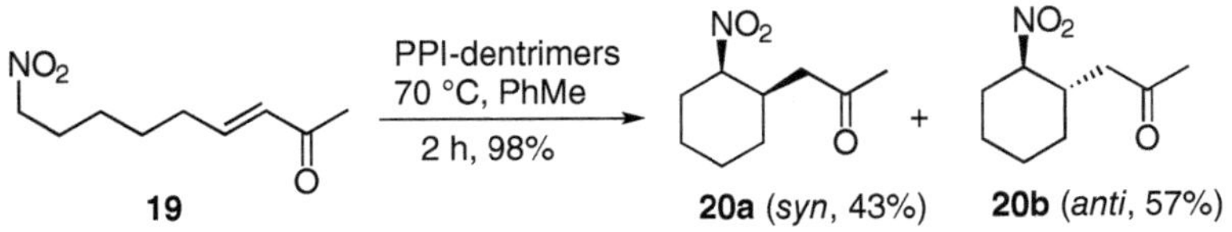

Scheme 5.7 Intramolecular Michael reaction.

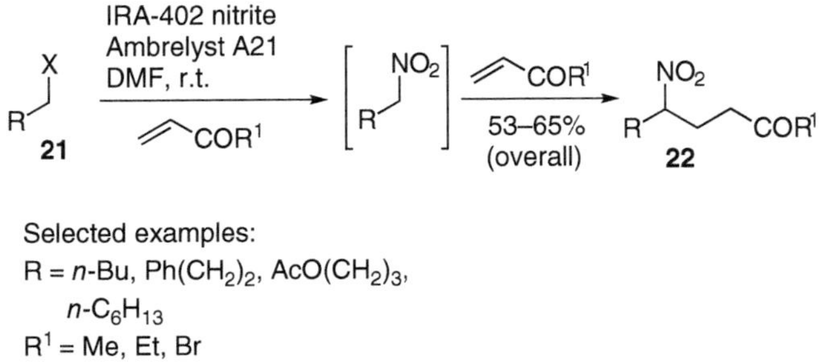

Scheme 5.8 One-pot Michael reaction from haloderivatives.

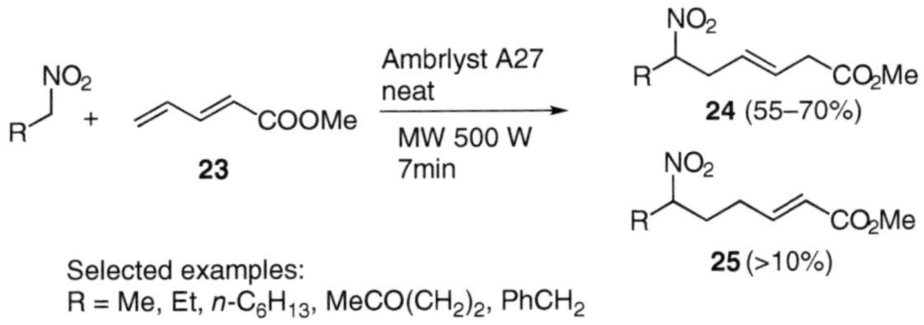

Scheme 5.9 Conjugate addition to the dienoic ester **23**.

5.3 Michael Reaction under Green Solvents

Due to the strategic role of the solvents in organic synthesis from sustainable point of view, a variety of procedures of Michael reaction under eco-friendly solvents have been reported in the last two decades.

MW1: The first report of conjugate addition of nitroalkanes to enones in water (without any organic solvents) was reported using NaOH 0.025 M, in the presence of cetyltrimethylammonium chloride (CTACl) as cationic surfactant (Table 5.3) [25].

Good yields (up to 93%) are obtained in short reaction times (one to two hours), even with hindered (entries 5–7) or non-polar (entry 3) starting molecules.

MW2: As extension of the aforementioned procedure, few years later the method has been successfully extended [26] to a variety of other electron-poor alkenes such as CH_2=CHCOOMe, CH_2=CHCN, CH_2=CHSO_2Ph.

MW3: As improvement of the previous procedure, the conjugate addition of primary nitroalkanes to electron-poor alkenes (EWG = COR, COOR, CN, Ts) is catalyzed with aqueous solution of cetyltrimethylammonium hydroxide (CTAOH) that acts

Table 5.3 Michael reaction promoted by NaOH/CTACl system (selected examples).

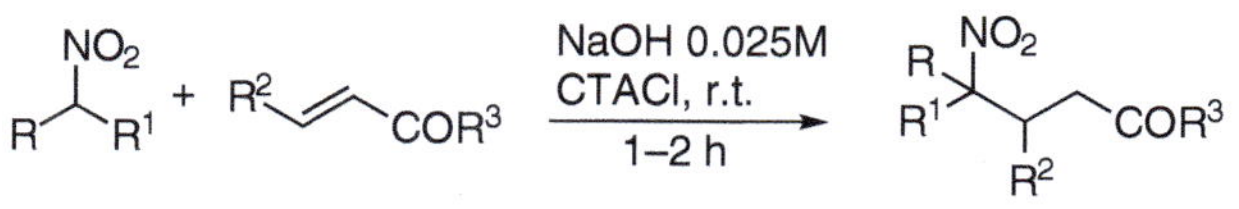

Entry	R	R^1	R^2	R^3	Yield (%)
1	CH_3	H	H	CH_3	76
2	$CH_3(CH_2)_3$	H	H	CH_3	83
3	$CH_3(CH_2)_9$	H	H	CH_3	74
4	CH_3	CH_3	H	$CH_3(CH_2)_2$	77
5	$-(CH_2)_5-$		H	CH_3	72
6	$CH_3(CH_2)_3$		H	$-(CH_2)_3-$	79
7	$CH_3(CH_2)_3$		H	$-(CH_2)_2-$	93

both as base and surfactant [27]. Good yields (70–90%) of adducts are observed at room temperature and under short reaction times (one to two hours).

MW4: Conjugate addition of nitroalkanes to enones, in aqueous medium, offers the opportunity to obtain 1,4-difunctionalized structures in a one-pot process (Scheme 5.10) [28], starting from the same molecules, by the appropriate choice of the sequence addition of the following reagents: (i) K_2CO_3, (ii) $NaBH_4$, and (iii) H_2O_2.

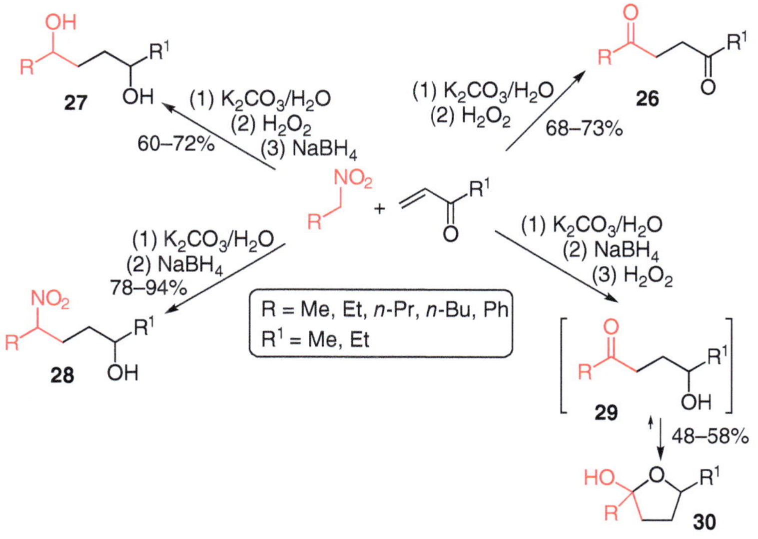

Scheme 5.10 One-pot synthesis of 1,4-difuntionalized molecules from nitroalkanes via the Michael reaction in water.

In fact, treating a mixture of nitroalkane and enone with aqueous K_2CO_3 (base for Michael reaction), and H_2O_2 oxidation (oxidant for Nef reaction) gives, directly, 1,4-diketones **26** in 68–73% yields.

Addition of $NaBH_4$ to the aforementioned procedure furnishes 1,4-diols **27** in 60–72% overall yields. If the starting molecules are treated with K_2CO_3/H_2O (Michael reaction and formation of γ-nitroketone) and $NaBH_4$ (reduction of carbonyl to hydroxyl), the nitroalcohols are prepared in 78–94% yield. Finally, adding the reagents in this sequence: K_2CO_3/H_2O (Michael reaction), $NaBH_4$ (reduction of carbonyl to hydroxyl), and H_2O_2 (Nef reaction), the hydroxy-carbonyls **29** (in equilibrium with the hemiketals **30**) are obtained in 48–58% yields.

Thus, by this one-pot procedure four different structures (**26–29**) can be prepared starting from common molecules. Moreover, the use of water as solvent, K_2CO_3 as base, and H_2O_2 as oxidant gives clear advantages from sustainable point of view.

MW5: The Michael reaction of nitroalkanes with chalcones **31**, in aqueous medium, is the key step for the one-pot synthesis of substituted Δ^1-pyrrolidines **33** (Scheme 5.11) [29].

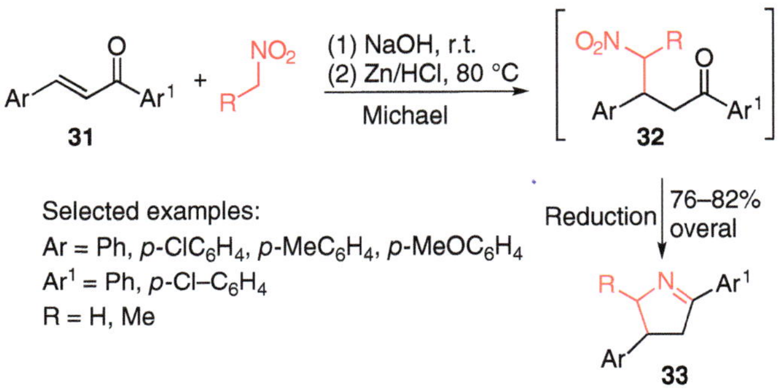

Scheme 5.11 Synthesis of substituted Δ^1-pyrrolines **33**.

Thus, $NaOH/H_2O$ promotes the Michael reaction, while Zn/HCl provides in situ reduction of the nitro group of **32**, with the consequent cyclization to the target structure **33**.

Ionic liquids (ILs) have been also studied as possible green solvents for the Michael reaction.

MIL1: During researches of the conjugate addition of active methylene compounds with electron-poor alkenes, in presence of basic ionic liquid,1-methyl-3-butylimidazonium hydroxide, [bmim]OH that acts both as solvent and as base [30], primary nitroalkanes have demonstrated to be highly reactive with enones and α,β-unsaturated esters.

MIL2: Few years later a more general procedure in IL was developed under base-free conditions [31]. As reported in Table 5.4, primary and secondary nitroalkanes, even functionalized ones, allow satisfactory to good yields with a variety of enones, in the presence of $BmimNTf_2$, as medium, that can be recycled several times without any appreciable loss of activity.

Table 5.4 Michael reaction performed by $BmimNTf_2$ system (selected examples).

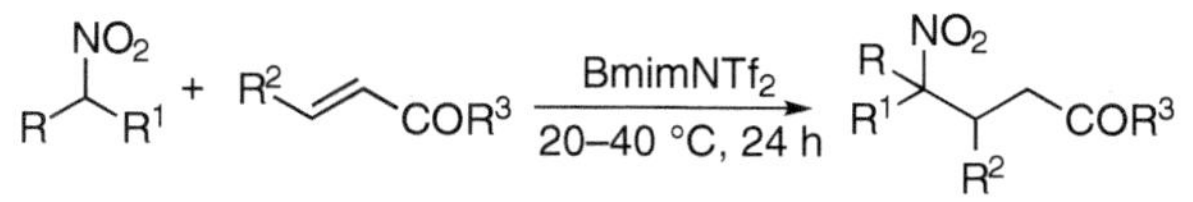

Entry	R	R[1]	R[2]	R[3]	Yield (%)
1	$CH_3(CH_2)_4$	H	H	CH_3	60
2	$Ph(CH_2)_2$	H	H	CH_3	53
3	$CH_2{=}CH(CH_2)_3$	H	H	CH_3CH_2	55
4	CH_3	$CO_2CH_2CH_3$	$-(CH_2)_3-$		72
5	H	$CO_2CH_2CH_3$	$-(CH_2)_3-$		60
6	PhCO	H	$-(CH_2)_3-$		68
7	$CH_3(OCH_2CH_2O)CCH_2$	H	$-(CH_2)_3-$		65
8	$NC(CH_2)_4$	H	$-(CH_2)_3-$		70

MIL3: Solvent-free synthesis of γ-nitro ketones and γ-nitro esters, promoted by 1-butyl-3-methylimidazolium-BF_4 ([bmim][BF_4])/K_2CO_3 has been developed for the Michael reaction of nitroalkanes with enones [32].

MIL4: During a general study of the Michael reaction of methylene active compounds to enones, it has been reported some examples of Michael reaction of primary nitroalkanes in IL and in presence of Me_2CO_3 as base [33].

5.4 Asymmetric Michael Reaction

Because the conjugate addition of nitroalkanes to electron-poor alkenes often involves the formation of stereogenic centers, a huge number of stereoselective methods have been reported [1].

An asymmetric Michael reaction can be performed following two different approaches: (i) asymmetric reaction using optically active alkenes or (ii) use of chiral catalysis. Thus, this chapter reports some of the most recent and significant applications of asymmetric Michael reaction of nitroalkanes.

5.4.1 Asymmetric Michael Reaction with Enones

MA1: Asymmetric Michael reaction of nitroalkanes with enones has been investigated using chiral imidazolidine-2-yltetrazole organocatalysis **35** [34].

The reaction is performed (Scheme 5.12) at room temperature in presence of catalyst **35** (0.1 mmol) without any solvents.

Despite the long reaction times (up to 300 hours), the results in terms of yields (up to 93%) and ee (up to 92%) are completely satisfactory.

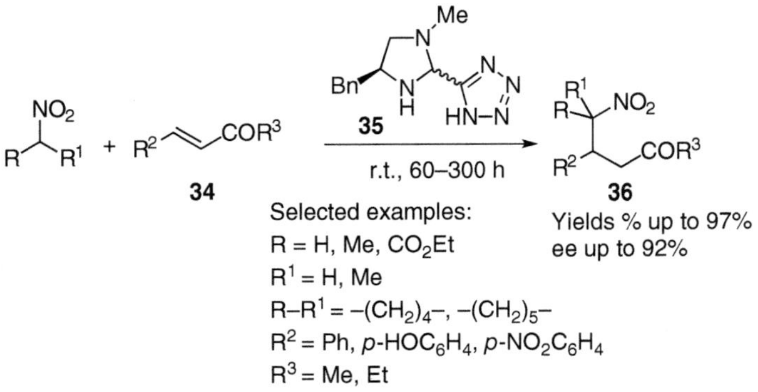

Scheme 5.12 Michael reaction of nitroalkanes under chiral imidazoline–tetrazole **35** catalysis.

MA2: 5-Pyrrolidin-2-yltetrazole is a strategic organocatalyst for the asymmetric conjugate addition of primary and secondary nitroalkanes to cyclic and acyclic enones, in the presence of *trans*-2,5-dimethylpiperazine as base (Scheme 5.13) [35].

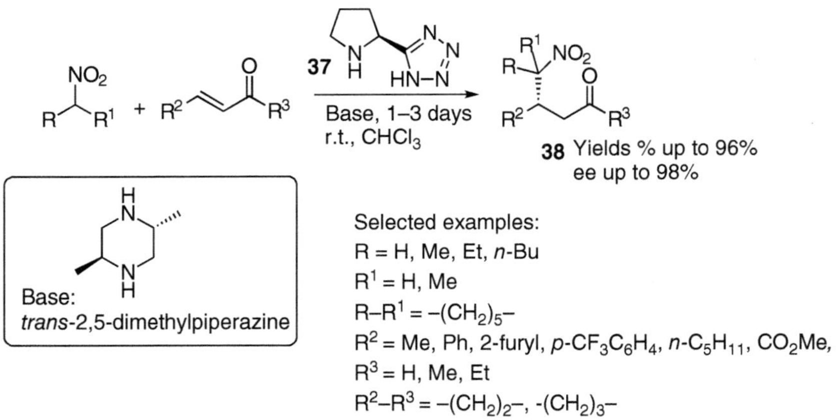

Scheme 5.13 Michael reaction of nitroalkanes under 5-pyrrolidin-2-yltetrazole **37** catalysis.

Under the catalyst **37** the reaction is scalable, providing enantiomeric excess of up to 98%, in one to three days, and using only 2 equiv of nitroalkane.

MA3: 4-*trans*-amino-Proline (di- and tetra-peptides, **39** and **40**) has been successfully applied as chiral organocatalyst for the enantioselective Michael reaction of primary and secondary nitroalkanes to cyclic enones (Scheme 5.14) [36].

This method was tested, separately, with both dipeptide **39** and Tetrapeptide **40** and for many substrates the results are comparable. Anyway, satisfactory to good yields of **41** are obtained (up to 100%) with ee up to 88%.

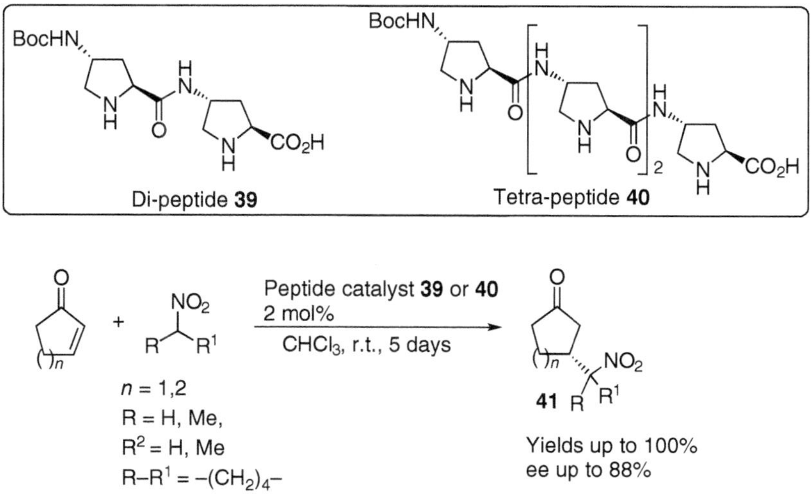

Scheme 5.14 Michael reaction of nitroalkanes catalyzed by *trans*-amino-prolines **39** or **40**.

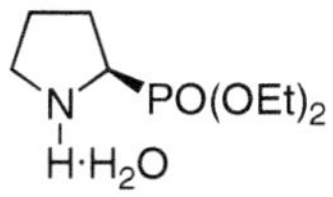

Figure 5.3 α-Aminophosphonate.

MA4: Asymmetric addition of nitroalkanes to enones was reported under chiral α-aminophosphonates (Figure 5.3) as a new class of catalysts for asymmetric Michael reaction in presence of *trans*-2,5-dimethylpiperazine as base [37].

The hydrate form of the catalyst seems important because it increases the catalytic activity. Thus, good to high yields and good enantioselectivities are achieved, in few days, with a range of substrates.

MA5: Asymmetric Michael reaction of nitroalkanes to enones, catalyzed by multifunctional organocatalyst **42** (Scheme 5.15), allows good yields (up to 98%) of adducts **43** (ee up to 98%) [38].

The organocatalyst, consisting of 1,2-diaminocyclohexane and 9-aminocinchona alkaloid derivatives, possesses three strategic functional groups: primary amine that activates the enone, tertiary amine, and thiourea that activate the nucleophiles.

MA6: Enantio- and diastereoselective conjugate addition of nitroalkanes to enones has been performed with thioureas derived from α-amino acids **44** (Figure 5.4) [39]. The method was successfully tested with chalcones and cyclic and acyclic enones.

The reaction was performed under neat conditions followed by computational study and the authors suggested that the reaction involves an interaction of the nitro group with thiourea while the ketone is activated in the protonated amine ($NHMe_2^+$).

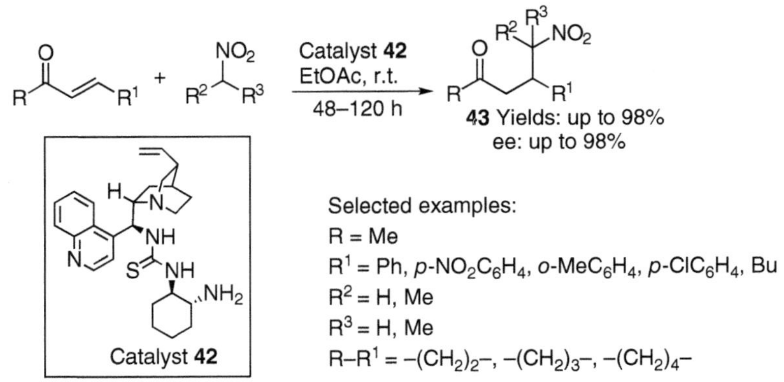

Scheme 5.15 Michael reaction of nitroalkanes catalyzed by *Cinchona* alkaloid derivatives **42**.

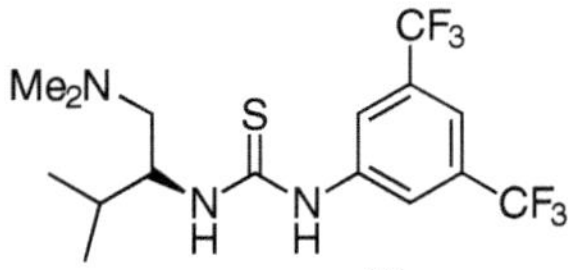

44

Figure 5.4 Catalyst **44**.

MA7: Cyclohexanediamine-derived primary amine thiourea **45** catalyzes the asymmetric conjugate addition of nitromethane to enones (Scheme 5.16) [40]. A wide range of enones have been tested and high to excellent enantioselectivities (92–99%) of products **46** are achieved.

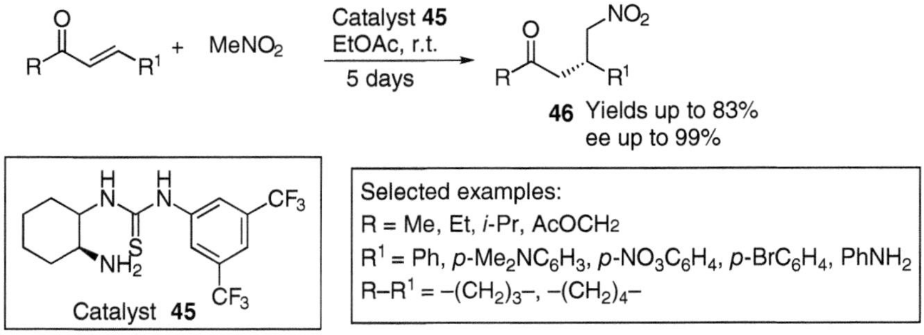

Scheme 5.16 Michael reaction catalyzed by the thiourea **45**.

The authors assert that, following their studies, it is possible to reveal that slight structural modification of an organocatalyst can create a more efficient catalytic system for improving reaction efficiency that cannot be achieved with the existing promoters.

MA8: Enantioselective conjugate addition of α-nitroacetate **47** to enones in aqueous medium and under chiral bifunctional organocatalyst **48** achieves the corresponding adducts **49** with high enantioselectivity (Scheme 5.17) [41]. Thus, the process

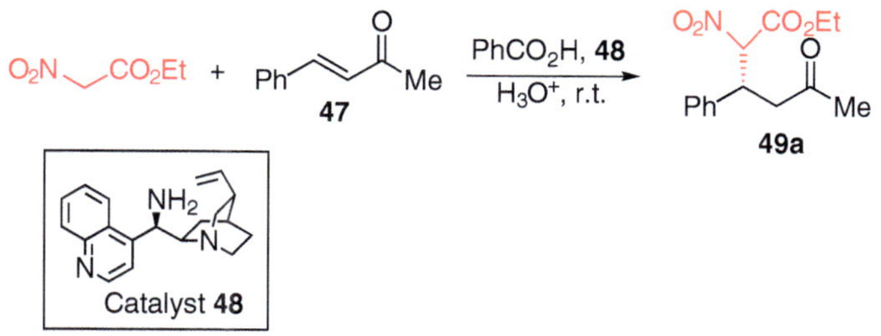

Scheme 5.17 Conjugate addition of α-nitroacetate to α,β-unsaturated ketones in water.

is efficiently catalyzed by *Cinchona* alkaloid derivative **48** containing chiral primary amine. Good yields (up to 98%) and excellent enantiomeric excess (up to 93% ee) are obtained.

The significance of polyfunctionalized compounds **49** is demonstrated by their possible transformations into γ-nitro ketones **50**, via decarboxylation, or into γ-keto esters **51**, via denitration (Scheme 5.18).

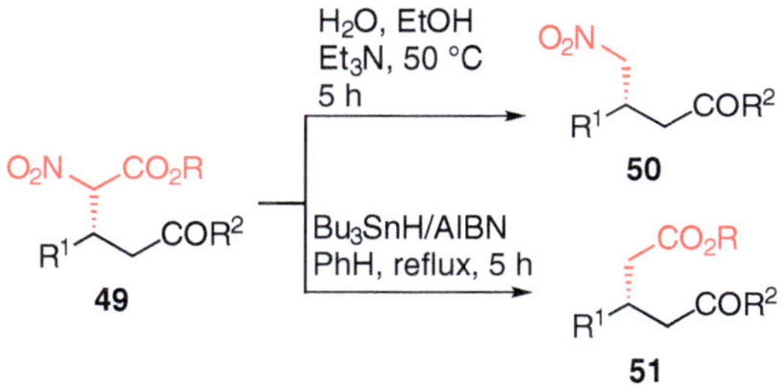

Scheme 5.18 Synthetic applications of **49**.

MA9: Camphor-derived C_1-symmetric chiral diamine **52**, in combination with acid additive Boc-L-PheOH shows high efficiency in promoting the asymmetric conjugate addition of nitroalkanes ($MeNO_2$, $EtNO_2$, *n*-$PrNO_2$, *i*-$PrNO_2$) to enones with high yields (up to 96%) and enantioselectivities (up to 98%), under mild conditions [42]. Scheme 5.19 reports some representative examples.

It is important to point out that the catalyst, due to its stability, could be recovered by simple aqueous acid/base work-up.

MA10: Conjugate addition of nitroalkanes to acrylate equivalent is of great interest because reduction of the nitro group to amine affords γ-amino acid (γ-AA) that are a strategic class of building blocks in biology and sciences. In this context, a decade ago Palomo and coworkers [43] published an efficient selective conjugate addition of prochiral nitroalkanes to acrylate equivalents **53** (Scheme 5.20) having an auxiliary (X_c, camphor) with high level of asymmetric induction.

The method employs the following tertiary catalytic system: $Cu(OTf)_2$ as Lewis acid, *N*-methyl piperidine (NMP) as Brønsted base and molecular sieves (MS). The chiral adducts **54** are obtained in good yields (57–89%), and excellent dr (up to 97/3)

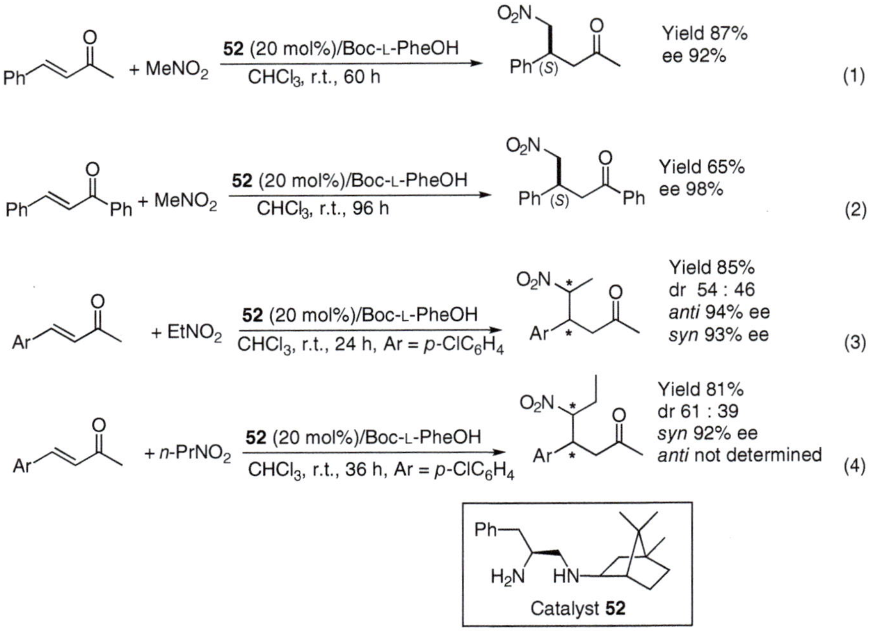

Scheme 5.19 Michael reaction catalyzed by diamine **52**.

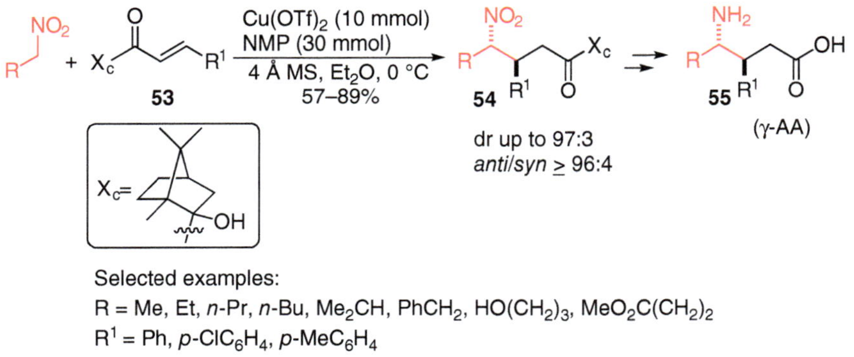

Scheme 5.20 Conjugate addition of prochiral nitroalkanes to acrylate equivalents **53**.

and *anti/syn ratio* (≥96/4). Moreover, **54** can be an easily source of enantioenriched γ-amino acids **55** with recovery of camphor auxiliary.

MA11: Enantioselective conjugate addition of nitroalkanes to 5-styrylisoxazoles **56**, as new acceptors, in presence of **57** as catalysis, was studied by Bernardi and coworkers [44] As a representative example we report the application of this method to the synthesis of (*S*)- and (*R*)-Baclofen **60**, a $GABA_B$-receptor agonist used in treatment of spasticity (Scheme 5.21).

The first step consists in the enantioselective Michael addition (ee 89–91%) of nitromethane to **56**. The latter is converted into nitro acid **59** that was reduced to the target compounds **60a,b** (*S*)-(+)-Baclofen-HCl and (*R*)-(−)-Baclofen-HCl.

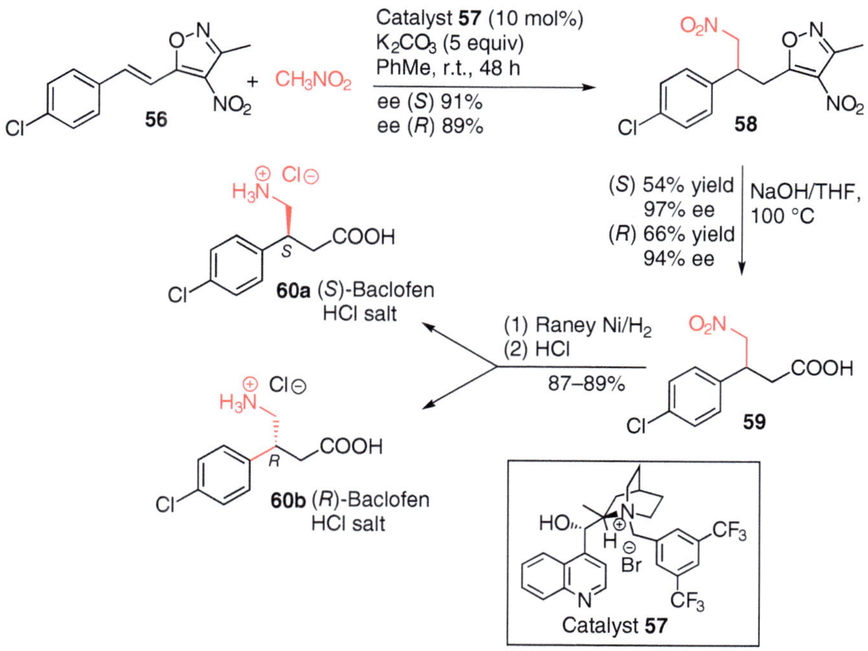

Scheme 5.21 Synthesis of (*S*)- and (*R*)-Baclofen **60a,b**.

5.4.2 Asymmetric Michael Reaction with Enals

Three possible competitive reactions could be observed during the conjugate addition of nitroalkanes to enals: Michael reaction (way a) or Henry reaction (way b), or both Michael and Henry reactions (way c), see Figure 5.5. Thus, very important are the reaction conditions and the *ratio* nitroalkane/enal.

MA12: Highly enantioselective organocatalytic Michael reaction of nitroalkanes to enals was developed under (*S*)-diphenylprolinol TMS ether **62** catalysis, affording conjugate adducts with good yields (up to 90%) and enantiocontrol (≥87–99% ee, Scheme 5.22) [45].

Authors applied this procedure to a convenient synthesis of Baclofen, as reported in Scheme 5.23.

Figure 5.5 Possible adducts (ways a–c) of nitroalkanes to enals.

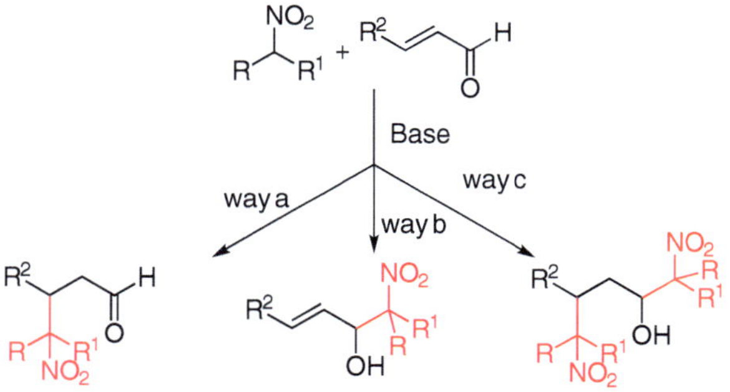

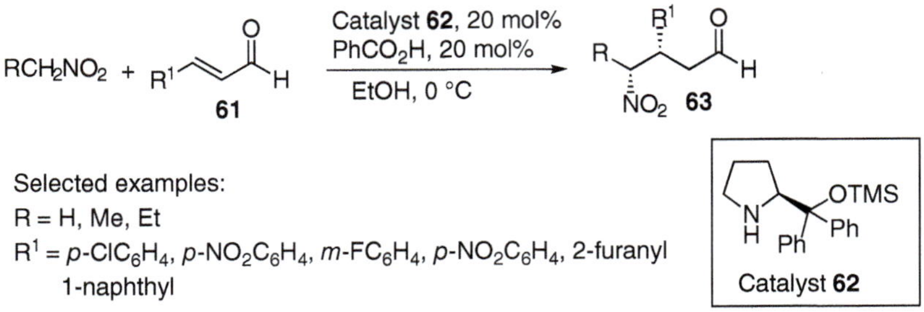

Scheme 5.22 Asymmetric Michael reaction promoted by the organocatalyst **62**.

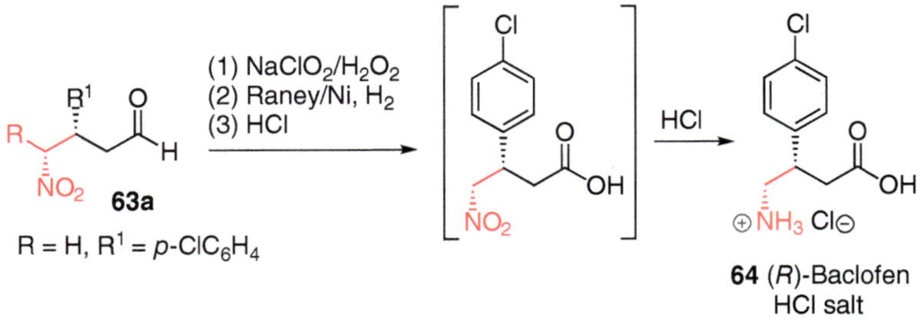

Scheme 5.23 Asymmetric synthesis of (*R*)-Baclofen·HCl salt **64**.

MA13: During studies of new family of prolinol-based catalysts, it has selected the triphenylsilyl ether derivative **65** as water compatible organocatalyst able to assure good enantioselectivity (ee up to 98%) in the Michael addition of nitromethane to a selection of enals (Scheme 5.24) [46].

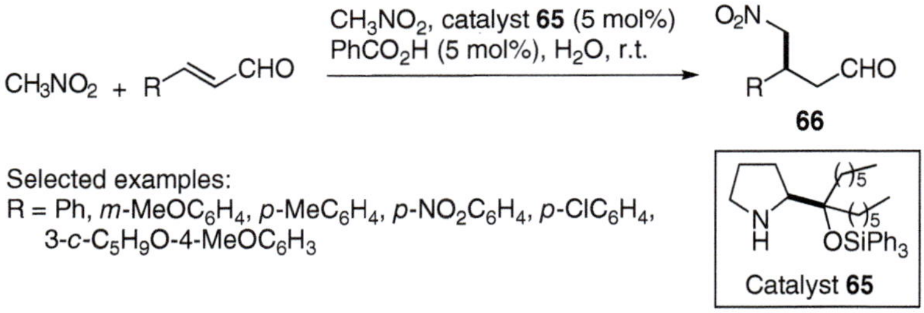

Scheme 5.24 Asymmetric Michael reaction promoted by catalyst **65** under aqueous conditions.

The adduct **66a** (R = 3-*c*-C_5H_9O-4-$MeOC_6H_3$), which has been obtained in very high enantioselectivity (ee 98%), has been employed as strategic precursor to the synthesis of (*S*)-Rolipram **69**, a type IV phosphodiesterase inhibitor (Scheme 5.25).

MA14: Diphenylprolinol silyl ether **62** was further successfully tested as organocatalyst of asymmetric Michael reaction of nitroalkanes (mainly nitromethane) with enals [47a]. As a representative example, the authors reported the synthesis of Pregabalin **73**, an important anticonvulsant drug (Scheme 5.26).

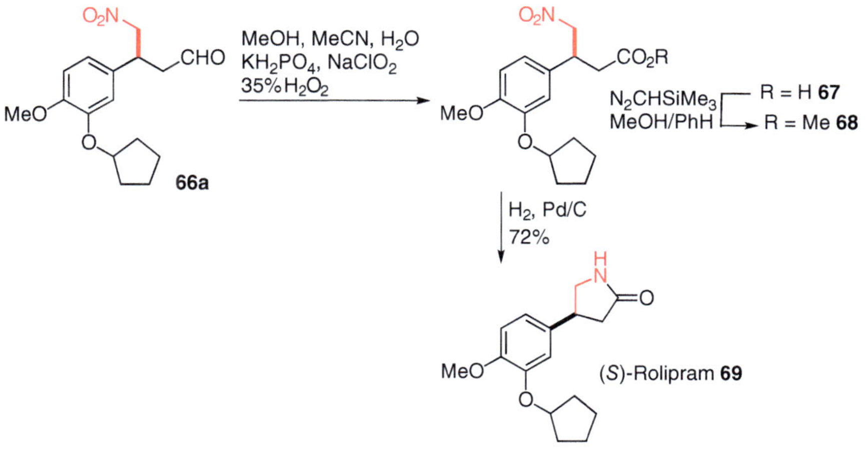

Scheme 5.25 Synthesis of (*S*)-Rolipram **69**.

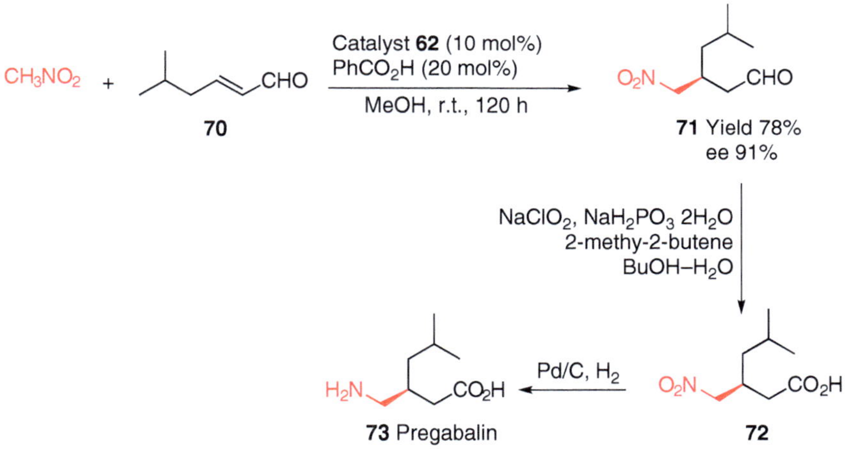

Scheme 5.26 Synthesis of Pregabalin **73**.

MA15: A new imidazole-containing imidazolidinone catalyst **74** was reported for enantioselective Michael reaction of nitroalkanes with enals (Scheme 5.27) [48]. The reaction is performed at room temperature, under inert atmosphere (N_2), and with nitroalkanes that act even as solvents. The method facilitates the reaction of straight-chain nitroalkanes and aromatic α,β-unsaturated aldehydes; however, good yields (up to 91%) and enantioselectivity (up to 92%) of the adducts **75** are achieved.

MA16: Further studies of Michael reaction under (*S*)-diphenylprolinol TMS ether **62** catalysis, in presence of lithium acetate as additive, demonstrated the efficiency of the system that was applied to the total synthesis of (*R*)-Baclofen·HCl, following the steps reported in Scheme 5.23 [49].

MA17: Catalyst **62** (see Schemes 5.22 and 5.26) has demonstrated to be useful in asymmetric Michael reaction of simple nitroalkanes enals. However, further

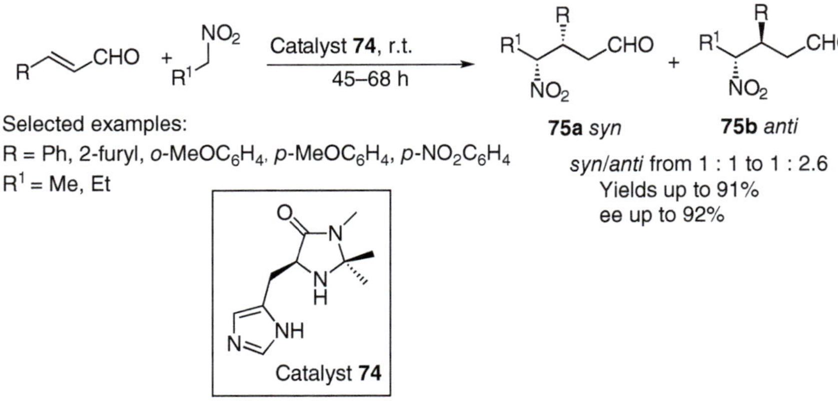

Scheme 5.27 Michael reaction promoted by the catalyst **74**.

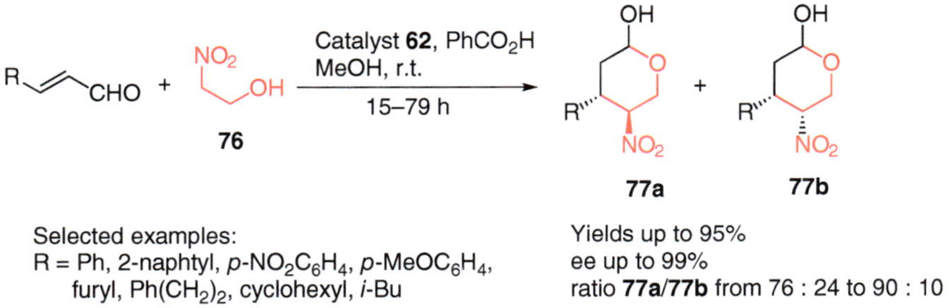

Scheme 5.28 Michael reaction promoted by the catalyst **62**.

application has been reported for the conjugate addition of nitroethanol to enals (Scheme 5.28) [50].

The process allows the Michael reaction and in situ cyclization of the obtained adducts to afford substituted tetrahydropyrans **77**.

Authors reported in the same paper the application of compound **77a** as key structure to the synthesis of a variety of important targets **78–82**, as reported in Scheme 5.29.

5.4.3 Asymmetric Michael Reaction with α,β-Unsaturated Esters

MA18: Conjugate addition to α,β-unsaturated esters **83** has been reported as internal asymmetric Michael reaction, catalyzed with organocatalyst **84** (Scheme 5.30) [51].

5.4.4 Asymmetric Michael Reaction with Conjugate Nitroalkenes

Conjugate addition of nitroalkanes to conjugate nitroalkenes is a fascinating process due to the opposite behavior of two nitro derivatives. In fact, nitroalkanes

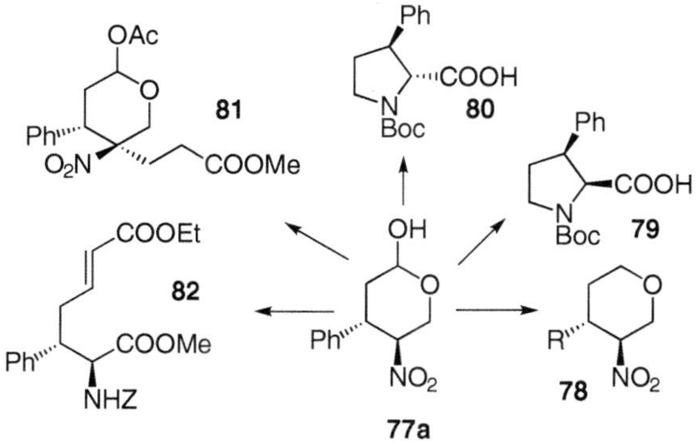

Scheme 5.29 Synthetic applications of compound **77a** (R = Ph).

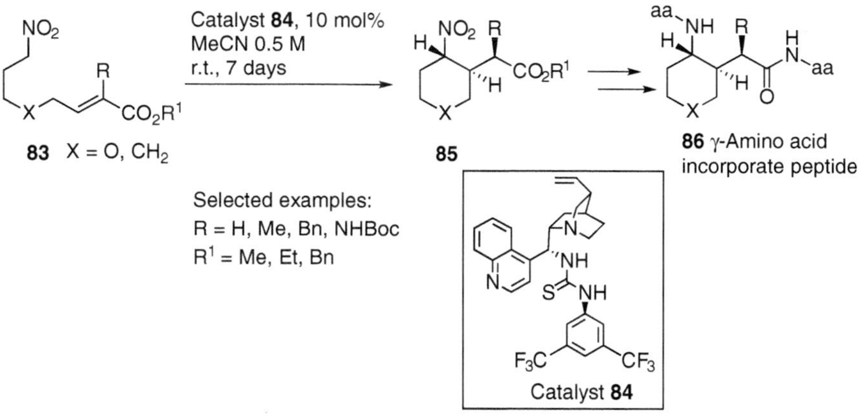

Scheme 5.30 Intramolecular reaction promoted by the catalyst **84**.

act as nucleophiles, while nitroalkenes as electrophiles. Thus, Michael reaction of nitroalkanes with nitroalkenes is a very important way to obtain 1,3-dinitroalkanes.

MA19: The first asymmetric Michael reaction of nitroalkanes to nitroalkenes was reported by Du and coworkers [52] using C_2-symmetric tridentate bis(oxazoline) **87** as organocatalyst (Scheme 5.31, eq. a).

The adducts **88** show high enantioselectivities with a series of substituted phenyl nitroalkenes bearing both electron-donating and electron-withdrawing substituents.

As application of this procedure, the authors reported (Scheme 5.31, eq. b) the use of adduct **88a** (R = p-MeC_6H_4, R^1 = Me, R^2 = H) to the synthesis of chiral cyclic thiourea derivative **90** in 53% yield and without loss of enantioselectivity. The compound **90** could be a valuable synthetic intermediate; moreover, chiral thoureas have attracted attention due to their *anti*-HIV activities.

MA20: In the same year, Maruoka and coworkers developed a high diastereo- and enantioselective conjugate addition of nitroalkanes (under silyl nitronates

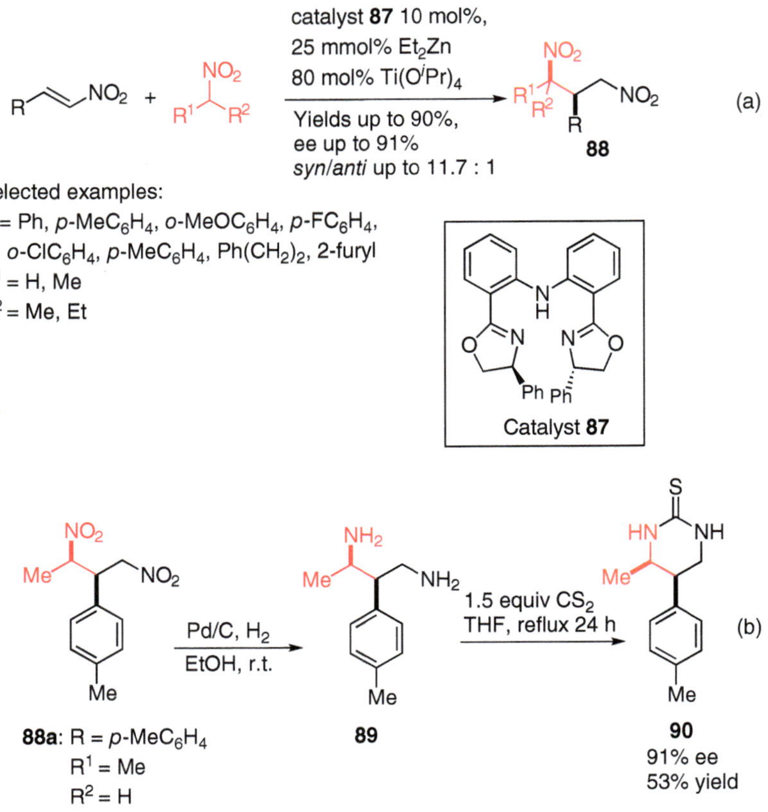

Scheme 5.31 Asymmetric Michael reaction with nitroalkenes promoted by the catalyst **87**.

derivative **91**) to nitroalkenes **92** by chiral ammonium bifluoride catalyst **93** (Scheme 5.32) [53].

The adducts **94**, with two consecutive stereocenters, are achieved in good yields and with excellent diastereo- and enantioselectivities employing 2 mol% of catalyst.

The resulting 1,3-dinitro compounds can be easily reduced to the corresponding chiral 1,3-diamines that could be versatile chiral building blocks in organic synthesis.

5.4.5 Asymmetric Michael Reaction with Vinyl Sulfones

Cinchona alkaloid-derived-thiourea catalyst **96** was found as efficient organocatalysis for the asymmetric conjugate addition of nitroalkanes to vinyl sulfone **95** [54]. A variety of adducts **97** have been prepared (Scheme 5.33) [54] in good yields (up to 87%) and enantioselectivities (ee up to 84%).

The adducts **97** are very useful intermediates in organic synthesis because the reduction the nitro group in combination with ready desulfonation provides an easy access to chiral amines. A very important application of the procedure is reported in Scheme 5.34, with the synthesis of **102**, a potent inhibitor of phenylethanolamine *N*-methyltransferase.

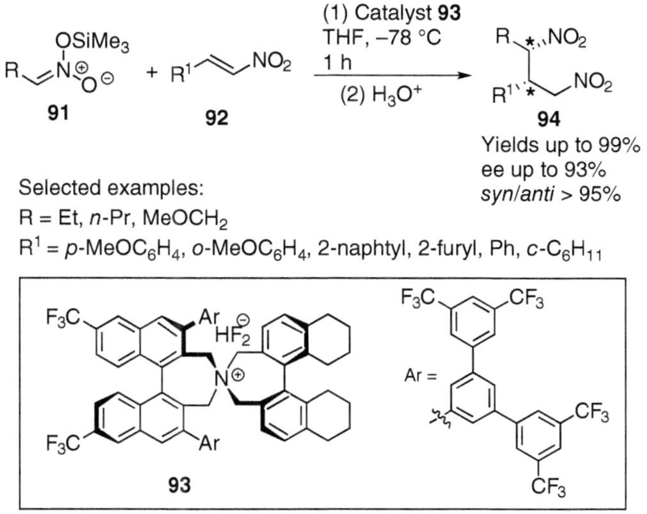

Scheme 5.32 Asymmetric Michael reaction with nitroalkenes promoted by the catalyst **93**.

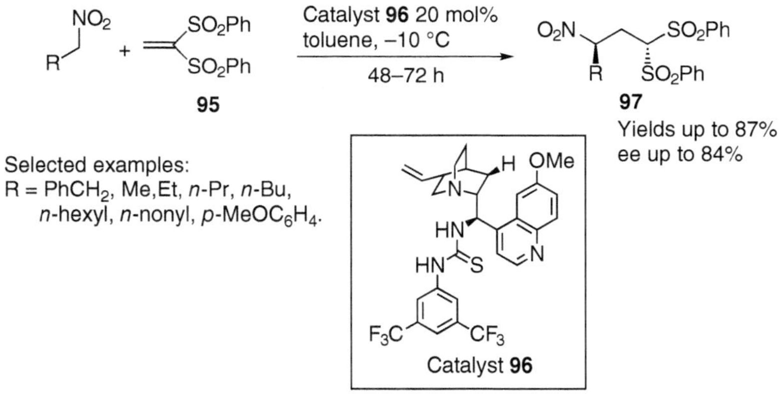

Scheme 5.33 Asymmetric Michael reaction with vinyl sulfone promoted by the catalyst **96**.

The adduct **97a** (R = Bn), reduced with Zn/AcOH, affords the amine **98** that was protected with Cbz. Successive desulfonation to **99** and reductive transformation gives **100**, the key intermediate to the target compound. In fact, treatment of the latter with polyphosphoric acid (PPA) affords the amide **101**, prone to be converted into **102** with $LiAlH_4$.

5.5 Synthetic Applications of Michael Reaction

Many of the topics of Chapters 7–13 will include the Michael reaction. However, here are reported some representative synthesis of important targets in which the conjugate addition of nitroalkanes to electron-poor alkenes plays a strategic role.

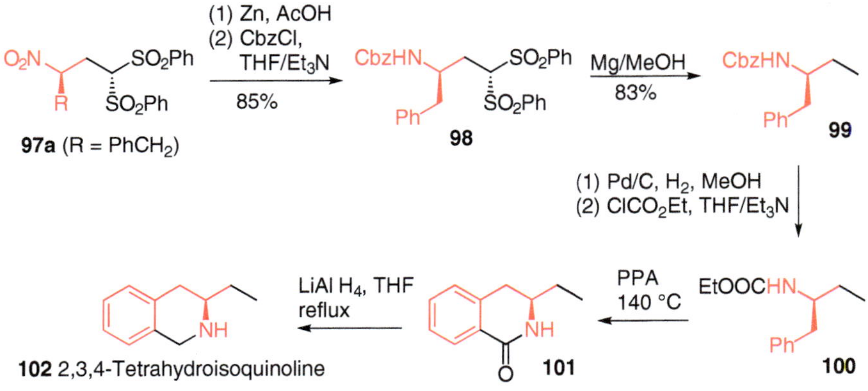

Scheme 5.34 Synthesis of 2,3,4-tetrahydroisoquinoline **102**.

(±)-Norsalanadione **105** is a terpenoid isolated from aged Burley, Turkish, and Greek tobacco leaves, and it is an attractant for *tabakoshibanmushi*, a pest insect that infests stored foods and tobacco. Moreover, **105** is a flavorant and flavor enhancer for tobacco.

The synthesis of **105** has been developed (Scheme 5.35) [55] by a double Michael reaction of *i*-$BuNO_2$ with 3-butyn-2-one and then with methyl vinyl ketone, in the presence of KF/*n*-Bu_4NCl/DMSO, with the formation (51% yield) of **104**. Successive reductive (NaHTe) denitration achieves Norsolanadione **105** in 64% yield.

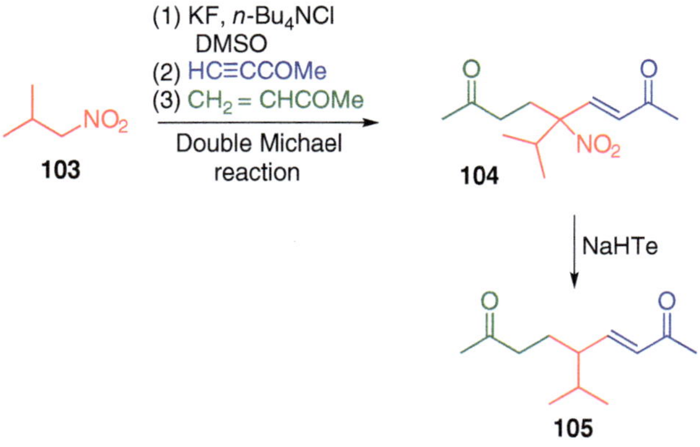

Scheme 5.35 Synthesis of (±)-Norsalanadione **105**.

Taxol (Figure 5.6) is a very important drug that possesses three main rings (**A–C**), and a big problem is procuring adequate strategies for the synthesis of this molecule and new contributes in this sense are of high importance.

In this context, a new procedure for the synthesis of **A** ring was realized starting from the Michael reaction of nitromethane with the enone **106** (Scheme 5.36) [56].

Figure 5.6 Taxol.

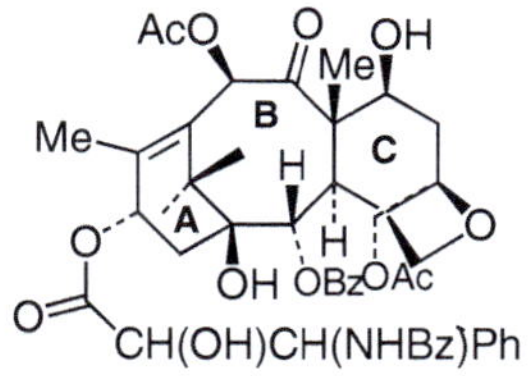

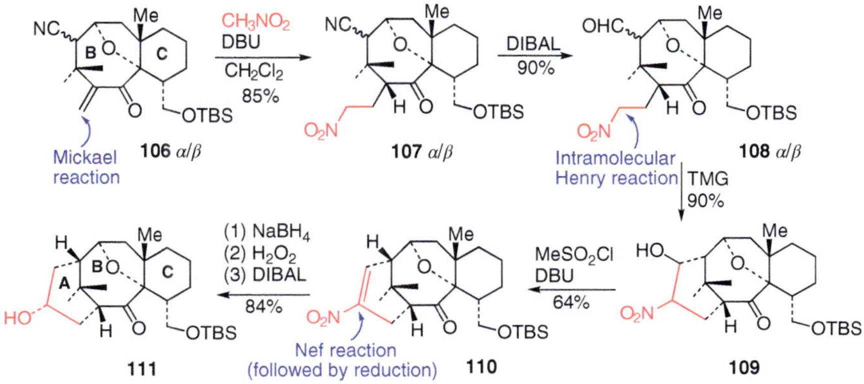

Scheme 5.36 Synthesis of **A** ring of Taxol **111**.

Thus, *exo*-methylene ketone **106** is treated with nitromethane under basic condition (DBU) allowing the adduct **107** in excellent yield. Treatment with diisobutylaluminium hydride (DIBAL), and intramolecular nitroaldol reaction of **108**, promoted by TMG as base, allows cyclic nitroalkanols **109**, precursor of the **A** ring.

Dehydration of the latter and successive Nef transformation ($NaBH_4/H_2O_2$), followed by reduction (DIBAL), gives the **A** ring **111**.

Although the method was proposed more than two decades ago, it remains highly instructive to understand the chemical versatility of nitroalkanes since the procedure involves the main reactions of these compounds: (i) Michael reaction (**106** to **107**), (ii) Henry reaction (**108** to **109**), and (iii) Nef reaction (**110** to **111**).

Protected nitro ketone **112** has been used, as key functionalized nitroalkane, in the synthesis of (±)-Trachelanthamidin **117**, an alkaloid isolated from *Eupatorium maculatum*, *Heliotropium strigosum*, and *Heliotropium*-Arten (Scheme 5.37) [57]. Thus, the Michael reaction of **112** with diethylfumarate affords the adduct **113** that, hydrogenated with H_2, Pd/C, allows pyrrolidine **114**. Successive reduction ($NaBH_4$) and double protection of the hydroxyl and NH groups as benzyl derivatives afford **115** in 69% yield. Reduction ($LiAlH_4$) of the amide and removal of the acetal gives the target's precursor **116**. Finally, hydrogenation (H_2, Pd/C) yields (±)-Trachelanthamidin **117** in 6% yield.

Isolactone **122** is an important member of alkaloids, detected in Greek tobacco, marijuana, and aroma of carrots. The Michael reaction of *i*-$BuNO_2$ **118** with acrylonitrile, under KF/EtOH, has been used as first step for the synthesis of **122** (Scheme 5.38) [58]. The adduct **119** yields keto-nitrile **120** through the Nef reaction

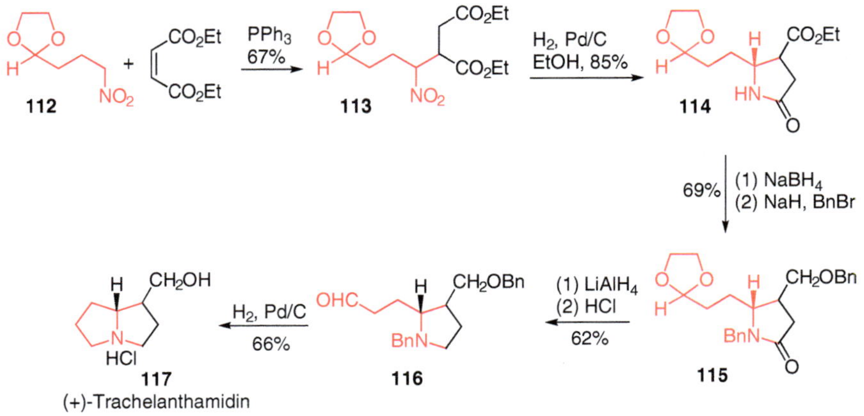

Scheme 5.37 Synthesis of (+)-Trachelanthamidin **117**.

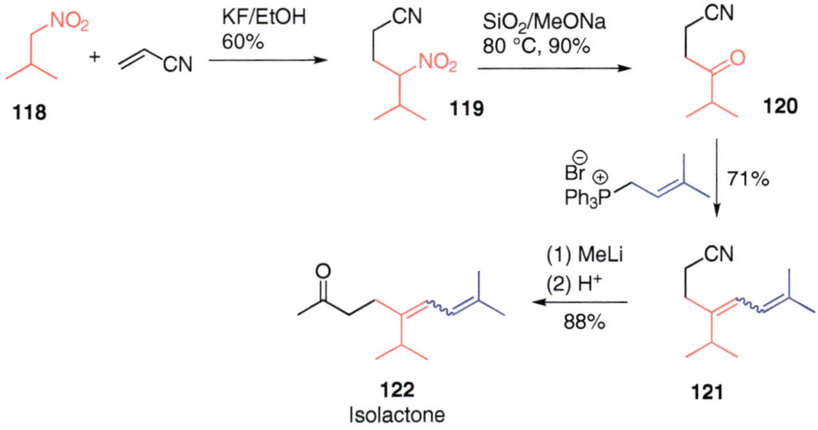

Scheme 5.38 Synthesis of Isolactone **122**.

(SiO_2/MeONa), successive Wittig olefination allows the formation of the dienic system **121**.

Finally, methylation of **121** achieves the target product **122** in 88% yield (overall yield: 44%).

Levoglucosenone **123** is a rigid bicyclic structure able to efficiently react with a large excess of nitromethane, in the presence of TMG as base, giving Michael and Henry reactions (in this order), with the formation of dinitro adduct **124** in good yields and diastereoselectivity (10 : 1 at C_2) (Scheme 5.39) [59].

Deoxygenation and denitration by Bu_3SnH in a radical process afford **126**, a key intermediate for the enantioselective synthesis of aggregation pheromone (−)-δ-Multistriatin **127**.

Nitromethane addition to α,β-unsaturated diester **128**, under DBU catalysis, produces the adduct **129** (Scheme 5.40) [60]. The latter, submitted to reduction (Pd/C,

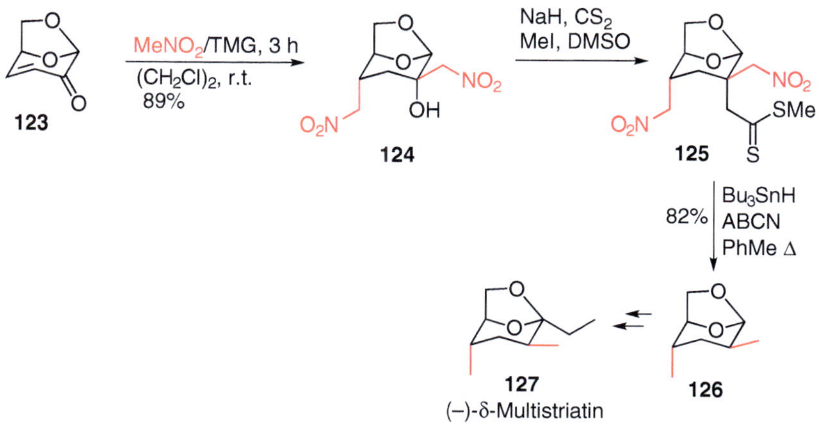

Scheme 5.39 Synthesis of (−)-δ-Multistriatin pheromone **127**.

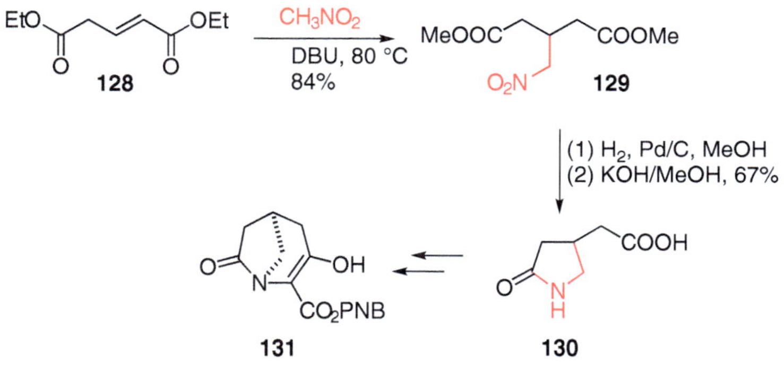

Scheme 5.40 Synthesis of bicyclic-γ-lactam proteins **131**.

H_2) of the nitro group and treatment with alcoholic KOH, gives pyrrolidinone **130** that can be easily converted, in few steps, into bicyclic-γ-lactam proteins **131**.

It is important to know that asymmetric hydrolysis of racemic **129**, using two enantiocomplementary enzymes, allows the preparation of optically active **130** in both enantiomeric forms.

(*R*)-Patulolide A and B are important natural macrolides, with *anti*-fungal, antibacterial, and *anti*-inflammatory activities, isolated from the culture broth of *Penicillium urticae*.

Michael reaction of optically active nitroacetate **132**, easily prepared by enzymatic resolution of the corresponding nitro alcohol, with methyl propiolate achieves the adduct (*E*)-**133** (Scheme 5.41) [61].

The nitro group is than submitted to the Nef reaction and the obtained keto ester **134**, after hydrolysis of the acetate, is converted into the macrolide (*R*)-Patulolide A **135** in 67% yield. Similar process can be employed to prepare Palulolide B, (*Z*)-**135**.

Paroxetine **140** is an important drug with important biological activities and an interesting synthesis of this target has been reported [62] starting from

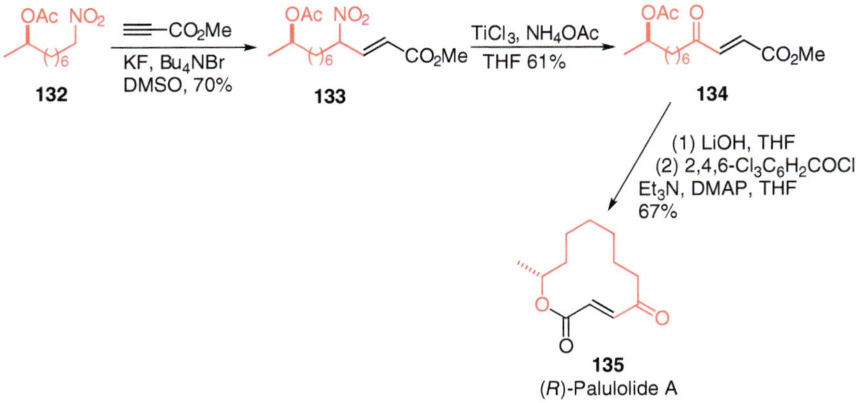

Scheme 5.41 Synthesis of (*R*)-Palulolide A **135**.

enantioselective conjugate addition of nitromethane to enal **136**, catalyzed with *O*-TMS-α,α-diphenylpropanol, and *N*-bromosuccinimide-induced oxidation of resulting **137**, emiacetalic form of the adduct, to (*R*)-nitroester **138**. The latter, via heterocyclization ($BnNH_2/CH_2O$), furnishes the piperidone **139**, (48% yield, 94% ee, 3 : 1 mixture of isomers [major *trans*-diastereomer]), key precursor of **140** in only few steps (Scheme 5.42).

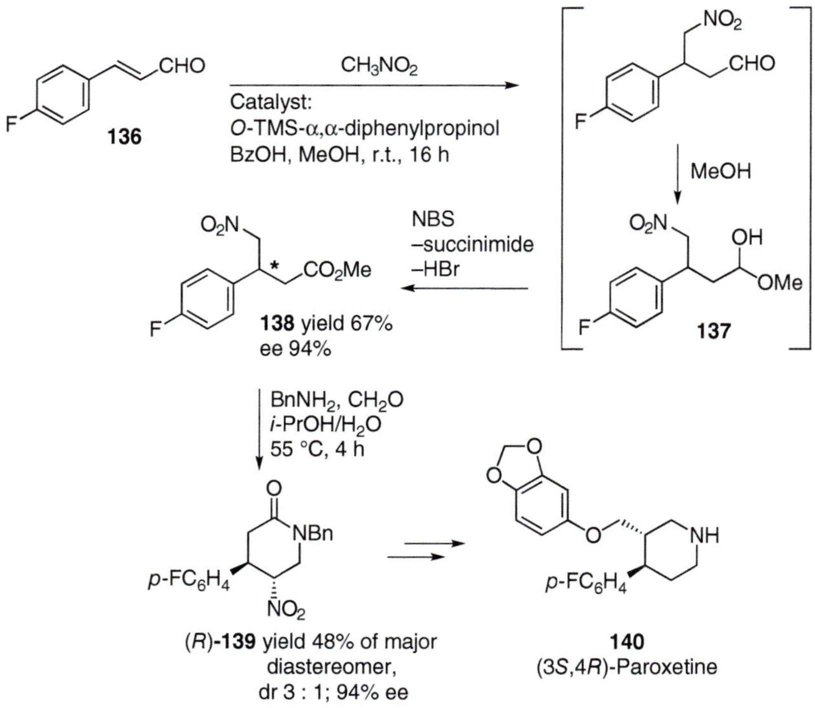

Scheme 5.42 Synthesis of (3*S*,4*R*)-Paroxetine A **140**.

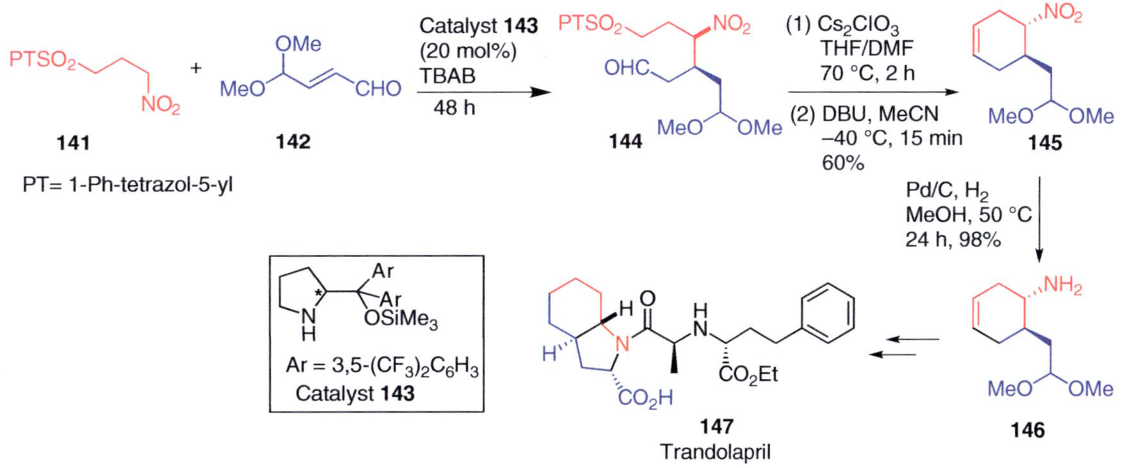

Scheme 5.43 Synthesis of Trandolapril **147**.

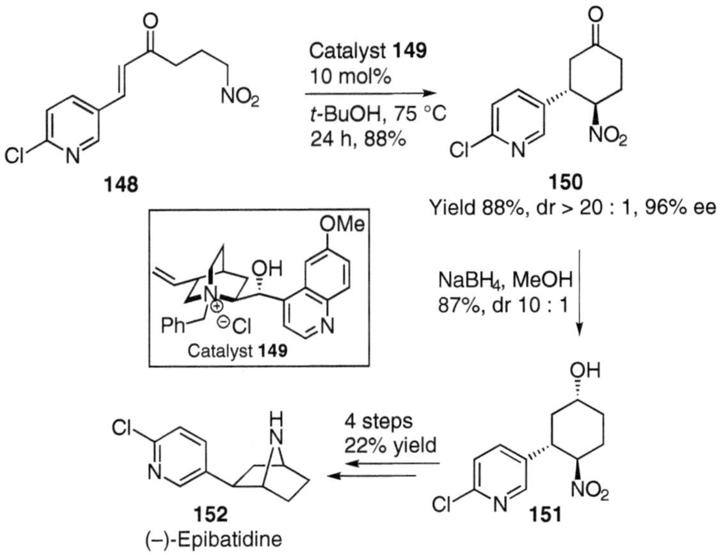

Scheme 5.44 Synthesis of (–)-Epibatidine **152**.

Trandolapril **147** is an angiotensin converting enzyme (ACE) inhibitor, cardiovascular drug for controlling hypertension and drug to treat epilepsy. This target has been synthesized (Scheme 5.43) [63] starting from Michael reaction of functionalized nitroalkane **141** and enal **142**, under organocatalyst **143**, with the formation of the adduct **144** as a mixture of diastereomers (dr 64:36).

The synthesis combined the organocatalytic Michael addition with intramolecular Julia–Kocienski olefination for the preparation of enantioenriched nitro cicloesene **145**, with good diastereo and enantioselectivity (dr 90 : 10, 94% ee). Successive reduction (Pd/C, H_2) of both C=C double bond and nitro group allows the amino cyclohexene **146** (38% overall yield), key precursor of Trandolapril **147**, in few steps.

(−)-Epibatidine **152**, isolated from frog *Epibatidores tricolor*, shows important bioactivities such as agonist of the nicotinic acetylcholine receptor and analgesic 200 times more potent than morphine (Scheme 5.44; [64]).

Thus, an important synthesis of **152** starts with intramolecular Michael reaction of nitro enone **148**, under organocatalysis (catalyst **149**), affording *trans*-isomer of the adduct **150** in high yield (88%) and high diastereo- and enantioselectivity (>20 : 1 dr, 96% ee). Reduction of carbonyl in **150** with $NaBH_4$ affords nitroalkanol **151** in 87% yield and high diastereoselectivity (dr 10:1). Finally, (−)-Epibatidine **152** can be prepared from **151** in only four steps, including mesylation, reduction/cyclization and epimerization.

(+)-Manzacidin **A** and **C** are two important alkaloids prepared in a multistage nitro-Michael/Henry cascade reaction (Scheme 5.45), through a three components system (enal **153**, nitroethane and formaldehyde) [65].

Michael addition of $EtNO_2$ to enal **153** furnishes the adduct **154** that, by successive nitroaldol reaction furnishes the nitroalkanol **155** prone to cyclize into the compound **156**, then oxidized (potassium dichromate, PDC) into the lactonic form **157**.

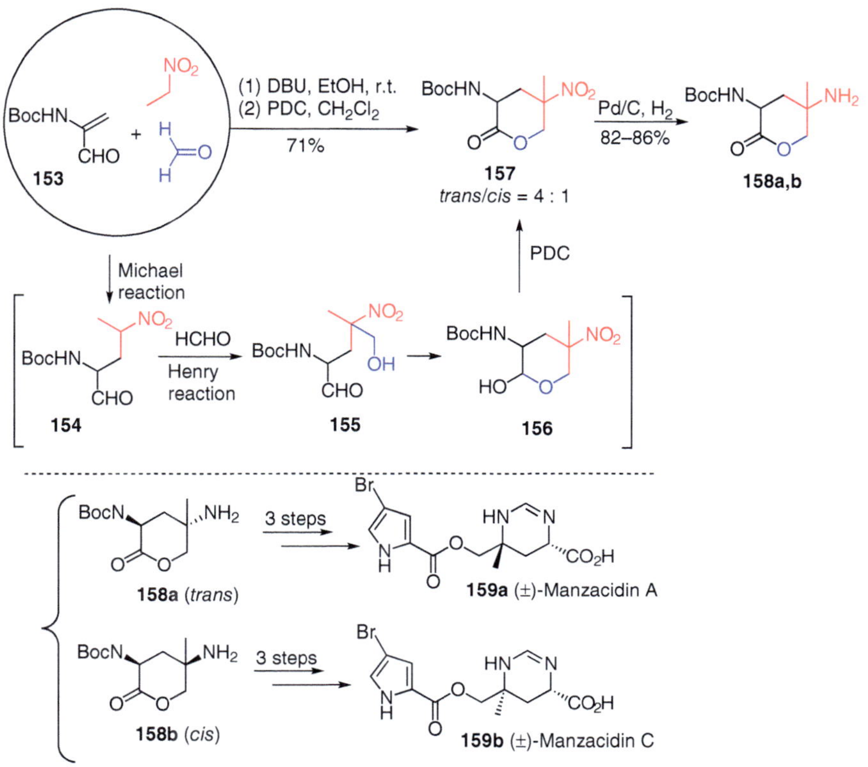

Scheme 5.45 Synthesis of (±)-Manzacidin **A** and **C**.

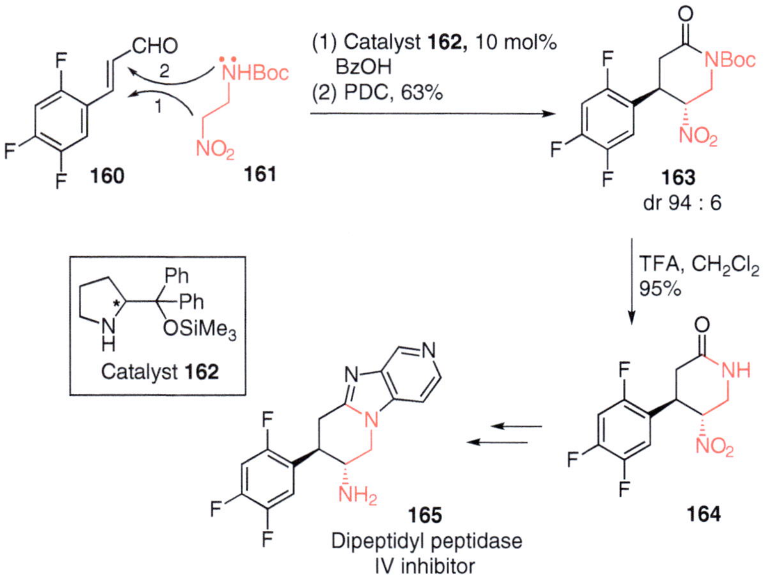

Scheme 5.46 Synthesis of Dipeptidyl peptidase IV inhibitor **165**.

Reduction of **157** gives the amine **158**, as *trans/cis* form. Thus, as reported in braces parenthesis, *trans* form of **158** is the precursor, in only three steps, of (±)-Manzacidin A, while *cis*-**158** gives (±)-Manzacidin C.

Dipeptidyl peptidase IV inhibitor **165**, selected as a drug candidate for treatment of type 2 diabetes mellitus, was prepared from conjugate addition of Boc-protected 2-amino-1-nitroethane **161** to 2,4,5-trifluorocinnamic aldehyde **160** (Scheme 5.46), under organocatalyst **162** [66]. The adduct **163** is obtained in 63% yield with excellent diastereoselectivity (dr 94:6). The pure *trans* isomer of **163**, isolated with column chromatography, is than converted into pyperidine-2-one **164** in 95% yield and 99.8% ee, by treatment with trifluoroacetic acid (TFA). Compound **164** is the key intermediate for a large-scale production of the target molecule **165**.

References

1 (a) Ballini, R., Bosica, G., Fiorini, D. et al. (2006). *Chem. Rev.* 105: 933–971. (b) Fujieda, N., Ichihashi, H., Yuasa, M. et al. (2020). *Angew. Chem. Int. Ed.* 59: 1–5.

2 Nakashita, Y. and Hesse, M. (1983). *Helv. Chim. Acta* 66: 845–860.

3 Derri Alcantara, M.P., Escribano, F.C., Gómez-Sánchez, A. et al. (1996). *Synthesis*: 64–70.

4 Clark, J.K. (1980). *Chem. Rev.* 80: 429–452.

5 Grimm, E.L., Zechiesche, R., and Reissig, H.U. (1985). *J. Org. Chem.* 50: 5543–5545.

6 (a) Chasar, D.W. (1982). *Synthesis*: 841–842. (b) Chetia, A., Saikia, C.J., Lekhok, K.C., and Boruah, R.C. (2004). *Tetrahedron Lett.* 45: 2649–2651.

7 (a) Ono, N., Kamimura, A., and Kaji, A. (1984). *Synthesis*: 226–227. (b) Pollini, G.P., Barco, A., and De Giuli, G. (1972). *Synthesis*: 44–45. (c) Chavan, S.P., Pasupathy, K., Venkatraman, M.S., and Kale, R.R. (2004). *Tetrahedron Lett.* 45: 6879–6882.

8 Peera, A. and Tomlinson, I.A. (2012). *Tetrahedron Lett.* 53: 3322–3324.

9 Ballini, R., Bosica, G., and Livi, D. (2001). *Synthesis*: 1519–1522.

10 Ballini, R., Fiorini, D., Gil, M.V., and Palmieri, A. (2003). *Tetrahedron Lett.* 44: 9033–9034.

11 Rosini, G., Marotta, E., Ballini, R., and Petrini, M. (1986). *Synthesis*: 237–238.

12 Bergbreiter, D.E. and Lalonde, J.J. (1987). *J. Org. Chem.* 52: 1601–1603.

13 Ballini, R., Petrini, M., and Rosini, G. (1987). *Synthesis*: 711–713.

14 Ballini, R., Marziali, P., and Mozzicafreddo, A. (1996). *J. Org. Chem.* 61: 3209–3211.

15 (a) Kisaga, P.B., Ilankumaran, P., Fetterly, B.M., and Verkade, J.G. (2002). *J. Org. Chem.* 67: 3555–3560. (b) Reddy, C., Reddy, V., and Verkade, J.G. (2007). *J. Org. Chem.* 72: 3093–3096.

16 Keller, E. and Feringa, B.L. (1997). *Synlett*: 842–844.

17 Ballini, R., Fiorini, D., Gil, M.V., and Palmieri, A. (2003). *Green Chem.* 5: 475–476.

18 Ballini, R., Bosica, G., Palmieri, A. et al. (2008). *Green Chem.* 10: 541–544.

19 Ballini, R., Barboni, L., Castrica, L. et al. (2008). *Adv. Synth. Catal.* 350: 1218–1224.

20 Motokura, K., Tanaka, S., Tada, M., and Iwasawa, Y. (2009). *Chem. Eur. J.* 15: 10871–10879.

21 Maeno, Z., Mitsudome, T., Mizugaki, T. et al. (2012). *Chem. Lett.* 41: 801–803.

22 Ballini, R., Barboni, L., and Palmieri, A. (2007). *Synlett* 19: 3019–3021.

23 (a) Mdoe, J.E.G., Clark, J.H., and Macquarrie, D.J. (1998). *Synlett*: 625–627. (b) Boruah, A., Baruah, M., Prajapati, D., and Sandhu, J.S. (1996). *Chem. Lett.*: 965–966. (c) Michaud, D., Abdallah-El Ayubi, S., Dozias, M.-J. et al. (1997). *Chem. Commun.*: 1613–1614. (d) Michaud, D., Texier-Boullet, F., and Hamelin, J. (1997). *Tetrahedron Lett.* 38: 7563–7564. (e) Escalante, J. and Diaz-Coutiño, F.D. (2009). *Molecules* 14: 1595–1604.

24 Ballini, R., Bosica, G., and Fiorini, D. (2001). *Tetrahedron Lett.* 42: 8471–8473.

25 Ballini, R. and Bosica, G. (1996). *Tetrahedron Lett.* 37: 8027–8030.

26 Ballini, R. and Bosica, G. (1998). *Eur. J. Org. Chem.*: 355–357.

27 Ballini, R., Fiorini, D., Gil, M.V., and Palmieri, A. (2004). *Tetrahedron* 60: 2799–2804.

28 Ballini, R., Barboni, L., and Giarlo, G. (2003). *J. Org. Chem.* 68: 9173–9176.

29 Liang, Y., Dong, D., Lu, Y. et al. (2006). *Synthesis* 19: 3301–3304.

30 Ranu, B.C., Banerjee, S., and Jana, R. (2007). *Tetrahedron* 63: 776–782.

31 Mancuso, R., Palmieri, A., Ballini, R., and Gabriele, B. (2012). *Tetrahedron* 68: 5852–5856.

32 Zlotin, S., Bogolyubov, A.V., Kryshtal, G.V. et al. (2006). *Synthesis* 22: 3849–3854.

33 Fabris, M., Lucchini, V., Noè, M. et al. (2009). *Chem. Eur. J.* 15: 12273–12282.

34 Prieto, A., Halland, N., and Jørgensen, K.A. (2005). *Org. Lett.* 7: 3897–3900.

35 (a) Mitchell, C.E.T., Brenner, S.E., and Ley, S.V. (2005). *Chem. Commun.*: 5346–5348. (b) Mitchell, C.E.T., Brenner, S.E., Garcia-Fortanet, J., and Ley, S.V. (2006). *Org. Biomol. Chem.* 4: 2039–2049.

36 Tsogoeva, S.B., Jagtap, S.B., and Ardemasova, Z.A. (2006). *Tetrahedron: Asymmetry* 17: 989–992.

37 Malmgren, M., Granander, J., and Amedijkouh, M. (2008). *Tetrahedron: Asymmetry* 19: 1934–1940.

38 Li, P., Wang, Y., Liang, X., and Ye, J. (2008). *Chem. Commun.*: 3302–3304.

39 Manzano, R., Andrés, J.M., Álvarez, R. et al. (2011). *Chem. Eur. J.* 17: 5931–5938.

40 Mei, K., Jin, M., Zhang, S. et al. (2009). *Org. Lett.* 11: 2864–2867.

41 Moon, H.W. and Kim, D.Y. (2010). *Tetrahedron Lett.* 51: 2906–2908.

42 Zhou, Y., Liu, Q., and Gong, Y. (2012). *Org. Biomol. Chem.* 10: 7618–7627.

43 Garcia, J.M., Maestro, M.A., Oiarbide, M. et al. (2009). *Org. Lett.* 11: 3826–3829.

44 Baschieri, A., Bernardi, L., Ricci, A. et al. (2009). *Angew. Chem. Int. Ed.* 48: 9342–9345.

45 Zu, L., Xie, H., Li, H. et al. (2007). *Adv. Synth. Catal.* 349: 2660–2664.

46 Palomo, C., Landa, A., Mielgo, A. et al. (2007). *Angew. Chem. Int. Ed.* 46: 8431–8435.

47 (a) Gotoh, H., Ishikawa, H., and Hayashi, Y. (2007). *Org. Lett.* 9: 5307–5309. (b) Wang, Y., Li, P., Liang, X. et al. (2006). *Chem. Commun.*: 1237–1234.
48 Hojabri, L., Hartikka, A., Moghaddam, F.M., and Arvidsson, P.I. (2007). *Adv. Synth. Catal.* 349: 740–748.
49 Wang, Y., Li, P., Liang, X. et al. (2008). *Chem. Commun.*: 1232–1234.
50 Gotoh, H., Okamura, H., Ishikawa, H., and Hayashi, Y. (2009). *Org. Lett.* 11: 4056–4059.
51 Noedes, W.J., Nutt, D.R., Chippindale, A.M., and Cobb, A.J.A. (2009). *J. Am. Chem. Soc.* 131: 16016–16017.
52 Lu, S.-F., Du, D.-M., Xu, J., and Zhang, S.-W. (2006). *J. Am. Chem. Soc.* 128: 7418–7419.
53 Ooi, T., Takada, S., Doda, K., and Maruoka, K. (2006). *Angew. Chem Int. Ed.* 45: 7606–7608.
54 Zhu, Q. and Lu, Y. (2009). *Org. Lett.* 11: 1721–1724.
55 Anderson, D.A. and Hwo, J.H. (1990). *J. Org. Chem.* 55: 511–516.
56 Magnus, P. and Pye, P. (1995). *J. Chem. Soc., Chem. Commun.*: 1933–1934.
57 Horni, A., Hubacek, I., and Hesse, M. (1994). *Helv. Chim. Acta* 77: 579–585.
58 Lee, W.Y., Jang, S.Y., Chae, W.K., and Park, O.S. (1993). *Synth. Commun.* 23: 3037–3046.
59 Forsyth, A.C., Gould, R.O., Paton, R.M. et al. (1993). *J. Chem. Soc., Perkin Trans. 1*: 2737–2741.
60 Aszodi, J., Rowlands, D.A., Mauvais, P. et al. (2004). *Bioorg. Med. Chem. Lett.* 14: 2489–2492.
61 Kalita, D., Khan, A.T., Barua, N.C., and Bez, G. (1999). *Tetrahedron* 55: 5177–5184.
62 Jensen, K.L., Poulsen, P.H., Donslund, B.S. et al. (2012). *Org. Lett.* 14: 1516–1518.
63 Rodrigo, E., Garcia Ruano, J.L., and Cid, M.B. (2013). *J. Org. Chem.* 78: 10737–10746.
64 Jensen, K.L., Weise, C.F., Dickmeiss, C. et al. (2012). *Chem. Eur. J.* 18: 11913–11918.
65 Xiong, G., Wei, M., Zhou, Y. et al. (2011). *Synthesis*: 3439–3446.
66 Huo, L., Ma, A., Zhang, Y., and Ma, D. (2012). *Adv. Synth. Catal.* 354: 991–994.

6

Formation of C—C Bond by Coupling Nitroalkanes with Aryl Halides

Although the Henry and Michael reactions are still nowadays the main synthetic approaches providing new C—C bond starting from nitroalkanes, the coupling reaction of these with aryl halides, under palladium catalysis, emerged in the last years as a practical and powerful procedure for the preparation of more elaborated nitroalkanes to be employed in organic synthesis. In this sense, a pioneering work was reported, two decades ago [1], combining simple nitroalkanes ($EtNO_2$ and n-$PrNO_2$) with bromo-p-$^tBuC_6H_4$ and Chloro-p-MeC_6H_4, respectively, in presence of $Pd(OAc)_2$ and phosphine-ligands **A** or **B** (Figure 6.1).

This synthetic potential of this approach for generate new C—C bond from nitroalkanes was successively developed in order to enlarge the applicability and the use of this new way to prepare new classes of nitroalkanes.

6.1 Main Procedures for Coupling Nitroalkanes with Aryl Halides

(i) Buchwald and coworker expanded the original method to a large variety of nitroalkanes and aryl halides, as reported in Scheme 6.1 [2].
Good yields of nitro-benzyl derivatives **3** are achieved in 8–30 hours. Moreover, the high chemoselectivity of the process grants the survival of a variety of functionalities such as ketones esters and C=C double bond.
This procedure was successively applied to the synthesis of compounds **NH-9** (Scheme 6.2), very important for its thyroid hormone agonist/antagonist activity [3].
The synthesis starts from the alkyne-precursor **4**, easily converted into its derivative **5**, prone to give arylation with nitroethane and formation of secondary nitroalkane **6**. Successive Nef reaction and in situ hydrolysis of both "MOMO" and "methyl ester" functionalities produce the target **NH-9** in 41% yield.

(ii) α-Arylation of aryl nitromethanes **7** with **8** is a valuable reaction due to the importance of the obtained diaryl nitromethanes **9**. This reaction was developed [4] under palladium catalysis, in the presence of t-BuXPhos as ligand and

Nitroalkanes: Synthesis, Reactivity, and Applications, First Edition.
Roberto Ballini and Alessandro Palmieri.

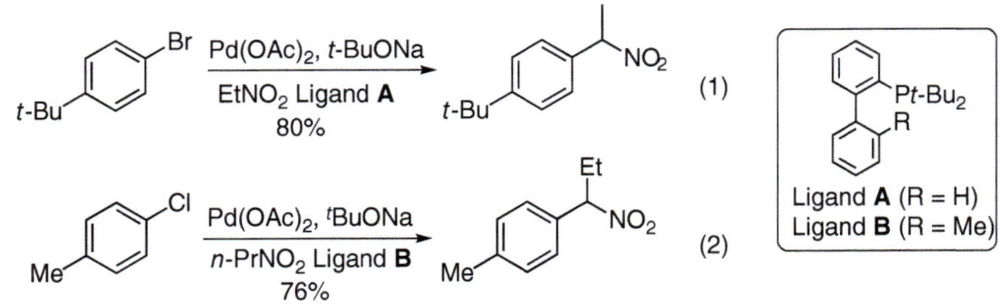

Figure 6.1 Pioneering coupling reactions.

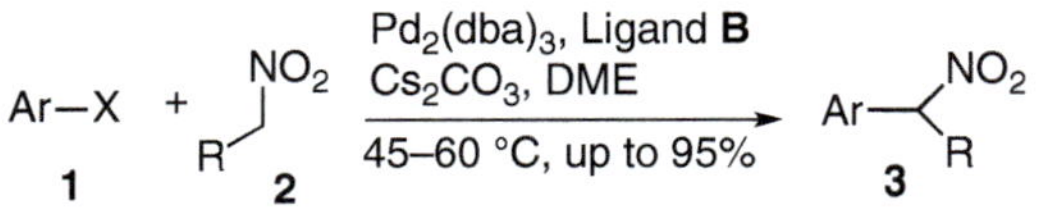

Selected examples:
X = Br, Cl
R = Me, Et, Bn, n-C_5H_{11}, $MeOCO(CH_2)_2$, $CH_2{=}CH(CH_2)_3$
Ar = Ph, m-EtC_6H_4, p-$MeOC_6H_4$, p-BrC_6H_4, p-MeCO, m-$Me_2NC_6H_4$

Scheme 6.1 Coupling of aryl halides **1** with nitroalkanes **2**.

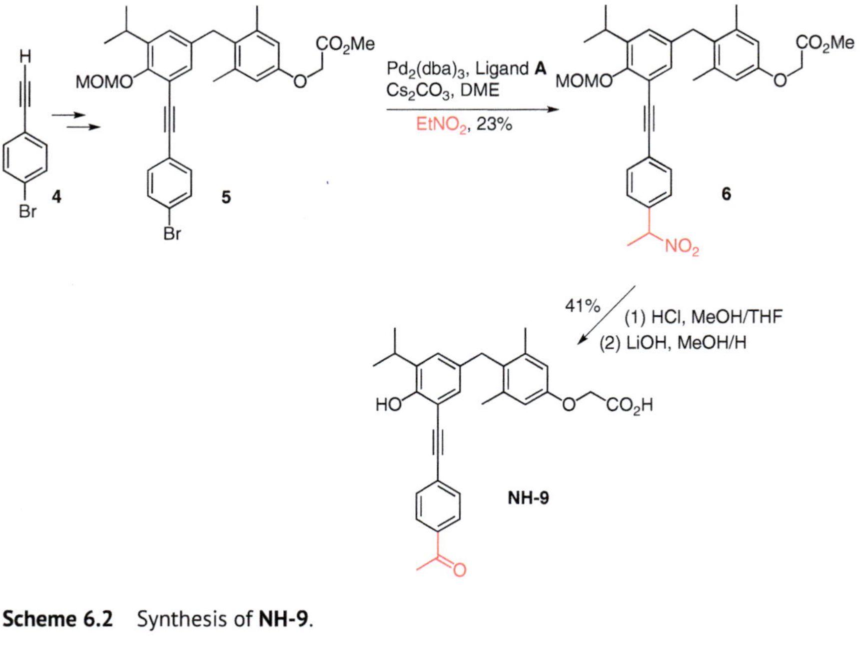

Scheme 6.2 Synthesis of **NH-9**.

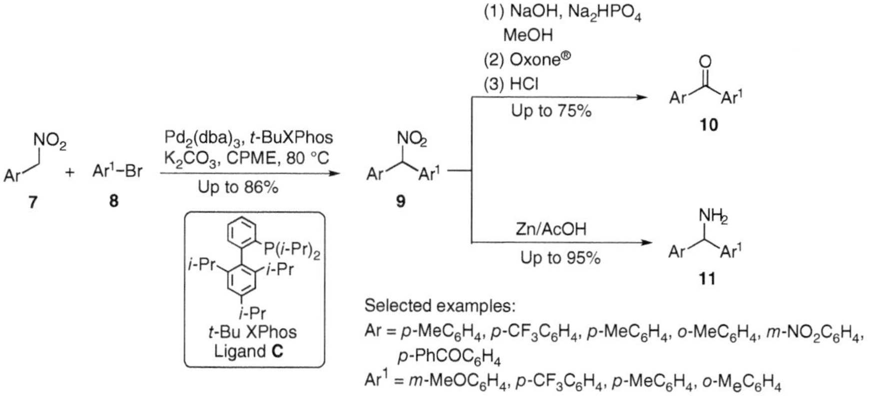

Scheme 6.3 α-Arylation of aryl nitromethanes **7**.

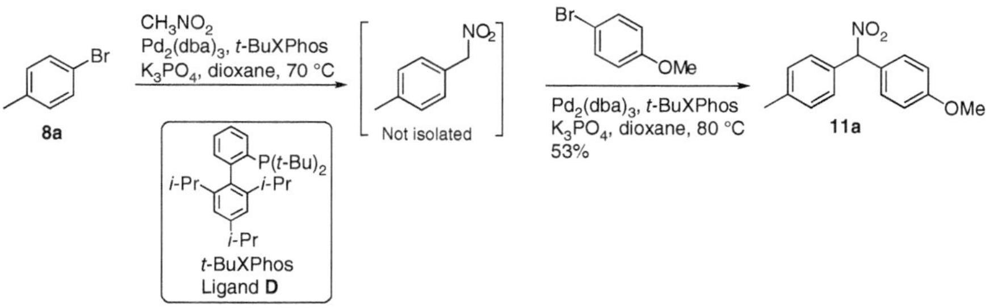

Scheme 6.4 α-One-pot diarylation of nitromethane.

cyclopentyl methyl ether (CPME) as green solvent (Scheme 6.3). The importance of diaryl nitromethanes is demonstrated by their easy conversion into acetophenones **10** through oxidative Nef reaction or into diarylamines **11** via reduction with Zn/glacial AcOH system.

As further improvement of this method, authors reported a one-pot diarylation of nitromethane, as illustrated in Scheme 6.4.

Arylation of nitroalkanes can be even performed using aryl chlorides or alkyl triflates, while iodides give low yields (Table 6.1) [5]. The method, which involves palladium as catalyst and XPhos as ligand at moderate temperature, works well with a large variety of substrates and tolerates an array of functionality.

(iii) Green procedure for sp^2–sp^3 coupling of nitroalkanes with aryl bromides, in aqueous medium and in presence of surfactant FI-750-M, was realized under palladium catalysis and *t*-BuXPhos (Ligand **D**, Figure 6.2) [6]. A number of polyfunctionalized nitroalkanes can be easily prepared under environmentally conditions, in the presence of benign amphiphilic molecules that mimics polar solvents.

Moreover, the procedure shows general applicability (28 examples are reported), low palladium loading was observed, and the surfactant can be recycled.

Table 6.1 Arylation of nitroalkanes minimizing their amount (selected examples).

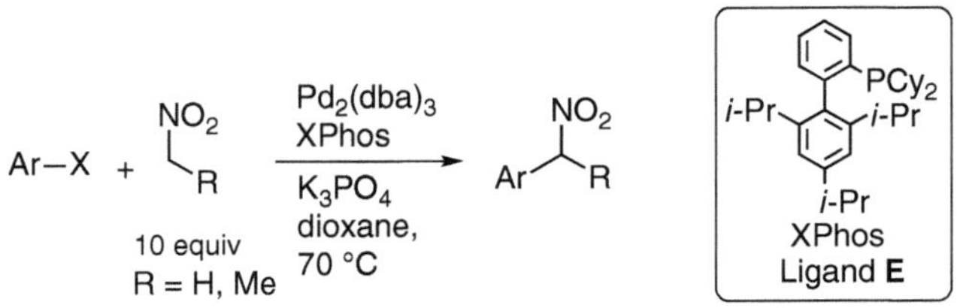

Entry	Ar–X	R	Product	Yield (%)
1	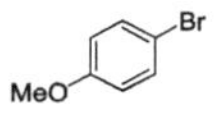	H	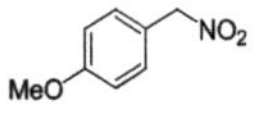	94
2	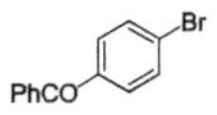	H	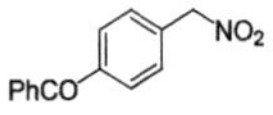	82
3	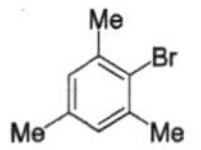	H	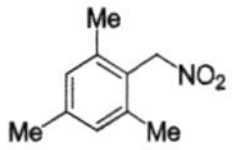	90
4	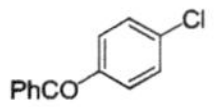	H	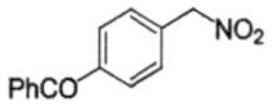	77
5	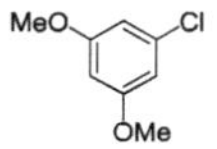	H	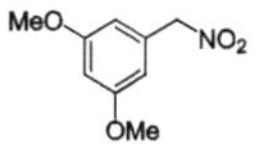	77
6	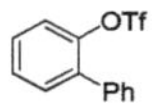	H	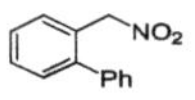	86
7	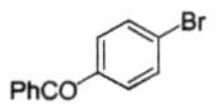	CH_3	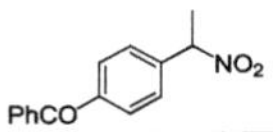	98

6.2 Application of C–C Coupling Nitroalkanes with Aryl Halides

The C—C coupling of aryl halides (or thiflates) with nitroalkanes would represent an ideal method to prepare a wide range of polyfunctionalized aryl nitroalkanes, precursors of an array of valuable compounds. Thus, Pd-catalyzed arylations of nitroalkanes have been used as first key step for the synthesis of a number of important targets.

(i) Orthogonal formylation equivalent of aryl halides was developed via their nitromethylation (Scheme 6.5) [7]. A large series of aryl derivatives **1** are nitro-methylated with success into **7** (20 examples are reported with good to high yields and high chemoselectivity) under Pd catalysis and in presence of Ligand **E**.
Then, reductive Nef reaction provides formylated aryls **13**, via (*E*)-oximes **12**, in satisfactory to good yields. The authors developed a direct, one-pot formylation process of bromo *p*-$MeOC_6H_4$ **1a** to *p*-MeO-benzaldehyde **13a** with excellent results since the process was realized in 76% overall yield. If necessary, the same

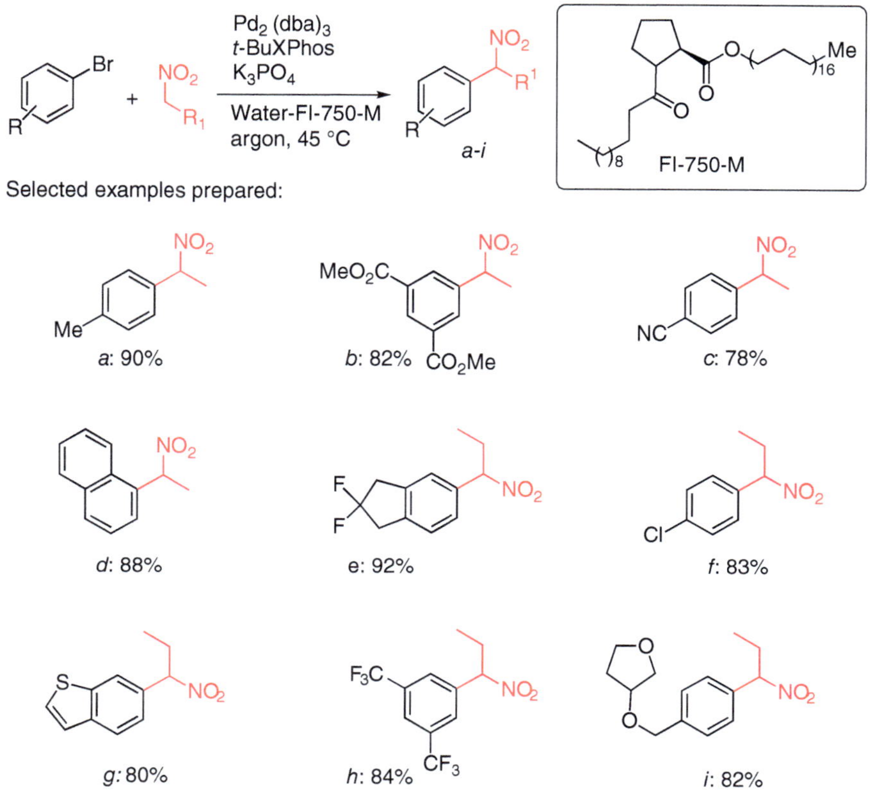

Figure 6.2 Arylation of nitroalkanes in water.

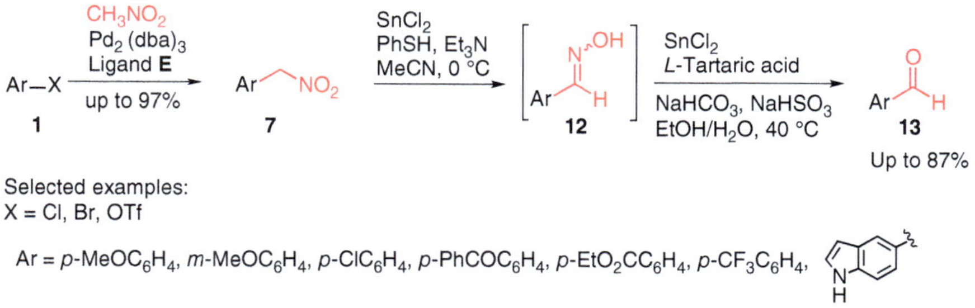

Scheme 6.5 Orthogonal formylation equivalent of aryl halides/triflates **1–13**.

one-pot procedure can be stopped to the synthesis of oxyme **12a**, in 74% overall yield.

(ii) Silyl nitronates **14**, prepared from arylnitromethane **7**, were found the right nucleophile for asymmetric 1,3-dipolar cycloaddition with 2-alkylacroleins **15**, key step for the preparation of isoxazolines **17**. Thus, the easy availability of **7** as precursors of **14**, independently from the chemical structures, is crucial. In this context, Jiao and coworkers [8] developed a new strategy for the synthesis of **17** starting from nitromethylation of aryl halides **8a** as first step (Scheme 6.6) [8].

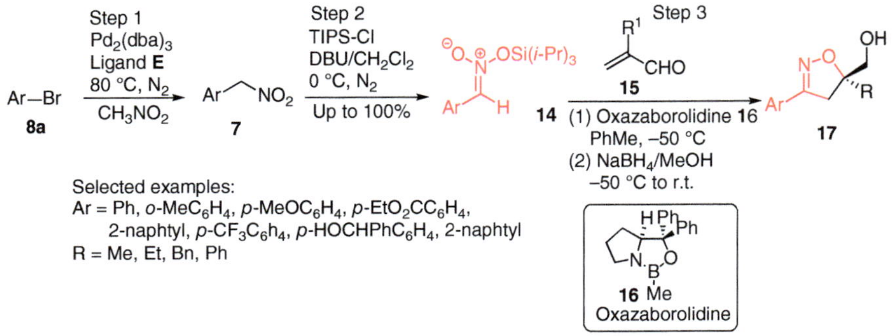

Scheme 6.6 Synthesis of 3-alkyl-2-isoxazolines **17**.

The obtained aryl nitromethanes **7** are converted into the corresponding triisopropylsilyl nitronates **14** that, by cycloaddition with **15** and with the aid of Corey's oxazaborolidine catalyst **16**, furnished 3-aryl-2-isoxazolines **17** in high enantioselectivities (up to 93% ee) and in moderate to excellent yields (up to 86%).

(iii) Tandem Pd-catalyzed C—C coupling (Scheme 6.7) [9] of 2-(2-bromoaryl) cyclopropnae-1,1-dicarboxylates **18** with primary nitroalkanes **2** in presence of JohnPhos ligand (Ligand **A** as bulky phosphine) and successive spontaneous in situ recyclization of **19** achieves 1-*H*-2,3-benzoxazine 3-oxides **20** in satisfactory to excellent yields (up to 82%).

(iv) A general method for the preparation of 5-alkyl-5-aryl-1-pyrroline *N*-oxides (AAPOs, **22**), a diffused and important heterocyclic structure, present in therapeutics, neurodegeneration, cardiovascular disease, and cancer, was realized (Scheme 6.8) [10] in three steps. (Step 1) Palladium-catalyzed arylation of

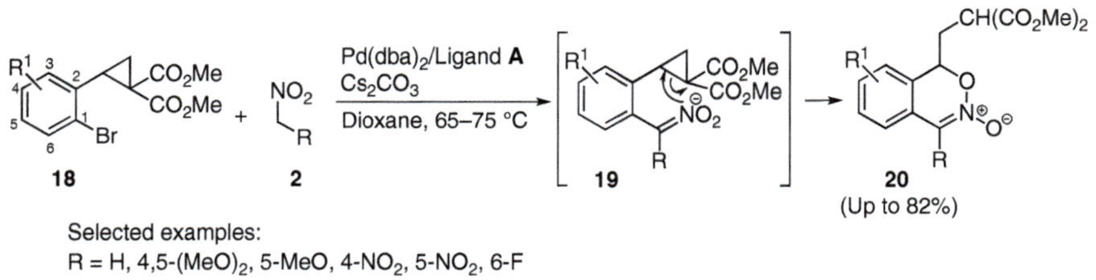

Scheme 6.7 One-pot synthesis of 1*H*-2,3-benzoxazine-3-oxide **20**.

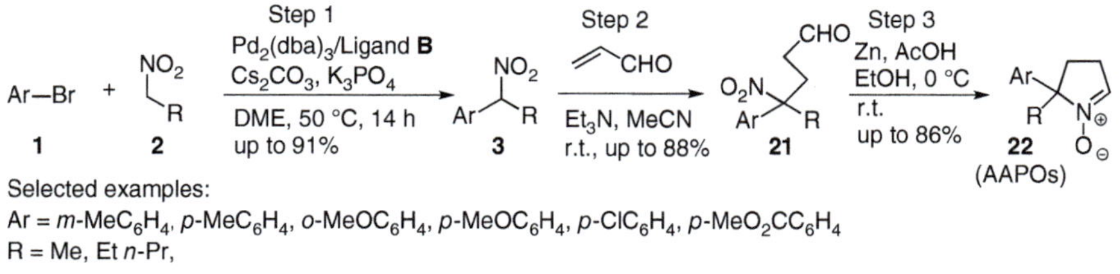

Scheme 6.8 Synthesis of 5-alkyl-5-aryl-1-pyrroline N-oxides (AAPOs) **22**.

nitroalkanes **2** into **3**, under Pd catalysis and in presence of Ligand **B**. (Step 2) Michael reaction of **3** with acrolein under basic conditions (Et_3N) and formation of the adduct **21**. (Step 3) Reductive cyclization of **21** (Zn/AcOH in EtOH at 0 °C) into the target heterocycles **22**.

(v) Aryl γ-lactam (pyrrolidin-2-one) structures have found in many natural products, thus their easy synthetic availability became very important. Few years ago, Dai and coworkers [11] developed a new synthesis of 5-alkyl-5-aryl-γ-lactams **24** using readily available aryl bromides 1, nitroalkanes 2, and methyl acrylate (MA). The process involves three steps and the first one concerns the arylation of nitroalkanes under palladium catalysis, in the presence of Ligand **B** (Scheme 6.8). Successive Michael reaction with MA, under DBU, furnishes the adducts **23** that, under $NaBH_4$-reduction, cyclize to 5-alkyl-5-aryl-γ-lactams **24** in good yields.

Since the palladium-catalyzed α-arylation of nitroalkanes (Step 1) tolerates a variety of functionalities, this method offers access to a diversity of functionalized γ-lactams.

6.3 Others

(i) Trifluoromethylation of secondary nitroalkanes **25** has been pointed out employing commercially available Umemoto's reagent **26** [12]. Although this approach is conceptually different from the arylation of nitroalkanes Pd-catalyzed, it represents a very important way for introduction of CF_3-group, via C—C coupling with nitroalkane. Thus, the method (Scheme 6.9) provides rapid access to highly complex α-(trifluoromethyl)nitroalkanes **27**, in good yields and diastereoselectivity. The wide functional group tolerance highlights the power of this procedure as a way for installation of trifluoromethyl group (Scheme 6.10).

In addition, compounds **27** can be converted into a large variety of other strategic molecular structures such as medicinally interesting α-trifluoromethylated amines and alkenes (via elimination of nitrous acid).

(ii) Direct coupling of allylic alcohols and secondary nitroalkanes can be obtained by the synergic action of palladium complexes and commercially available titanium(IV)isopropoxide (Table 6.2).

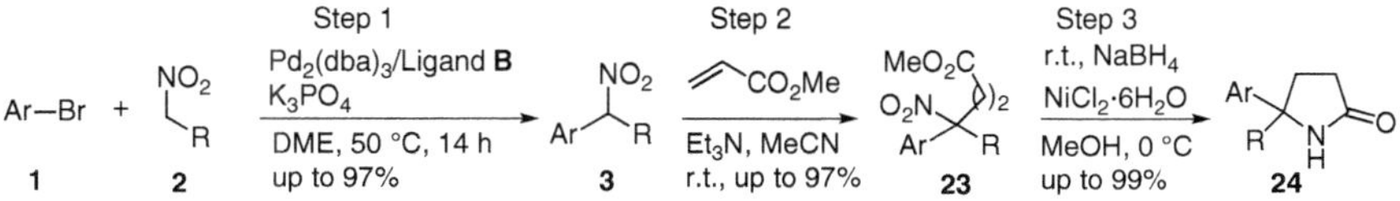

Selected examples:
Ar = o-MeC_6H_4, m-MeC_6H_4, o-$MeOC_6H_4$, p-$MeOC_6H_4$, p-ClC_6H_4, p-$MeO_2CC_6H_4$
R = Me, Et.

Scheme 6.9 Synthesis of γ-lactams **24**.

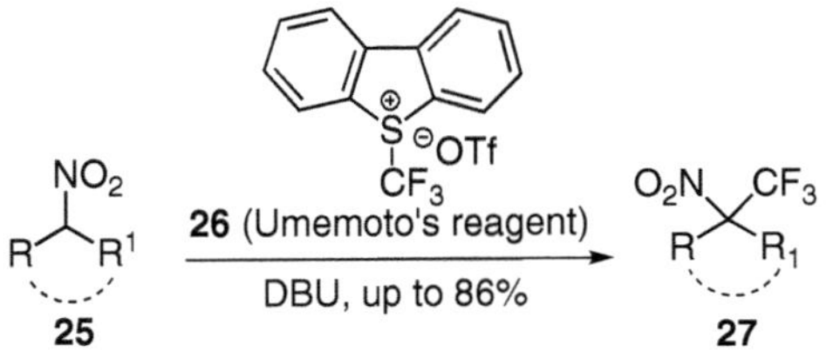

Selected examples

R^1 = CH_2=$CHCH_2$, *p*-$MeOC_6H_4$, *p*-$MeCOC_6H_4$, $Me_2NCO(CH_2)_2$, 3,4-$Cl_2C_6H_3CH_2NH$
R–R^1 = cycloalkyl derivatives

Scheme 6.10 Trifluoromethylation of secondary nitroalkanes into **27**.

Table 6.2 α-Arylation of secondary nitroalkanes **25** into **29** (selected examples).

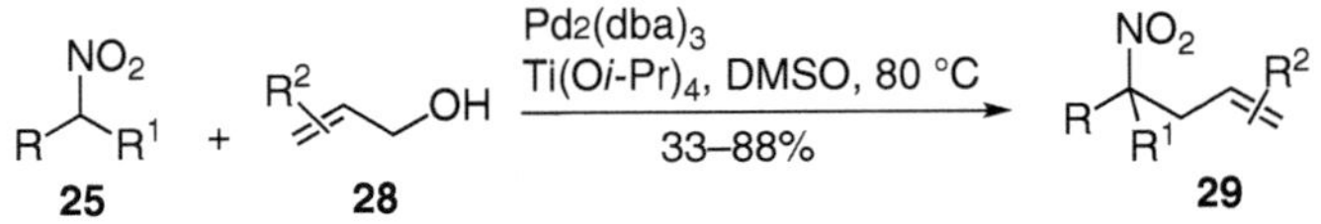

Nitroalkanes 25	Allylated products 29	Yield (%)
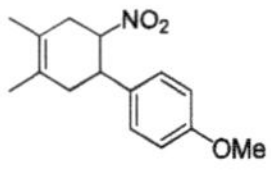	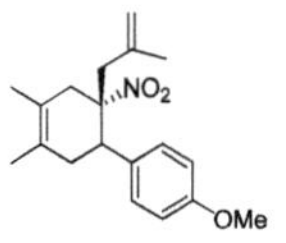	88
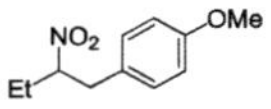	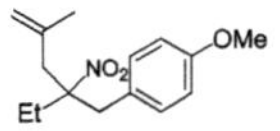	65
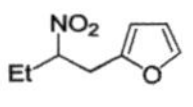	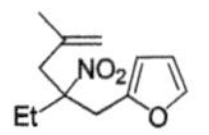	62
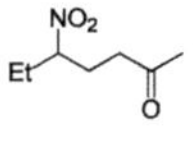	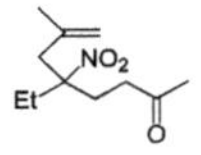	54
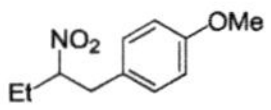	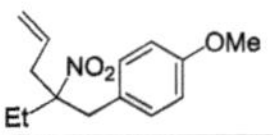	76

Authors applied the aforementioned method to the synthesis of (±)-Adalinine **34** (Scheme 6.11), a piperidine alkaloid isolated from chemical defense secretion of the European two-spotted ladybird beetle *Adalia bipunctata*. Secondary nitroalkane **30** (Seebach's reagent) is coupled with β-methallyl alcohol **28a** to the adduct **33** in 81% yield. The latter is ozonolyzed into **32**, which, by the Fischer esterification, converts into **33** that, under Zn/AcOH reduction, affords Adalinine **34** in 34% overall yield from the starting secondary nitroalkane **30** [13, 14].

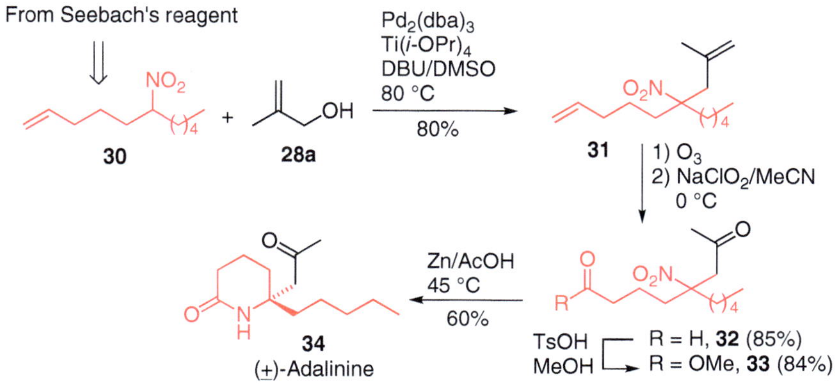

Scheme 6.11 Synthesis of (±)-Adalinine **34**.

References

1 Fox, J.M., Huang, X., Chieffi, A., and Buchwald, S.I. (2000). *J. Am. Chem. Soc.* 122: 1360–1370.
2 Vogl, E.M. and Buckwald, S.L. (2002). *J. Org. Chem.* 67: 106–111.
3 Nguyen, N.-H., Apriletti, J.W., Baxter, J.D., and Scanlan, T.S. (2005). *J. Am. Chem. Soc.* 127: 4599–4608.
4 VanGelder, K.F. and Kozlowski, M.C. (2015). *Org. Lett.* 17: 5748–5751.
5 Walvoord, R.R. and Kozlowski, M.C. (2013). *J. Org. Chem.* 78: 8859–8864.
6 Brals, J., Smith, J.D., Ibrahim, F. et al. (2017). *ACS Catal.* 7: 7245–7250.
7 Wolvoord, R.R., Berritt, S., and Kozlowski, M.C. (2012). *Org. Lett.* 14: 4086–4089.
8 Jiang, M., Feng, L., Feng, J., and Jiao, P. (2017). *Org. Lett.* 19: 2210–2213.
9 Mikhaylov, A.A., Dilman, A.D., Novikov, R.A. et al. (2016). *Tetrahedron Lett.* 57: 11–14.
10 Xu, J., Li, X., Wu, J., and Dai, W.-M. (2014). *Tetrahedron* 70: 6384–6391.
11 Xu, J., Li, X., Wu, J., and Dai, W.-M. (2014). *Tetrahedron* 70: 3839–3846.
12 Gietter-Burch, A.A.S., Devannah, V., and Watson, D.A. (2017). *Org. Lett.* 19: 2957–2960.
13 Chang, C.-Y. and Wu, Y.-K. (2018). *J. Org. Chem.* 83: 6217–6224.
14 Seebach, D. and Knochel, P. (1984). *Helv. Chim. Acta* 67: 261–283.

7

Synthesis and Reactivity of 1,3-Dinitroalkanes

1,3-Dinitroalkanes are an emerging class of nitroalkanes prone to be employed (Scheme 7.1) as (i) bidentate nucleophiles, (ii) dicarbonyls, and (iii) diamines. Moreover, 1,3-dinitroalkanes have, recently, demonstrated to be the key staring materials for the synthesis of many important targets [1].

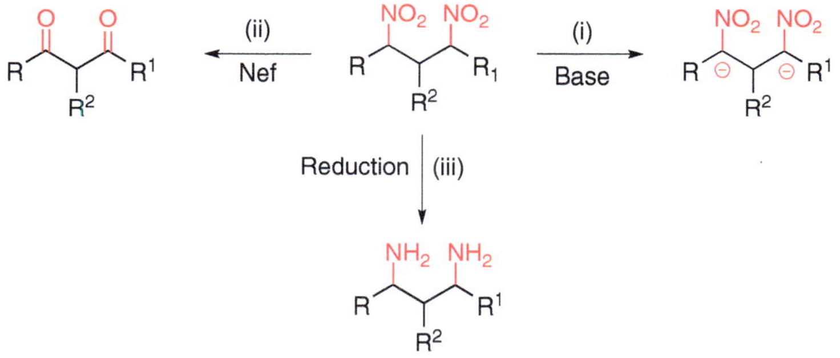

Scheme 7.1 1,3-Dinitroalkanes transformations.

7.1 Synthesis of 1,3-Dinitroalkanes

Two main procedures have been proposed for the synthesis of 1,3-dinitroalkanes **1**: Michael addition of nitroalkane **2** to pre-prepared conjugated nitroalkenes **3** (Scheme 7.2, pathway a) or nitroaldol condensation of an aldehyde **4** with nitroalkanes **2** followed by in situ Michael reaction of nitroalkane with the formed nitroalkene (pathway b).

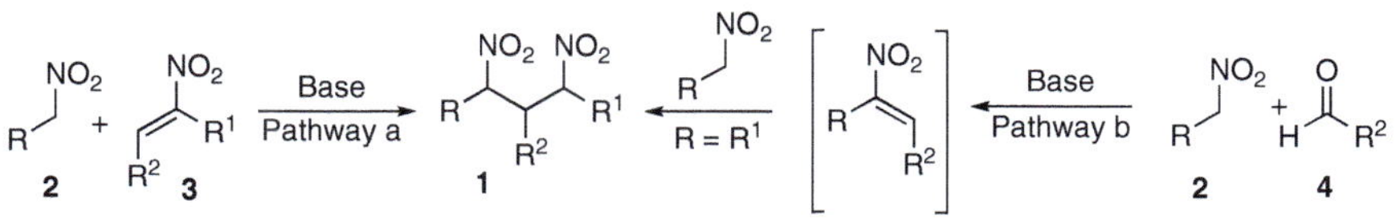

Scheme 7.2 Synthesis of 1,3-dinitroalkanes (pathway a or b).

Nitroalkanes: Synthesis, Reactivity, and Applications, First Edition.
Roberto Ballini and Alessandro Palmieri.

Michael reaction has been the topic of Chapter 6 that includes the conjugate addition of carbanions derived from nitroalkanes to electron-poor nitroalkenes, with the formation of 1,3-dinitroalkanes. For the efficiency of this procedure, the right choice of the catalytic system appears crucial. A pioneering method [2] proposed for primary nitroalkanes under Et_3N solvent free catalysis gave good yields of 1,3-dinitrocompounds. More recently, an efficient method applied to nitrostyrenes, under aqueous Na_2CO_3, achieved good yields (up to 86%) of products **1**, and in some cases with good *syn/anti* diastereoselectivity (up to 5.3 : 1) [3]. 1,8-Diazabicyclo[5.4.0]undec-7-ene (DBU) or TBAF have been proposed [4] as bases for the conjugate addition of CH_3NO_2 to 3-aminonitro alkenes in a study concerning the preparation of 1,3-nitroamines and 1,3-diamines.

7.1.1 Asymmetric Synthesis of 1,3-Dinitroalkanes

Conjugate addition of nitroalkanes to nitroalkenes allows the formation of stereogenic centers (for R and $R^1 \neq H$); thus, in the last years, interesting investigations of asymmetric procedures have been developed (Scheme 7.3).

Firstly, Wang et al. [5] reported the enantioselective conjugate addition catalyzed with modified *Cinchona* alkaloid **5a** under neat conditions, affording 1,3-dinitro compounds **10** in high yields (up to 82%) and good enantioselectivity (up to 88% ee). The same authors reported that when primary nitroalkanes are used (R = H), two new stereogenic centers are generated with 2 : 1 dr and up to 81% ee for the major isomer (65% ee for the minor) of adducts **10**.

Further studies [6] were successively reported involving *Cinchona* alkaloid **5b**. The procedure allows good results of adducts **11** (yields up to 85%, *syn/anti* = 67 : 33 and ee of *syn* 81%).

An efficient asymmetric organocatalysis was then reported [7] using tridentate catalyst (bis-oxazoline) **6**, with the help of Et_2Zn and Ti(O*i*Pr)$_4$. High enantioselectivity and acceptable *syn* diastereoselectivity of products **12** are reported. The presence of Ti(O*i*Pr)$_4$ seems crucial for the reaction because it might energize the Et_2N that acts as activator of nitroalkene.

A simple procedure was found in presence of thiourea catalyst **7** (2 mol%), Et_3N (2 mol%) as base, and PhMe as solvent [8]. The adducts **13** are obtained in good yields, mainly as *syn*-form (up to 92%) and dr up to 90 : 10.

A highly enantioselective version of the aforementioned procedure, based on bifunctional thiourea catalyst **8**, has been designed for this reaction [9]. Asymmetric induction increases with decreasing catalyst loading, with the optimal compromise between rate and induction being at a loading of 2 mol% of catalyst in PhH. Excellent enantioselectivities of **13** are observed over a range of electron-rich and electron-poor nitro styrenes.

Michael addition of $EtNO_2$ to a variety of nitroalkenes, in presence of lanthanum(III) triflate and in combination with *N,N'*-dioxide ligand **9** was reported [10]. It has been observed that the base dramatically influences the formation of the adducts **14**, and imidazole resulted the best. Thus, high yields, good diastereoselectivities (up to dr 93 : 7), and excellent ee value (up to 97%) are achieved.

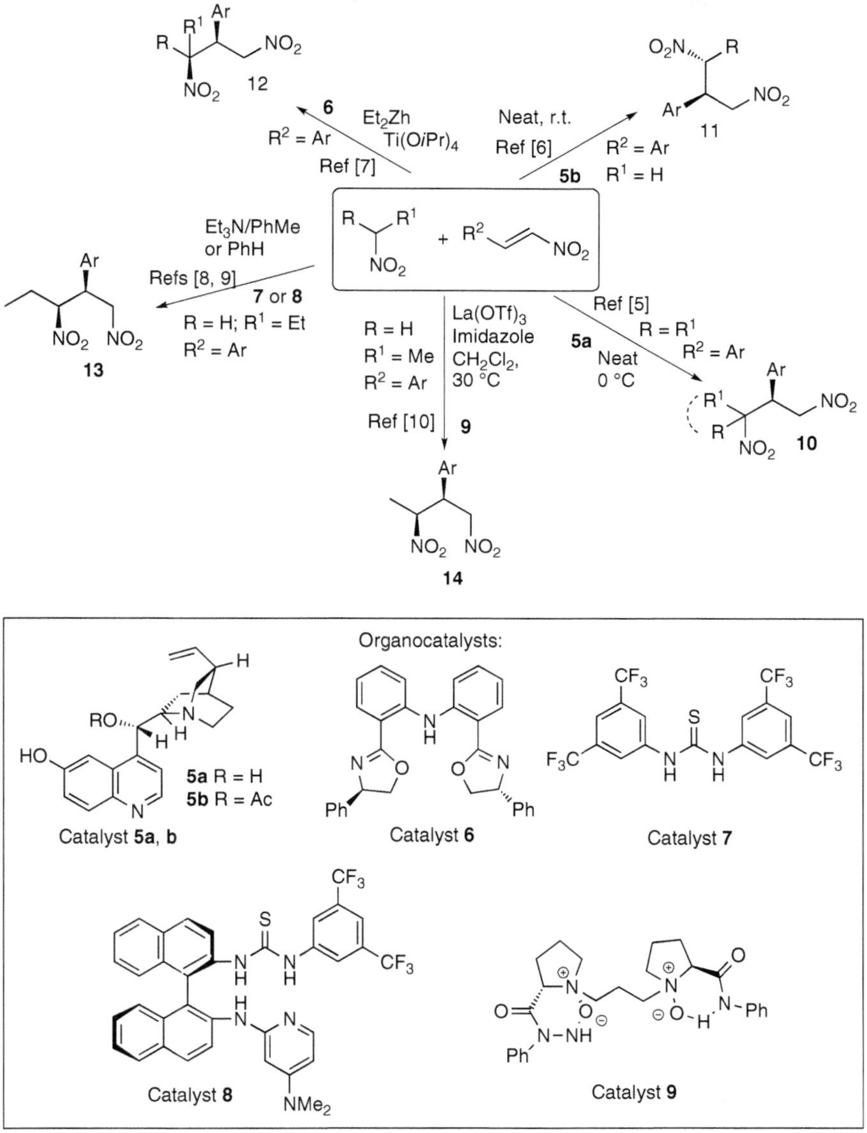

Scheme 7.3 Asymmetric conjugate addition of nitroalkanes to nitroalkenes.

One more asymmetric strategy, with improved stereo-control of the reaction, was realized using chiral ammonium bifluoride **16** in combination with silylated nitronates **15** (Scheme 7.4) [11]. With aromatic nitroalkenes the efficiency of the reaction and the diastereoselectivity are insensitive to the electron properties of the aromatic ring. However, the presence of electron-withdrawing substituents seems to be associated with slightly lower enantioselectivity.

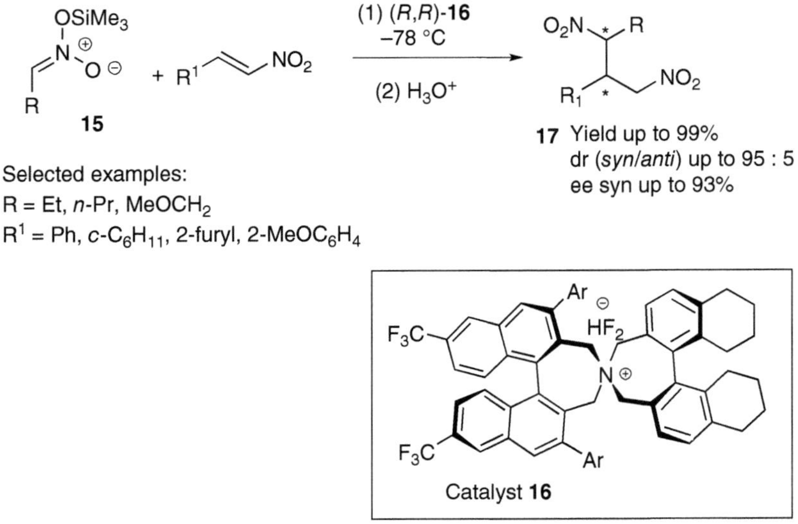

Scheme 7.4 Asymmetric conjugate addition of silyl nitronates promoted by the catalyst **16**.

7.1.2 Synthesis of Symmetric 1,3-Dinitroalkanes

Symmetric 1,3-dinitroalkanes (compound **1**, Scheme 7.2, R = R^1) are very interesting bidentate structures prone to be employed as key starting molecules for the synthesis of important targets, thus a convenient way for their preparations appears of great interest.

An electrochemical procedure was originally reported [12], reacting an aldehyde with CH_3NO_2 (Scheme 7.5).

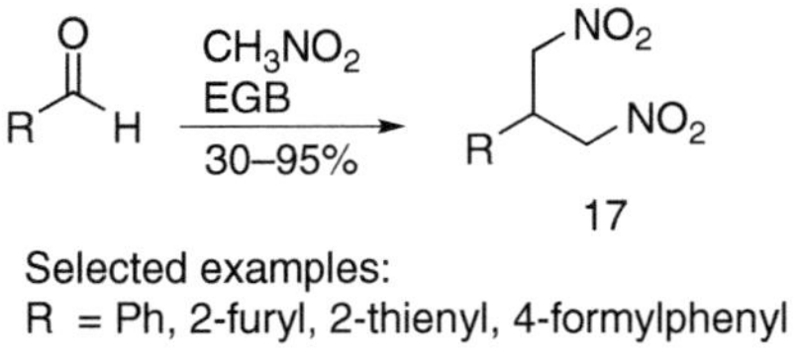

Scheme 7.5 Electrochemical synthesis of 1,3-dinitroalkanes.

Firstly, an electrogenerated base (EGB) is prepared via cathodic reduction of air-saturated $MeNO_2$ solution. The aldehyde is then added and the reaction proceeds in three base-catalyzed steps: (i) addition of $MeNO_2$ to the aldehyde (Henry reaction), with the formation of nitroalcohol, (ii) dehydration, with the generation of nitroalkene (nitroaldol condensation), and (iii) Michael addition of nitromethane to the formed nitroalkene (Michael reaction). Good yields can be achieved when the aryl group is of electron-donating character.

A more general and efficient one-pot synthetic process, under heterogeneous catalysis, was developed few years later [13]. Thus, treatment of an aldehyde with excess of $MeNO_2$ and in presence of basic alumina as solid catalyst gives the one-pot synthesis of 1,3-dinitroalkanes **17** (Scheme 7.6).

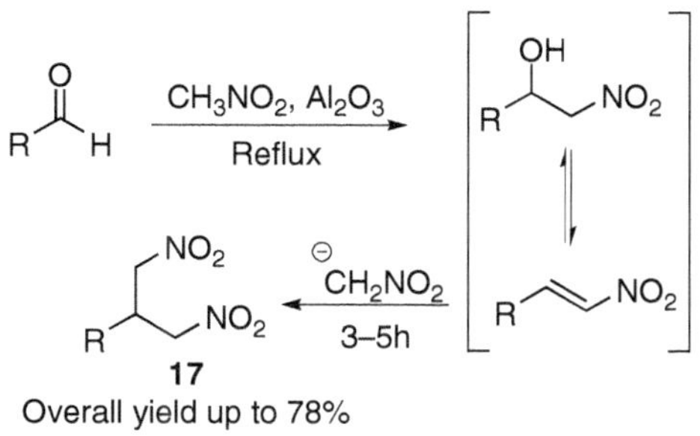

Selected examples:
R = Ph, $Ph(CH_2)_2$, *i*-Pr, *n*-C_8H_{17}, *p*-$NO_2C_6H_4$, *p*-FC_6H_4, 2-furyl, 3-pyridyl

Scheme 7.6 Synthesis of 1,3-dinitroalkanes promoted by basic Al_2O_3.

As development of this procedure, the use of calcium vanadade apatite (VAp) in water was successively reported [14]. Although the method produces high yields of compounds **17** (77–92%), it must be performed into a sealed vessel for 12 hours at 110 °C.

Further results, under heterogeneous catalysis, were reported [15] using acid-base bifunctional catalytic surfaces for nucleophilic addition reactions. In fact, the reaction is performed in the presence of SA-NH_2-NEt_2 (silica/alumina-immobilized primary and tertiary amines) as catalyst at 100 °C, yielding (up to 93%) 1,3-dinitroalkanes **17**.

An interesting procedure was reported under nickel species catalysis (Scheme 7.7) [16]. Good yields are acquired; however, the method needs the presence of 4 equiv of both Zn and Na_2CO_3 and appears to be restricted to aromatic aldehydes.

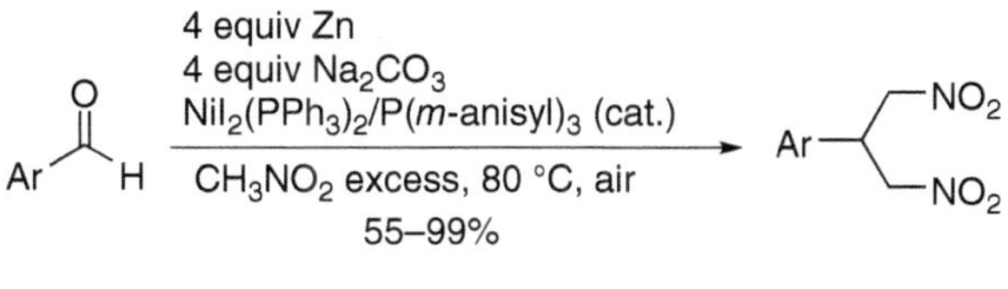

Scheme 7.7 Synthesis of 1,3-dinitroalkanes under Ni-species.

Linear 1,3-dinitro compounds of type **20** were easily prepared by condensation between lithium salt of an appropriate nitroalkane **18** and nitroalkene **19** (Scheme 7.8) [17].

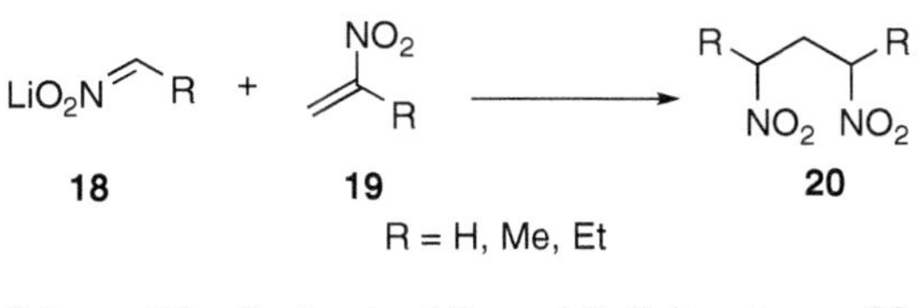

Scheme 7.8 Synthesis of linear 1,3-dinitroalkanes **20**.

The method needs an excess of **18** in order to avoid the formation of oligomers and appears efficient only for symmetric 1,3-dinitroalkanes.

7.2 Synthetic Applications of 1,3-Dinitroalkanes

The bidentate behavior of 1,3-dinitroalkanes makes these compounds strategic for the synthesis of a variety of targets. However, just in the last years there has been an increasing interest on the chemistry of these compounds.

7.2.1 Synthesis of 1,3-Diamines

The easy reduction of the nitro to amino group has applied to the conversion of chiral 1,3-dinitroalkanes into 1,3-diamines that are ubiquitous in a wide range of biologically active compounds or key starting compounds for the synthesis of agrochemicals and pharmaceuticals. This reduction has been proposed following a variety of methods (Scheme 7.9) [11, 18, 19].

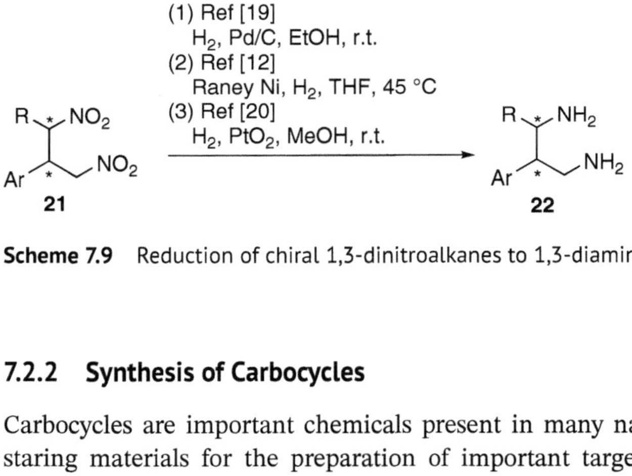

Scheme 7.9 Reduction of chiral 1,3-dinitroalkanes to 1,3-diamines **22**.

7.2.2 Synthesis of Carbocycles

Carbocycles are important chemicals present in many natural products and key staring materials for the preparation of important targets, thus their synthetic availability appears of great interest.

7.2.2.1 Synthesis of Dinitrocyclohexanols

One-pot synthesis of 2,4-dinitrocyclohexanols, important building blocks for several targets, is obtained in a one-pot procedure adding of 1,3-dinitroalkanes to acrolein under basic (Al_2O_3) neat conditions (Scheme 7.10) [20].

The first step consists in a Michael reaction followed by internal Henry reaction with the formation of **24** in good yields and dr up to 15 : 85.

Asymmetric version of the previous process was developed [21] using (*S*)-2-[bis(3,5-bis-trifluoromethylphenyl)(trimethylsilyloxymethyl)]pyrrolidine **25** as organocatalyst (Table 7.1).

During studies of the reaction it has been found that the best enantioselectivity was achieved with diazabicyclo[2.2.2]octane (DABCO) (10%) as additive base. A variety

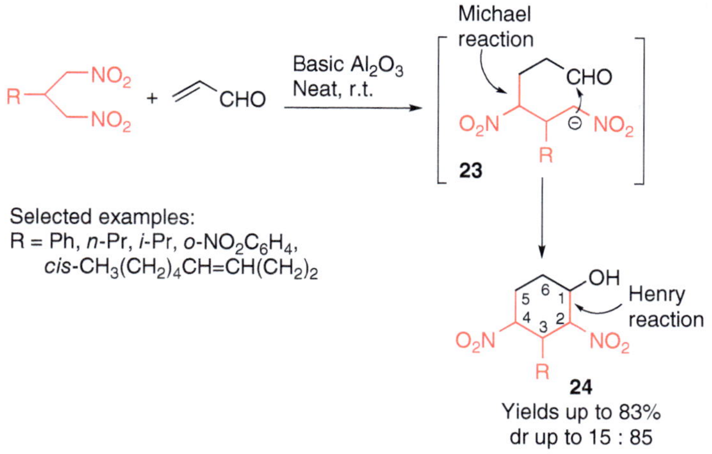

Scheme 7.10 One-pot synthesis of 2,4-dinitrocyclohexanols **24**.

Table 7.1 Asymmetric synthesis of tetrasubstituted cyclohexanols **26**.

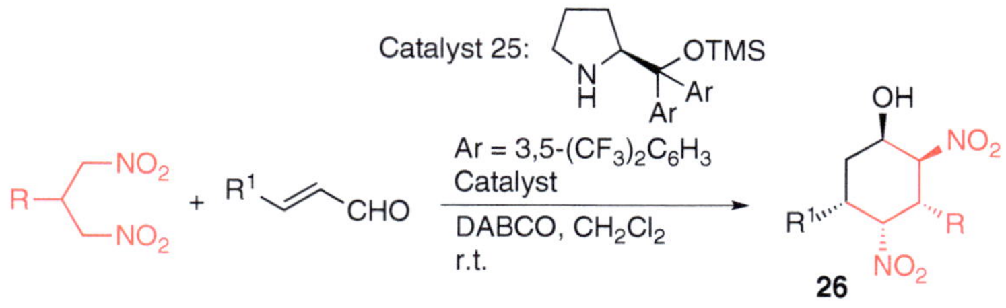

Entry	R	R^1	Yield (%)	dr	ee (%)
1	Ph	CH_3CH_2	45	4 : 2 : 1	90
2	Ph	$(CH_3)_2CH$	38	3 : 1 : 1	90
3	*p*-$CH_3OC_4H_6$	$(CH_3)_2CH$	53	5 : 1 : 1	89
4	2-Furanyl	$(CH_3)_2CH$	43	6 : 1 : 0	88
5	2-Thiophenyl	$(CH_3)_2CH$	43	5 : 1 : 1	90
6	Ph	$CH_3(CH)_3$	43	4 : 2 : 1	87

of enals and 1,3-dinitrocompounds can be employed; however positive effects on the stereoselectivity can be enhanced in the presence of bulkier alkyl substituents.

Thus, this domino process, catalyzed by commercially **25**, proceeds with interesting overall yields and high diastereo- and enantioselectivities.

7.2.2.2 Synthesis of Bicyclo[3.3.1]nonanes

Bicyclo[3.3.1]nonanes are important molecules from both theoretical and synthetic studies. In fact, this framework is common in nature as constituent of a variety of natural compounds and biologically active products or their metabolites. Moreover, these molecules are of interest because of their relation to alkaloids and adamantanoids. For these reasons, an easy procedure for the synthesis of bicyclo[3.3.1]nonanes appears of great interest. In this context, 1,3-dinitroalkanes have been found the key starting molecules for one-pot synthesis of bicyclo[3.3.1]nonanes **29** (Scheme 7.11) [22].

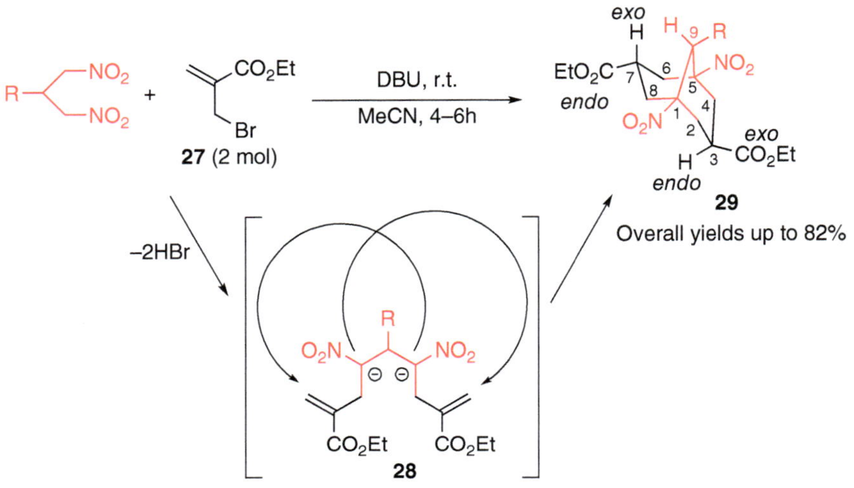

Selected examples:
R = Ph, *n*-Pr, $Ph(CH_2)_2$, *o*-MeC_6H_4, *p*-PhC_6H_4, *c*-C_6H_{11}, *n*-C_5H_{11},

Scheme 7.11 One-pot synthesis of bicyclo[3.3.1]nonanes **29**.

The process starts with a double Michael reaction of 1,3-dinitroalkanes with 2 mol of commercially available bromomethylacrylate **27**, under DBU catalysis and in presence of MeCN as solvent, with consequent elimination of 2 mol of HBr. The formed intermediate **28** is prone to a double internal Michael reaction that quickly allows the cyclization to the target bicyclic structures **29**, as *exo*-3, *endo*-7 diastereoisomers, in good overall yields and in very short reaction times (four to six hours).

A large number of alkyl or aryl groups can be easily introduced at 9-position with the right choice of the starting dinitro derivatives.

It is important to observe that this method generates four new C—C bond (C_1—C_2, C_1—C_6, C_5—C_4, and C_5—C_6) in a one-pot process.

The same authors reported the one-pot synthesis of bis-bicyclic structures **31** (Scheme 7.12) by reaction of tetranitro compound **30** (easily prepared under condensation of $MeNO_2$ with terephthalaldehyde) and 4 equiv of **27**. It is important to point out that the one-pot synthesis of **31** involves the formation of eight new C—C bonds in a single step.

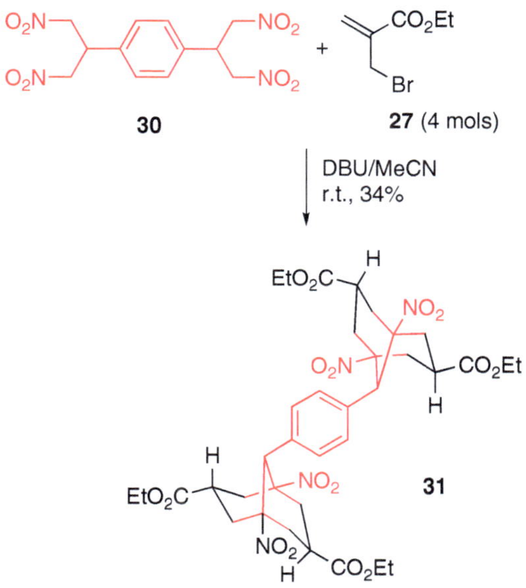

Scheme 7.12 One-pot synthesis of bis-bicyclic compounds **31**.

7.2.3 Synthesis of Benzene Derivatives

Benzenes are important molecules in organic synthesis and aromatization of precursors is an useful and convenient way to prepare these compounds. In this context, 1,3-dinitroalkanes have demonstrated to be crucial for this purpose.

7.2.3.1 Synthesis of Acetophenones and Benzoates

Alkylated acetophenones and methyl benzoates **36** are important building blocks for the synthesis of pharmaceuticals and other targets. An easy preparation of these molecules has been realized in a one-pot process that involves the Michael addition of 1,3-dinitroalkanes to enediones **32** in presence of DBU as base (Scheme 7.13) [23].

Thus, the first step consists in the Michael addition of dinitroalkane to **32** with consequent elimination of nitrous acid from the adduct **33**, allowing the polyfunctionalized enones **34**. Successive internal nitroaldol reaction produces the cyclic structure **35**. Quenching of the reaction with 4 N HCl favors the elimination of both water and nitrous acid, with the one-pot formation of **36** in good overall yields (up to 82%).

Several advantages can be obtained following this approach, including the avoidance of *ortho/meta/para* mixture formation commonly observed in conventional electrophilic aromatic substitutions.

In fact, this region-defined synthesis for acetophenone (R^1 = alkyl or aryl, R^2 = alkyl or aryl) and benzoate (R^1 = OMe, R^2 = alkyl or aryl) derivatives is very difficult to undertake via electrophilic substitution of pre-formed benzenes. Moreover,

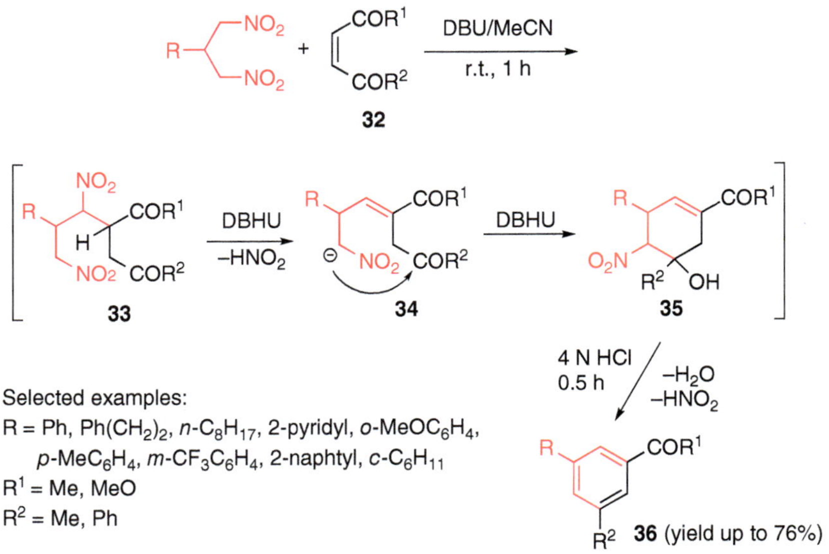

Scheme 7.13 One-pot synthesis of benzoyl derivatives **36**.

by this procedure, a large variety of alkyl or aryl groups can be introduced at the 9-position, independently from their nature (acyclic, cyclic, aromatic, polyaromatic, and heterocyclic).

As an extension of this procedure, the authors reported the one-pot synthesis of **38**, a key building blocks for the preparation of farnesyl-protein transferase inhibitor, achieved in 76% overall yields (Scheme 7.14).

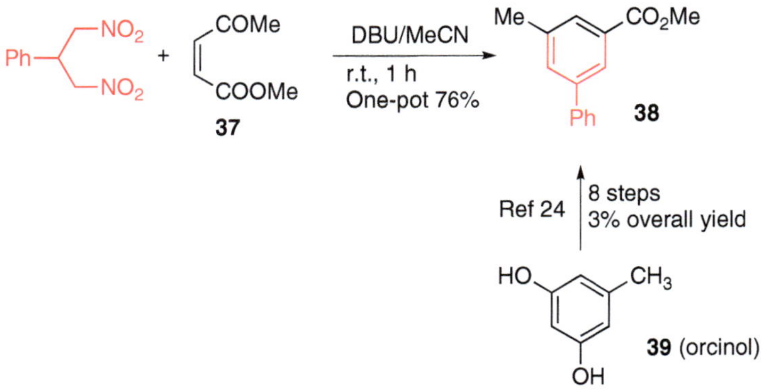

Scheme 7.14 One-pot synthesis of **38**.

In is important to note that **38** was previously synthesized in eight steps, from Orcinol **39**, in only 3% yield [24].

7.2.3.2 Synthesis of Arylamines

Arylamines are important molecules found in pharmaceuticals, materials, and ligand catalysis. 1,3-Dinitroalkanes have demonstrated to be versatile precursors for the preparation of symmetrical diarylamines **41** (Scheme 7.15) in a two-step process [25].

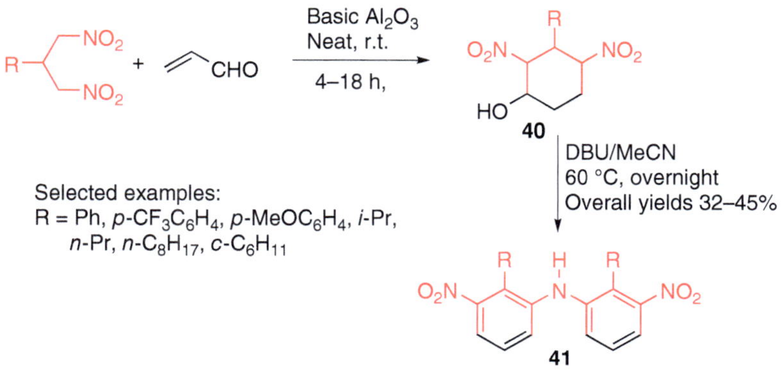

Scheme 7.15 Synthesis of diarylamines **41**.

In fact, reaction of 1,3-dinitroalkanes with acrolein, under basic Al_2O_3 and neat conditions, affords cyclohexanol **40** in a one-pot process. Successive treatment of **40** with 4 equiv of DBU, in MeCN as solvent and warming overnight at 60 °C, gives the direct formation of **41** in satisfactory overall yields (32–45%).

The mechanism of conversion **40** into **41** is not clear and a possible pathway could be the reductive coupling reaction (with loss of NO_2) that leads diarylamines **41**.

7.2.3.3 Synthesis of Polyfunctionalized Phenols

Phenols, in particular polyfunctionalized derivatives, are of special interest in organic synthesis since they are found in a large number of natural products and are the key building blocks for the preparations of important targets. In this context, the versatility of 1,3-dinitroalkanes has been exploited for a new procedure devoted to the synthesis of polyfunctionalized phenols. Thus, as reported in Scheme 7.16, a three-step sequence was developed starting from acrolein and dinitrolalkanes [26].

Cyclohexanols **40** are obtained following the previous procedure of Scheme 7.15. Then, the formed **40** is treated with H_2O/MeOH solution of K_2CO_3, followed by acidic work-up. Under these reaction conditions two transformations are observed: (i) Nef reaction and (ii) dehydration of the β-nitroalkanol. The formed nitroalkene **42** is brominated ($PhNMe_3Br{\cdot}Br_2$) giving the direct formation of polyfunctionalized phenols **43** in satisfactory to good overall yields (37–68%). Thus, the reaction represents a useful procedure for the preparation of a variety of polyfunctionalized phenols from acyclic precursors. Moreover, the appropriate choice of starting dinitroalkanes allows a large variety of substituents in *ortho*-position can be easily introduced.

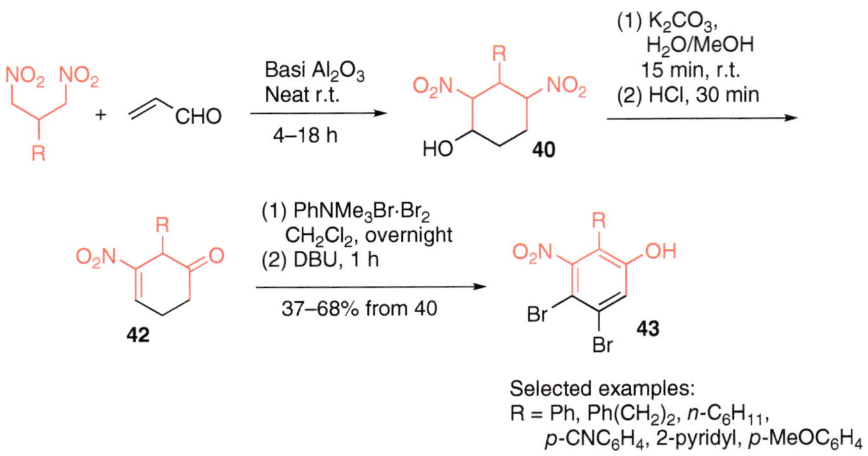

Scheme 7.16 Synthesis of phenols **43**.

An elegant synthesis of polysubstituted phenols **45** and **46** was elaborated from Baylis–Hillman adducts (as 1,3-dielectrophilic three-carbon components) and 1,3-dinitroalkanes as 1,3-dinucleophilic molecules, by a [3+3] annulation, with the help of DABCO (Scheme 7.17) [27].

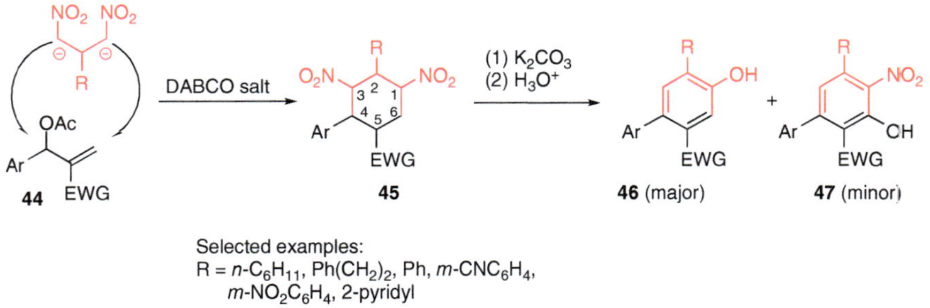

Scheme 7.17 Synthesis of phenols **46** and **47**.

A sequential S_N'/S_N2'/Michael addition produces a diastereomeric mixture of dinitrocyclohexane **45** that, after usage of K_2CO_3 in dimethylformamide (DMF) and acidic work-up, provides the formation of two new polysubstituted phenols **46** and **47**, difficult to obtain following other ways.

The formation of structure **46** can be explained with the elimination of nitrous acid from C_3 to C_4 and the simultaneous Nef reaction at C_1, with subsequent enolization of the formed ketone and dehydrogenation from C_5 to C_6 favored by the presence of H_3O^+ during the work-up. The formation of **47** follows the same steps as for **46** but without the Nef reaction.

7.2.3.4 Synthesis of Nitrobenzenes

The same authors found a new approach [28] for the synthesis one more category of polysubstituted phenols **50** (Scheme 7.18) from Baylis–Hillman adducts **44a**.

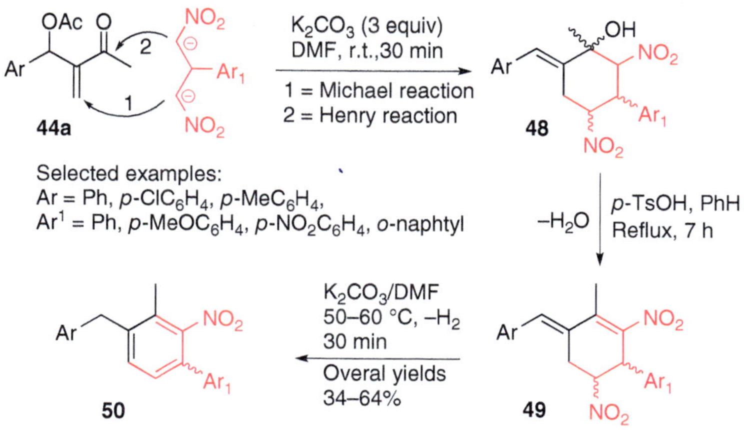

Scheme 7.18 Synthesis of phenols **50**.

The method starts with a Michael reaction (way 1) and successive Henry reaction (way 2), under basic conditions (K_2CO_3), with the formation of dinitrocyclohexanols **48**. Elimination of water (**48–49**), followed by nitrous acid elimination, achieves the targets molecules **50** in satisfactory to good overall yields (34–64%).

References

1 Ballini, R., Gabrielli, S., and Palmieri, A. (2014). *Eur. J. Org. Chem.*: 1805–1816.

2 Alcántara, M.-P.D., Escribano, F.C., Gómez-Sanchez, A. et al. (1996). *Synthesis*: 64–70.

3 Tulam, V.K., Kotte, S.C.B., Chinta, R.K.R. et al. (2012). *Int. J. Res. Pharm. Chem.* 2: 254–257.

4 Patrocinio Pereira, V.L.P., Silva Moura, A.L., Pires Vieira, D.P. et al. (2013). *Beilstein J. Org. Chem.* 9: 832–837.

5 Wang, J., Li, H., Zu, L. et al. (2006). *Adv. Synth. Catal.* 348: 2047–2050.

6 Deng, Y.-Q., Zhang, Z.-W., Feng, Y.-H. et al. (2012). *Tetrahedron: Asymmetry* 23: 1647–1652.

7 Lu, S.-F., Du, D.-M., Xu, L., and Zhang, S.W. (2006). *J. Am. Chem. Soc.* 128: 7418–7419.

8 Rabalakos, C. and Wulff, W.D. (2008). *Synlett*: 2826–2830.

9 Rabalakos, C. and Wulff, W.D. (2006). *J. Am. Chem. Soc.* 130: 13524–13525.

10 Yang, X., Zhou, X., Lin, L. et al. (2008). *Angew. Chem. Int. Ed.* 47: 7079–7081.

11 Ooi, T., Takada, S., Doda, K., and Maruoka, K. (2006). *Angew. Chem. Int. Ed.* 45: 7606–7608.

12 Niazimbetova, Z.I., Evans, D.H., Liable-Sands, L.M., and Rheingold, A.L. (2000). *J. Electrochem. Soc.* 147: 256–259.

13 Ballini, R., Bosica, G., Fiorini, D., and Palmieri, A. (2004). *Synthesis*: 1938–1940.

14 Hara, T., Kanai, S., Mori, K. et al. (2006). *J. Org. Chem.* 71: 7455–7462.

15 Motokura, K., Tada, M., and Iwasawa, Y. (2008). *Angew. Chem. Int. Ed.* 47: 9230–9235.

16 Gao, M. and Wei, V.-P. (2013). *J. Chem. Res.* 37: 146–148.

17 Wade, P.A., Dailey, W.P., and Carroll, P.J. (1987). *J. Am. Chem. Soc.* 109: 5452–5456.

18 Rios-Lomardia, N., Busto, E., Gotor-Fernández, V., and Gotor, V. (2010). *Eur. J. Org. Chem.*: 484–493.

19 Lu, S.-F., Du, D.-M., Xu, J., and Zhang, S.-W. (2004). *J. Am. Chem. Soc.* 128: 7418–7419.

20 Ballini, R., Barboni, L., Fiorini, D. et al. (2005). *Green Chem.* 7: 828–829.

21 Reyes, E., Jiang, H., Milelli, A. et al. (2007). *Angew. Chem. Int. Ed.* 46: 9202–9205.

22 Barboni, L., Gabrielli, S., Palmieri, A. et al. (2009). *Chem. Eur. J.* 15: 7867–7870.

23 Ballini, R., Barboni, L., Fiorini, D. et al. (2005). *Chem. Commun.*: 2633–2634.

24 Boyle, F.T., Davies, G.M., Wardleworth, J.M., and Arnould, J.C. 2002). Imidazolyl compounds as inhibitors of farnesyl-protein transferase. US Patent 6 414 145 B1, 02 July 2002.

25 Ballini, R., Barboni, L., Femoni, C. et al. (2006). *Tetrahedron Lett.* 47: 2295–2297.

26 Ballini, R., Barboni, L., Giarlo, G., and Palmieri, A. (2006). *Synlett*: 1956–1958.

27 Park, D.Y., Lee, K.Y., Gowrisankar, S., and Kim, J.N. (2008). *Bull. Korean Chem. Soc.* 29: 701–704.

28 Park, D.Y., Lee, K.Y., and Kim, J.N. (2007). *Tetrahedron Lett.* 48: 1633–1636.

8

Formation of Carbon=Carbon Double Bonds via Nitrous Acid Elimination (NAE)

The nitro group is a potent electron-withdrawing group; in addition it has demonstrated to be a valuable leaving group. In this context, the simultaneous presence of a nitro group and of an acidic hydrogen (often due to the presence of a geminal EWG) in the vicinal position strongly favors the elimination of nitrous acid (Figure 8.1) [1]. Thus, the sequential behavior of the nitro group as activator for the generation of C—C single bond and as leaving group is strategic for the preparation of alkenes, via β-elimination of nitrous acid.

Consequently, this opportunity was studied (mainly in the last two decades) and applied to the synthesis of a large variety of targets.

8.1 Synthesis of α,β-Unsaturated Carbonyl Derivatives

Michael addition of primary and secondary nitroalkanes to dimethyl maleate **1a**, in presence of DBU (diazabicyclo[5.4.0]undec-7-ene) as base and MeCN as solvent (Scheme 8.1) allows the formation of α,β-unsaturated esters **1**, via elimination of nitrous acid, in a one-pot process [2a, b].

The synthesis needs short reaction times allowing good overall yields (up to 96%) of **3**; see selected examples in Table 8.1.

The mild reaction conditions required allow high selectivity as demonstrated by the survival of several functionalities.

An important extension of the aforementioned method consists in the direct formation of 2-substituted succinate diesters **4a** and *N*-ethylsuccinate **4d**, by in situ reduction of the intermediate **3** (Scheme 8.2) [3].

In fact, the crude enone **3**, without any purification step, is treated with nickel boride ($NiCl_2/NaBH_4$), in MeOH/THF, affording **4a,d** in good overall yields (up to 94%).

It is important of note that, in this procedure, the nitroalkane acts as equivalent of the alkyl anion synthon **I** (Figure 8.2).

Although the synthetic utility of the obtained structures via Michael addition-elimination of nitrous acid (Schemes 8.1 and 8.2), the disappearance of the nitro

Nitroalkanes: Synthesis, Reactivity, and Applications, First Edition.
Roberto Ballini and Alessandro Palmieri.

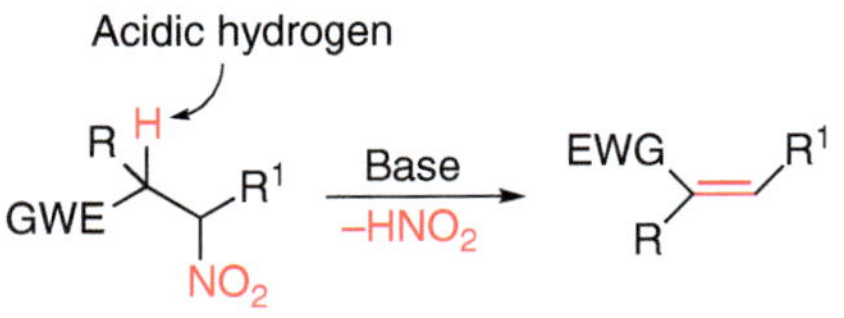

Figure 8.1 Base-induced elimination of nitrous acid.

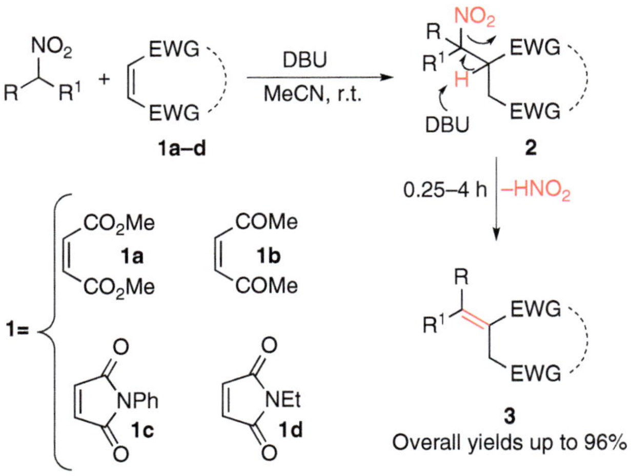

Scheme 8.1 Synthesis of α,β-unsaturated carbonyl derivatives **3**.

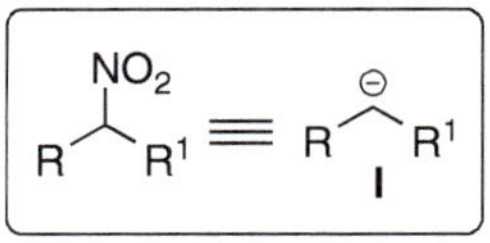

Figure 8.2 Nitroalkane as synthetic equivalent of the alkyl anion synthon **I**.

group in the products **3** and **4** represents an important loss in terms of reactivity for further synthetic applications, due to the great chemical versatility of the nitro functionality. Thus, a new chemoselective catalytic procedure, in which the temperature plays a key role, was elaborated using KF/basic alumina as solid base, under neat conditions (Scheme 8.3) [4]. In fact, the modulation of the temperature (r.t. or 55 °C) allows the chemoselective formation of the adducts **6**, with retention of the nitro group (r.t.) or generation of α,β-unsaturated enones **7** (55 °C), through elimination of nitrous acid, respectively. So, this procedure is the first versatile method for the chemoselective Michael addition of primary and secondary nitroalkanes to enediones **5**, in which the nitro group can be preserved or eliminated, simply, with the appropriate choice of the temperature.

Further results were obtained employing the Michael addition of primary nitroalkanes to trimethyl *trans*-aconitate **8**, in acetonitrile and in presence of DBU as base. The method achieves the one-pot formation of polyfunctionalized α,β-unsaturated carbonyl derivatives, such as methyl tri- and tetra-carboxylates **10** via base-induced nitrous acid elimination, from the adduct **9** (Scheme 8.4) [5].

The presence of double bond in compounds **10** results in the formation of α,β-, β,γ-, and γ,δ-unsaturated esters moiety being available in one molecule.

Table 8.1 Selected examples of α,β-unsaturated carbonyls **3**.

Entry	Enedione	Nitroalkane	Product 3	Yield (%)	Time (h)
1	**1b**	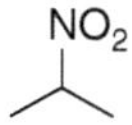	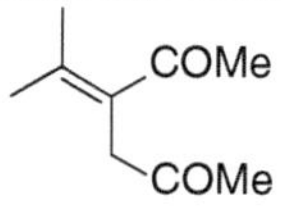	65	4
2	**1b**	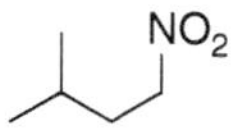	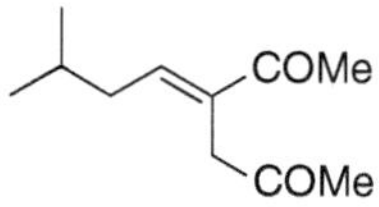	96	1
3	**1b**	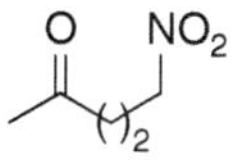	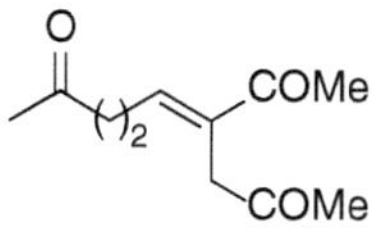	95	2
4	**1b**	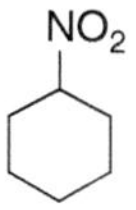	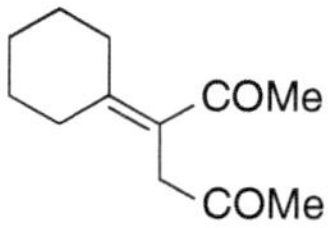	85	3
5	**1c**	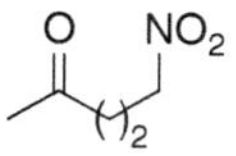	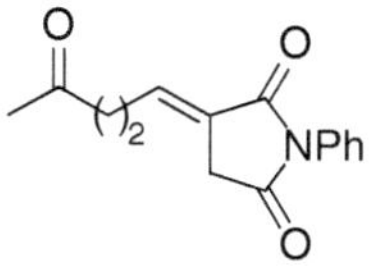	73	3
6	**1d**	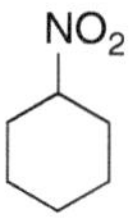	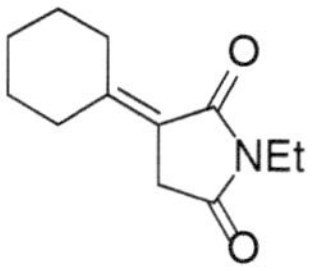	80	0.25
7	**1d**	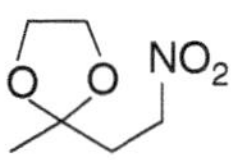	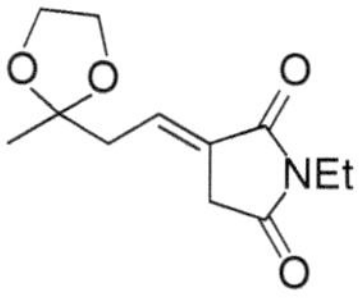	88	1
8	**1c**	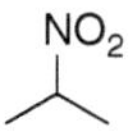	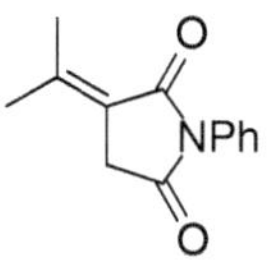	70	0.25
9	**1a**	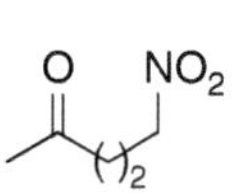	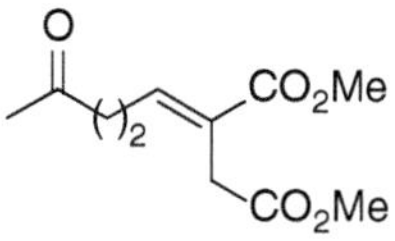	90	0.25
10	**1a**	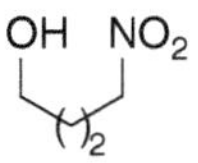	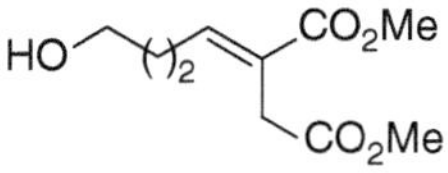	70	0.25

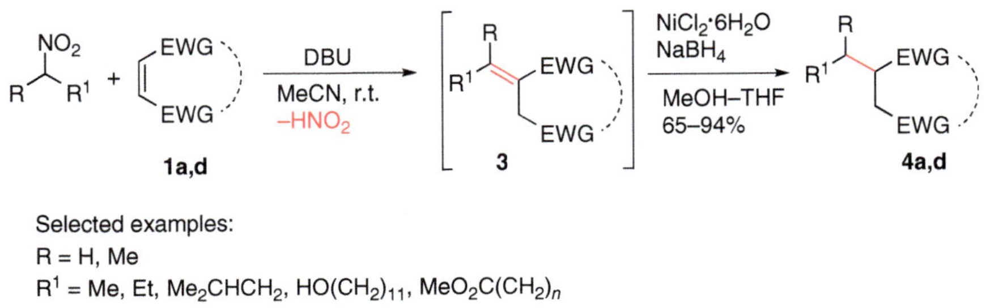

Scheme 8.2 Synthesis of succinate derivatives **4a** and *N*-ethylsuccinimmides **4d**.

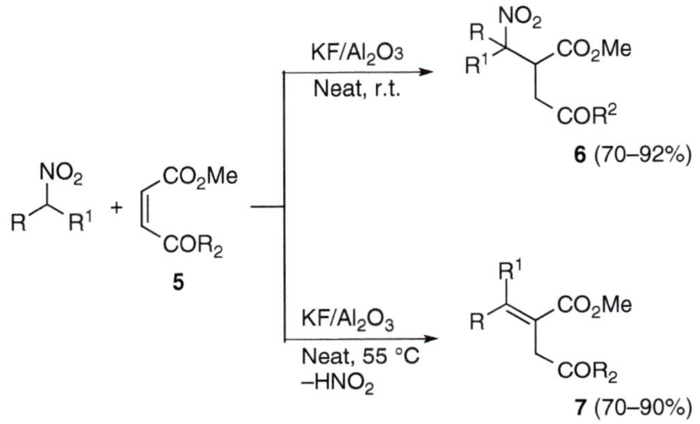

Scheme 8.3 Chemoselective formation of adducts **6** or **7**.

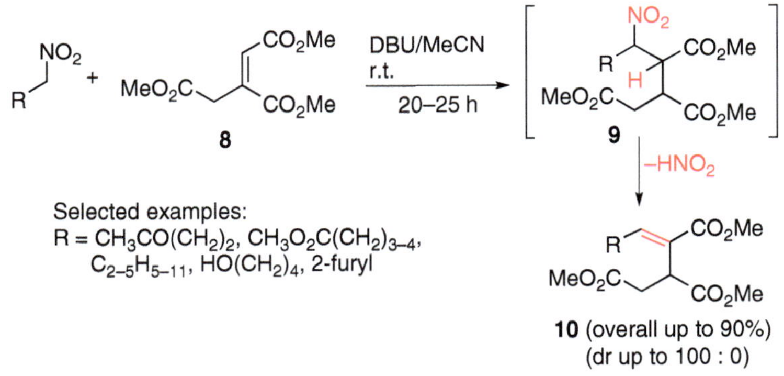

Scheme 8.4 Chemoselective one-pot formation of adducts **10**.

The same authors reported an interesting application starting from δ-nitroalkanol **11** and enone **8**, prone to achieve the one-pot formation of polyfunctionalized tetrahydrofurans **12** in good overall yields (Scheme 8.5).

The same paper reported the reduction of **10** into triols **13** in good yields (Scheme 8.6). It is important to observe that in compound **13** three different hydroxyls are simultaneously exhibited: allylic, homoallylic, and bis-homoallylic.

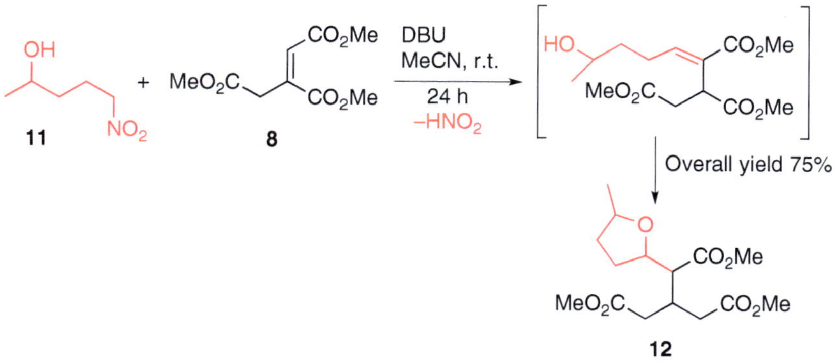

Scheme 8.5 One-pot formation of tetrahydrofuran **12**.

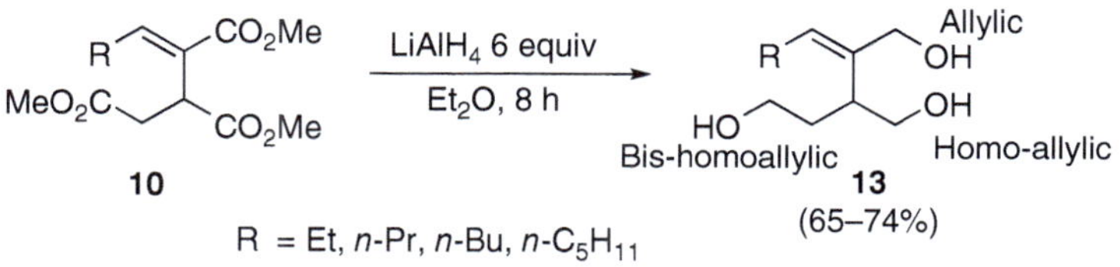

Scheme 8.6 Reduction of **10** into triols **13**.

Reduction of β-nitroacrylates **14** using $NaBH_4$ proceeds with the formation of saturated β-nitroesters **15** that quickly convert into α,β-unsaturated esters **16** (Scheme 8.7), by in situ elimination of nitrous acid, in very good overall yields (71–91%) [6].

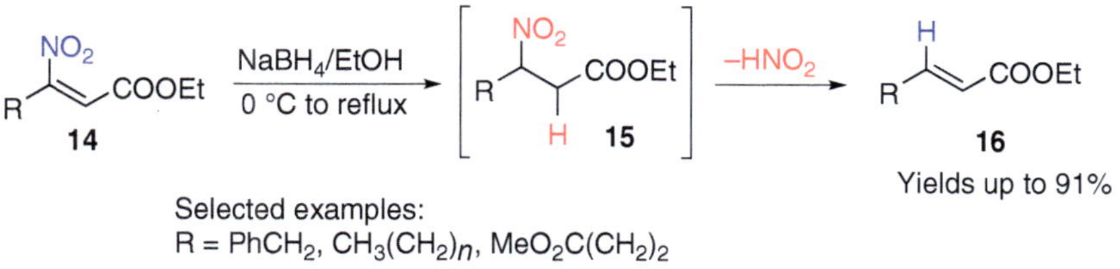

Scheme 8.7 Synthesis of α,β-unsaturated esters **16**.

It is important to observe that the conversion of **14** into **16** appears as a formal substitution of the vinyl nitro group with a hydrogen.

Conjugate addition of nitroalkanes to β-nitroacrylates **14**, promoted by carbonate on polymer, allows the one-pot formation of a diastereomeric mixture of a new class of nitro-functionalized α,β-unsaturated esters **18** and **19** (Scheme 8.8) with high prevalence of *E*-isomers [7].

Both isomers are very important from synthetic point of view because the presence of a secondary nitro group can favor the formation of stabilized carbanions prone to give the construction of new C—C bond. Moreover, for R = R^1 only a type of molecules is obtained.

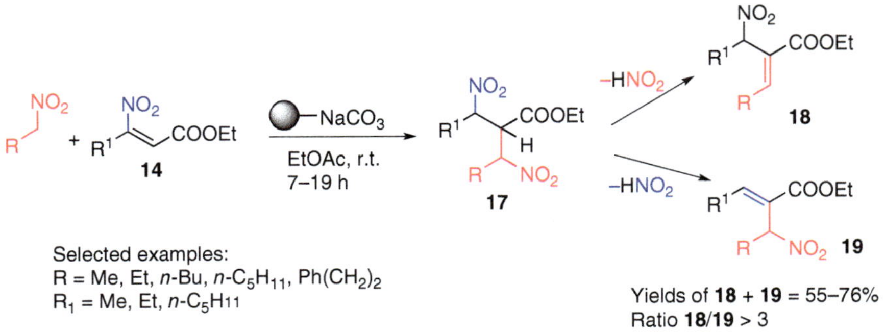

Scheme 8.8 Synthesis of nitro-functionalized α,β-unsaturated esters **18** and **19**.

(*E*)-Enediones **23** are valuable intermediates for the synthesis of important molecules such as prostaglandin, perfumes, pheromones, and other natural products.

Compounds **23** can be easily prepared (Scheme 8.9) [8] in good yields (up to 85%) from Henry reaction of an aldehydes with β-nitroalkanol **20**, under neat condition and in the presence of Amberlyst A21. The obtained nitro-diols **21** are oxidized with $K_2Cr_2O_7$, under phase-transfer conditions, yielding the nitro dione **22**.

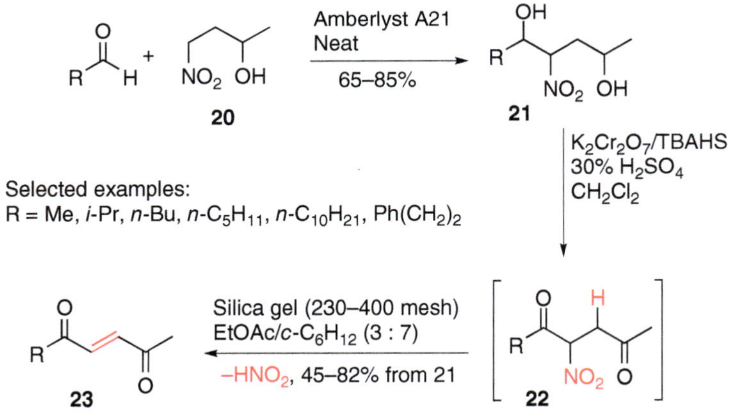

Scheme 8.9 Synthesis of enediones **23**.

Robust stirring (20 hours) of the latter in cyclohexane/EtOAc and final flash chromatography achieves the formation of the target enediones **23** in satisfactory to good yields, via elimination of nitrous acid.

8.2 Nitroaldol Reaction, Nitrous Acid Elimination vs Water Elimination

The Henry reaction of primary nitroalkanes and 2-quinolinecarboxaldehyde **24**, in presence of tetramethyl guanidine (TMG) as base and in PhMe as solvent, at 110 °C, yields (Scheme 8.10) [9] the unexpected ketone **27** due to the elimination of nitrous acid instead of the expected water elimination.

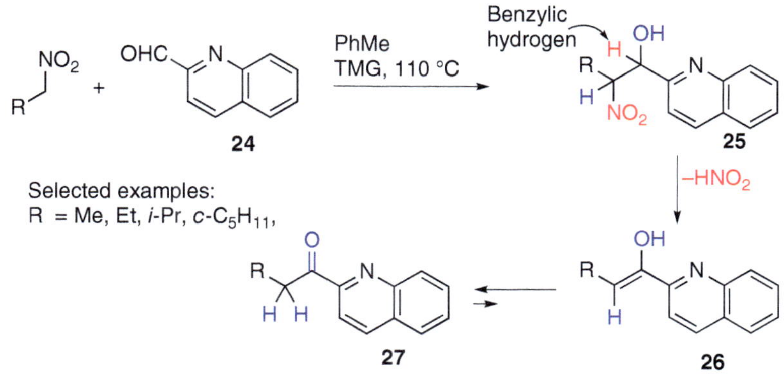

Scheme 8.10 Synthesis of ketones **27** via HNO_2 elimination vs. H_2O.

Probably, the acidity of the benzylic hydrogen in the intermediate **25** strongly favors HNO_2 elimination vs. dehydration.

8.3 Synthesis of Cyclic Compounds

Cyclic compounds are the main molecular structures in organic and industrial chemistry and efficient procedures for their preparation are of great interest. Thus, some of the more representative synthetic procedures by elimination of nitrous acid will be reported.

8.3.1 Synthesis of Aromatic Rings

Aromatic structures are very important for their synthetic utility and because belong to a great number of compounds with biological activities. The use of nitroalkanes as starting molecules together with the possibility to generate C=C double bond, through elimination of nitrous acid, was successfully used or the preparation of a variety of aromatic rings.

8.3.1.1 Synthesis of Benzene Ring

In this context, it has been discovered (Scheme 8.11) [7] that double Michael addition of nitroalkanes to enone **28**, under basic conditions and in aqueous medium (K_2CO_3/H_2O), affords the double adducts **29** that quickly cyclize with an intramolecular Henry reaction, into cyclohexanols **30**. The latter, under refluxing toluene, in the presence of TsOH, and with injection of air, gives (i) dehydration, (ii) elimination of nitrous acid, and (iii) dehydrogenation, with the direct synthesis of acetophenones **33** in good overall yields.

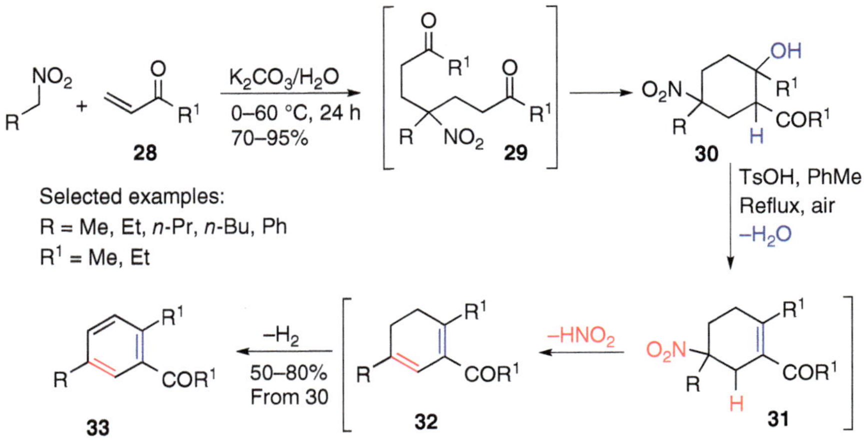

Scheme 8.11 Synthesis of acetophenones **33**.

This method achieves the preparation of aromatic systems from open-chain molecules. Moreover, trisubstituted benzenes are obtained with the avoidance of the formation of *ortho–meta–para* mixtures commonly observed in conventional aromatic electrophilic substitution.

Carbazoles are tricyclic heteroaromatic molecules that show a pyrrole linked to a benzenic ring. Thus their synthesis from pyrroles needs the construction of a benzene ring that can be performed [7] via cyclization of the appropriate nitroalkylindole **34** (easily obtained from the corresponding sulfonylindoles) [10], followed by water and nitrous acid eliminations from the tricyclic structure **35** and consequent formation of the target structure **36** in good overall yields (Scheme 8.12).

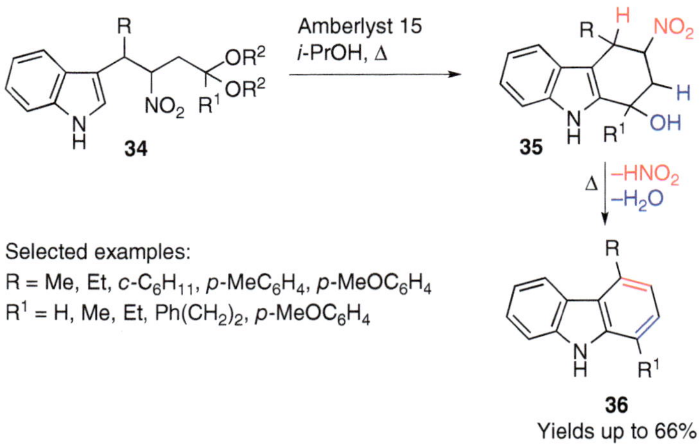

Scheme 8.12 Synthesis of carbazoles **36**.

Naphthyl derivatives have been prepared from the Michael reaction of primary nitroalkanes to Baylis–Hillman adducts **37**, under basic conditions (K_2CO_3), and successive addition–elimination protocol affording **38** (Scheme 8.13) [11]. Cyclization of the latter gives the cyclic nitrocompound **39** prone to eliminate nitrous acid with the consequent formation of the naphthalenes **40** in excellent overall yields (up to 89%).

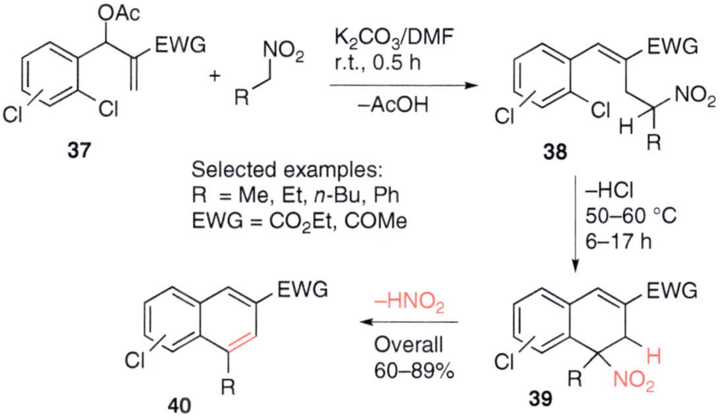

Scheme 8.13 Synthesis of naphthyl derivatives **40**.

Baylis–Hillman adducts and nitroalkanes have been even used for the preparation of tetrasubstituted benzenes, as reported in Scheme 8.14 [12].

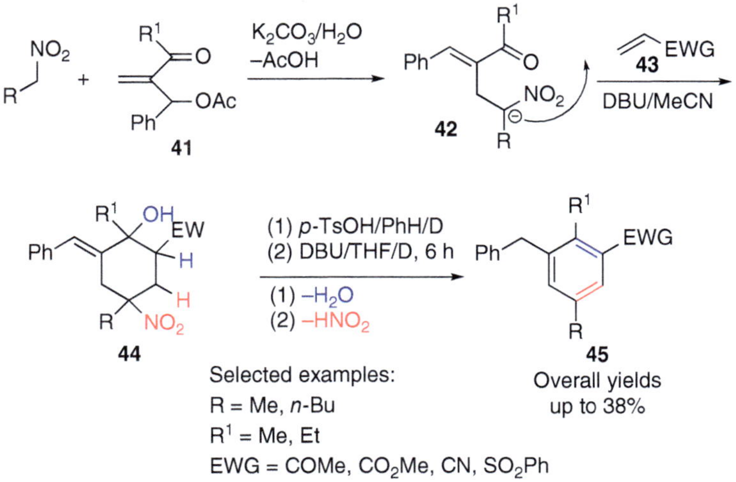

Scheme 8.14 Synthesis of tetrasubstituted benzenes **45**.

The method starts with the conjugate addition of primary nitroalkanes to the electrophilic Baylis–Hillman structure **41** (four carbon unit for benzene ring), with in situ elimination of AcOH, under basic conditions (aqueous K_2CO_3). The successive step consists in the consecutive (i) conjugate addition of the adduct **42–43** (two carbon unit for benzene ring), in presence of DBU as base, and in situ aldol-type cyclization affording the nitro cyclohexanols **44**. The next step consists in the (ii) aromatization of **44** through the consecutive Michael addition of **42–43,** then aldol cyclization in presence of DBU/MeCN. The intermediate **44** was not separated due to the complex mixture of diastereoisomers; however, the crude mixture of **44** via dehydration and nitrous acid elimination, helped by TsOH and DBU, gives the formation of tetrasubstituted benzenes **45**.

8.3.1.2 Synthesis of Furan Ring

Furans are probably the most studied heteroaromatic structures and their synthetic availability is essential. Thus, nitroalkanes have demonstrated to be strategic for the synthesis of furans, often via elimination of nitrous acid.

An interesting procedure was developed from the Michael addition of primary nitroalkanes to enedione **46**, obtained by oxidation of 2,5-dimethyl furan (Scheme 8.15) [13]. Thus, conjugate addition of nitroalkane to **46**, in presence of DBU, allows the adduct **47** prone to give elimination of nitrous acid and formation of enone **48**.

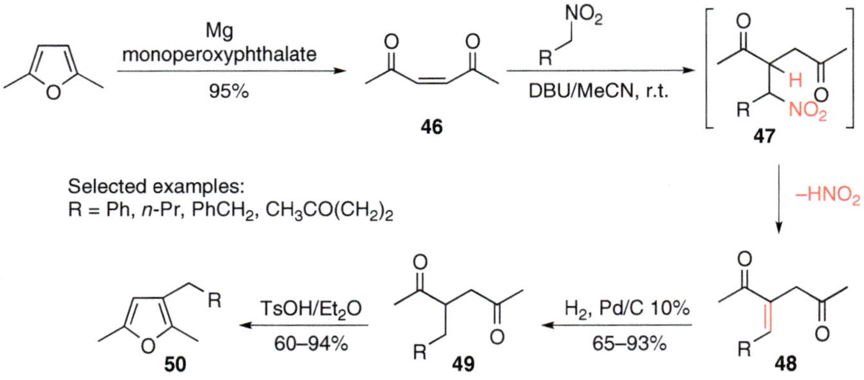

Scheme 8.15 Synthesis of trisubstituted furans **50**.

Successive hydrogenation of **48**, and acidic cyclization of **49**, achieves the formation of trialkylated furans **50** in good overall yields.

Thus, primary nitroalkanes acting as an indirect alkylating in 3-position of 2,5-dimethyl furan can be considered as an equivalent of the alkyl cation synthon **II** (Figure 8.3).

Nitro ketal **51** is the key starting molecule for the preparation of 2,5-disubstituted furans via a two-step sequence, as reported in Scheme 8.16 [14].

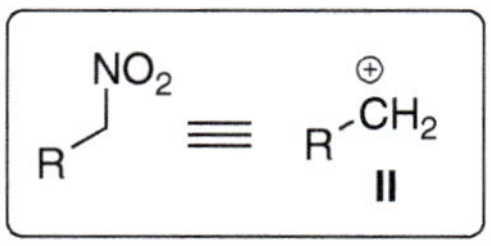

Figure 8.3 Nitroalkane as synthetic equivalent of the alkyl cation synthon **II**.

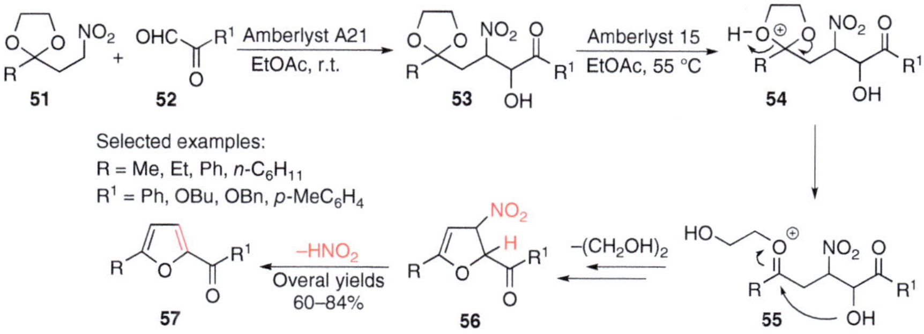

Scheme 8.16 Synthesis of disubstituted furans **57**.

The process starts with the Henry reaction of **51** with **52** catalyzed with Amberlyst A21 to form nitroalkanols **53**. The latter, in a one-pot process, under Amberlyst 15, converts into the cyclic structure **56** that, through spontaneous elimination of nitrous acid, gives 2,5-disubstituted furan **57**.

This procedure is an unprecedented and regiodefined synthesis of functionalized 2,5-disubstituted furan derivatives in good overall yields (up to 84%).

The applicability of this method has reported in the same paper with the synthesis of furan ring of 1-benzyl-3-(5′-hydroxymethyl-2′-furyl)-indazole (**YC-1**) (Scheme 8.17), a very important pharmaceutical target that exhibits significant inhibitory effects against thrombin-AA, -collagen, and PFA-induced platelet aggregation.

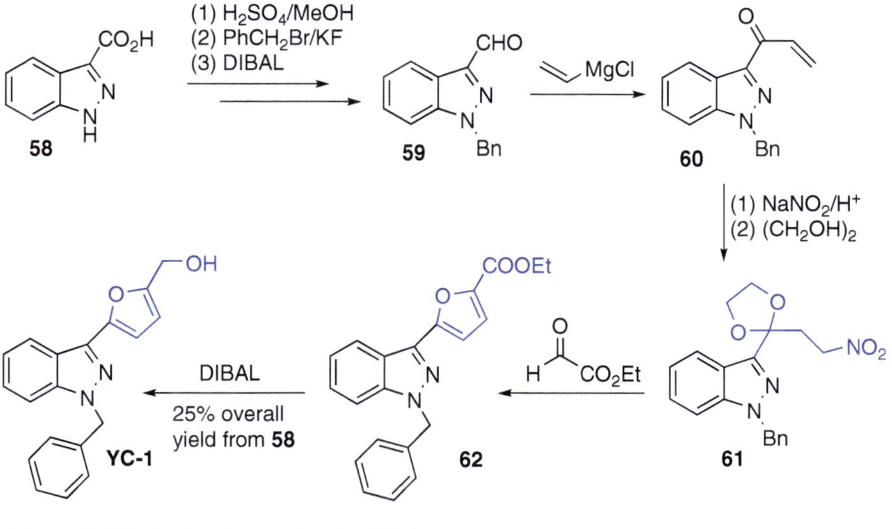

Scheme 8.17 Synthesis of furan ring **YC-1**.

The synthesis starts from commercially available indazole-3-carboxylic acid **58**, then converted into aldehyde **59** that, by Grignard reaction with vinyl-MgCl, allows the α,β-unsaturated structure **60**, prone to be nitrated and protected as ketal **61**, the key structure for the formation of furan ring of the adduct **62**, easily reducible to **YC-1**. The process achieves the formation of **YC-1** from **58** in 25% overall yield.

8.3.1.3 Synthesis of Pyrrole Ring

A valuable procedure was found from functionalized nitroalkane **51** and α-amido sulfones **63** (Scheme 8.18) [15]. Their reaction, under basic conditions (KF/Al_2O_3), produces the simultaneous (i) elimination of *p*-toluenesulfonic acid and (ii) addition of the nitronate derived from **51**, with the formation of nitrocarbamate adduct **64** in good yields (up to 89%).

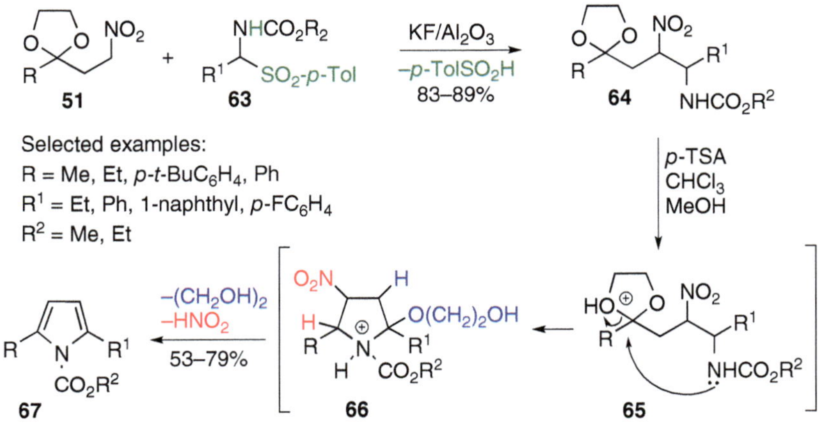

Scheme 8.18 Synthesis of pyrroles **67**.

Then, acid-promoted (*p*-TsOH) cascade process, involving ketal protonation **65** and ring closure, furnishes pyrrolidine **66** that is prone to aromatization via elimination of both ethylene glycol and nitrous acid, with the formation of polysubstituted pyrroles **67** in excellent overall yields from **64** (up to 79%).

8.3.1.4 Synthesis of Isoxazole Ring

Isoxazoles are attractive targets due to their biological activities and their availability is relevant. In this context, nitroalkanes have been studied as possible synthetic precursors of these molecules, by elimination of nitrous acid [16].

As reported in Scheme 8.19, a nitroaldol condensation of arylglyoxal **68** with nitroacetate **69** affords the adduct **70** prone to be submitted to a Michael reaction with an excess of nitro compound **69** and formation of **71**, predisposed to nitrous acid elimination and consequent cyclization to **72**.

Successive conversion of **72** into **73** and dehydration affords isoxazoles **74** in good overall yields (up to 85%).

8.3.2 Synthesis of Heterocyclic (Non-Aromatic) Rings

Non-aromatic heterocycles are maybe of lower interest respect to the aromatic ones; however they possess important biological properties and for this reason their synthetic availability is very valuable. In this context, the combination of nitroalkanes

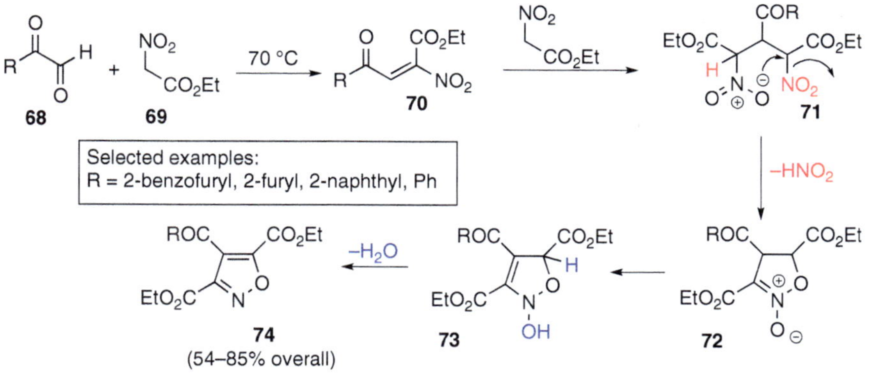

Scheme 8.19 Synthesis of isoxazoles **74**.

and elimination of nitrous acid remain as great opportunities for the synthesis of some of them.

8.3.2.1 Synthesis of Dihydropyranol Ring

As illustrated in Scheme 8.20 [17], dihydropyranols **78** can be prepared in a one-pot process via Michael reaction of β-nitroalkanols **75** with *cis*-3-hexen-2,4-enedione **1b**, in DBU/MeCN.

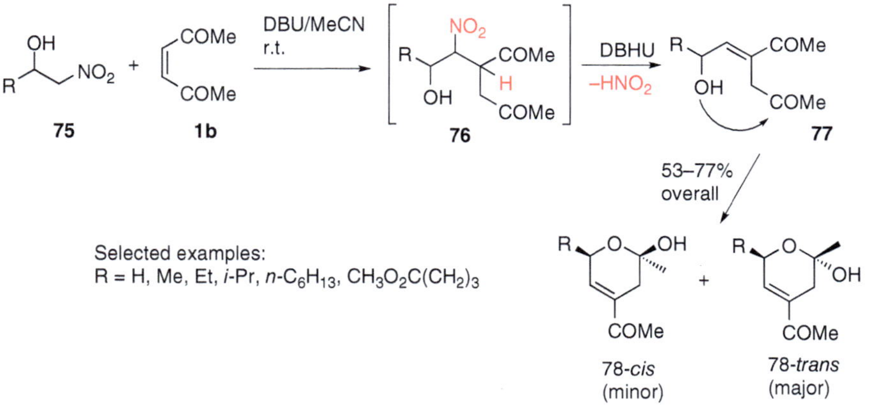

Scheme 8.20 Synthesis of dihydropyranols **78**.

The formed structure **77** (acquired by nitrous acid elimination) converts, via internal *emi*-ketal formation, into a *trans/cis* mixture of dihydropyranols **78** in excellent overall yields (up to 77%).

8.3.2.2 Synthesis of Butyrolactone Ring

The great chemical versatility of nitroalkanes has been demonstrated even in the synthesis of butyrolactones in only two steps. The first process involves the regiospecificxs Michael addition of primary nitroalkanes to enedione **82** with the direct formation of α,β-unsaturated ester **81**, via elimination of nitrous acid from the adduct **80** (Scheme 8.21) [18].

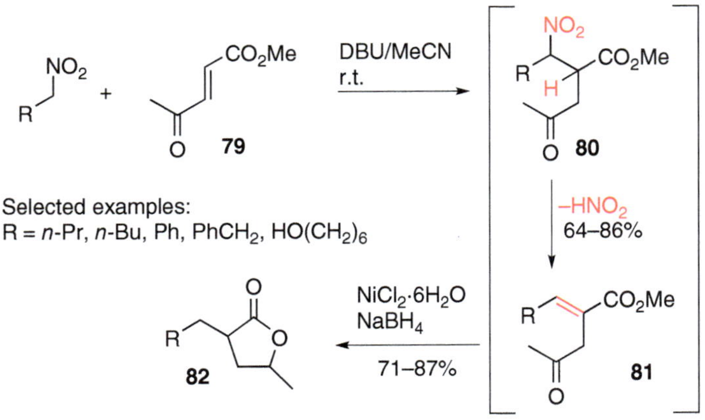

Scheme 8.21 Synthesis of butyrolactones **82**.

Successive $NiCl_2/NaBH_4$ reduction of **81** affords the direct lactonization to **82** (71–87%) under reduction of the C=C double bond. Good yields are observed in both the steps.

As improvement of the aforementioned method, it was reported the synthesis of α-methylene-γ-butyrolactones **84** and **85** from primary and secondary nitroalkanes (Scheme 8.22) [19]. The first step is similar to that reported in Scheme 8.21, with the formation of the α,β-unsaturated esters **83**. The latter has been employed as key structure for the synthesis of mono- or dialkylated α-methylene-γ-butyrolactones following two different ways.

Thus, reduction of carbonyl in **83** with $NaBH_4$ (Path a), and acidic work-up, allows monoalkylated α-methylenelactone **84**, while (Path b) treatment of **83** with a Grignard reagent, in presence of $CeCl_3$, and acidic work-up, affords the dialkylated lactones **85**.

8.3.2.3 Synthesis of Pyrrolidine Ring

Michael addition of nitroalkanes to enone **79** is the starting step even for the synthesis of pyrrolidines **86**, as reported in Scheme 8.23 [20]. In fact, the formed molecular structures **83** are strategic for this purpose because, after reduction ($LiAlH_4$) to diol, they can be converted into mesylate forms, propense to afford pyrrolidines **86** by reaction with benzylamine (R^2 = Bn) or methylbenzenesulfonamide (R^2 = *p*-Ts).

It is important to point out that this three-step sequence shows important advantages such as good yields for each step and the possibility to insert a variety of substituents in the heterocyclic ring.

8.3.2.4 Synthesis of Succinic Anhydride Ring

Primary and secondary nitroalkanes and dimethyl maleate **1a** have been employed even as precursors for the synthesis of succinic anhydrides **89** and **92** (Scheme 8.24) [21].

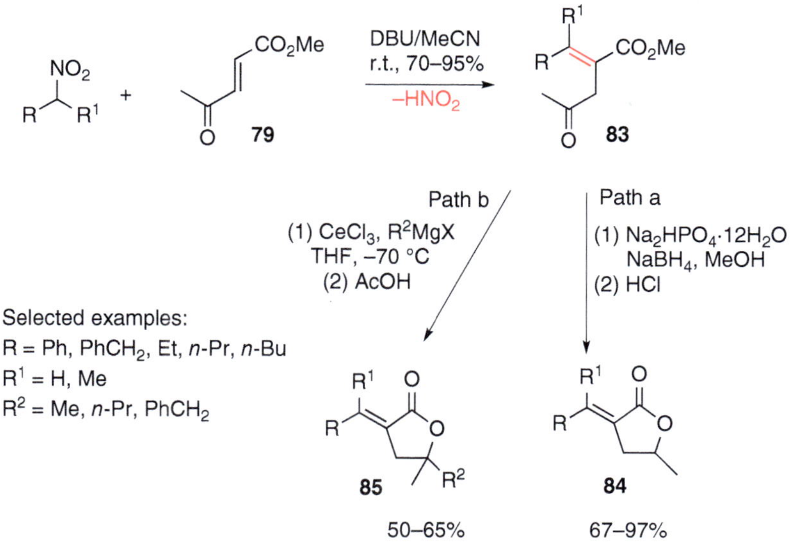

Selected examples:
R = Ph, $PhCH_2$, Et, *n*-Pr, *n*-Bu
R^1 = H, Me
R^2 = Me, *n*-Pr, $PhCH_2$

Scheme 8.22 Synthesis of α-methylene-γ-butyrolactones **84** and **85**.

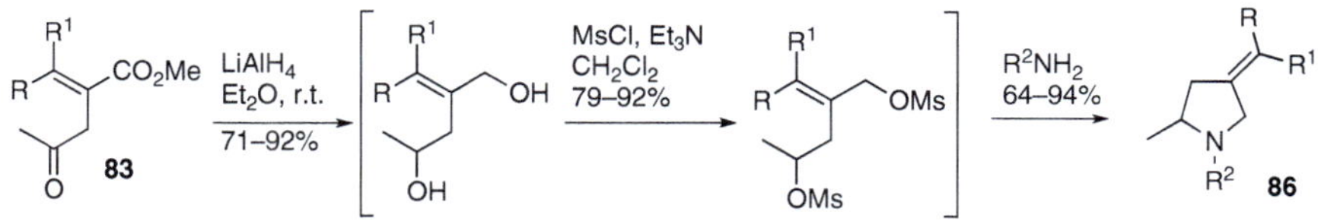

Selected examples:
R = H, Me, cycloalkyl
R^1 = Me, Et, *n*-Pr, *n*-C_5H_{22}, Ph
R^2 = Bn, *p*-Ts

Scheme 8.23 Synthesis of pyrrolidines **86**.

In fact, the formed (via tandem addition nitrous acid elimination) adduct **3a** converts (hydrolysis) into dicarboxylic acid **87** predisposed to produce succinic anhydride **88** with AcCl, under reflux.

Moreover, the adduct **3a** can be reduced (H_2, Pd/C) into saturated derivative **89**, then hydrolyzed to dicarboxylic acid **90** that, by the help of AcCl, allows **91** in good yields.

8.3.3 Synthesis of Cyclopentenone Ring

Cyclopentenone ring is one of the most recurring structural units in many targets of interest, including perfumes and nitroalkanes, which have demonstrated to play an important role in their synthesis (Scheme 8.25) [22].

The method consists in a one-pot process that includes three different steps: (i) chemoselective aldol condensation of **92–93**, (ii) Michael addition of nitroalkane

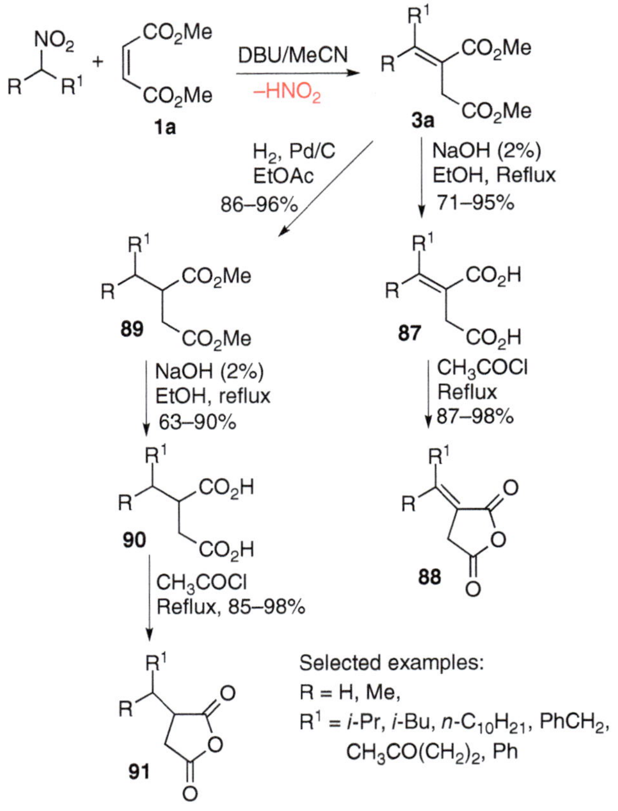

Scheme 8.24 Synthesis of succinic anhydrides **88** and **91**.

to **93** with the formation of adduct **94**, and (iii) base (DBU) assisted-elimination of nitrous acid yielding **95** in very good overall yields (up to 93%).

8.4 Synthesis of Polyenes

Conjugated polyenes are an important class of molecules due to their versatility as building blocks for the construction of a multitude of targets. Nitroalkanes have been successfully employed as key starting molecules for the one-pot synthesis of 1,3-diene-tricarboxylates **99**, as reported in Scheme 8.26 [23].

Double conjugate addition of nitro ester **96** to phenyl propiolates **97**, in presence of Ph_3P as base and under refluxing conditions (PhH), affords the intermediate **98**. Successive spontaneous elimination of 2 mol of nitrous acid produces the one-pot formation of diene **99** in good overall yields (up to 80%).

The conjugate addition of nitroalkanes to β-nitroacrylates was previously reported in this Chapter (Scheme 8.8, [7]) for the synthesis of α,β-unsaturated esters **18** and **19**. However, a small modification of the reaction conditions achieves the formation

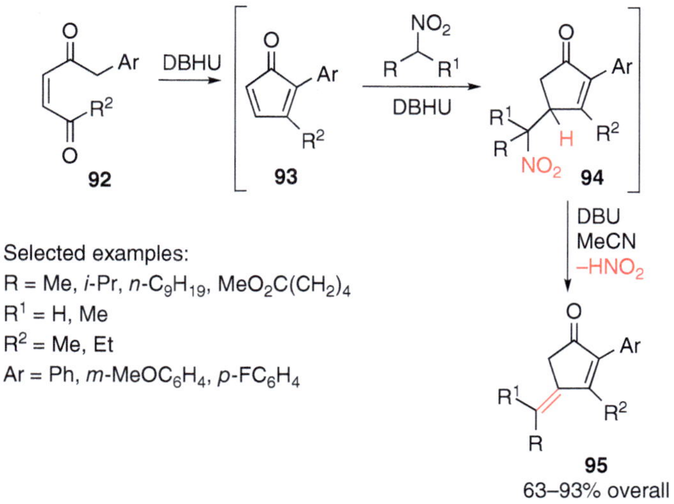

Scheme 8.25 Synthesis of cyclopentenones **96**.

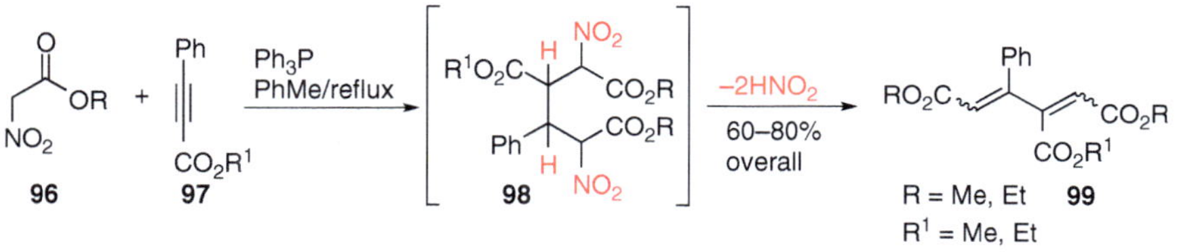

Scheme 8.26 Synthesis of 1,3-diene-tricarboxylates **99**.

of 1,3-butadienes, as reported in Scheme 8.27 [7], through elimination of 2 mol of nitrous acid, favored by the presence of supported $NaHCO_3$.

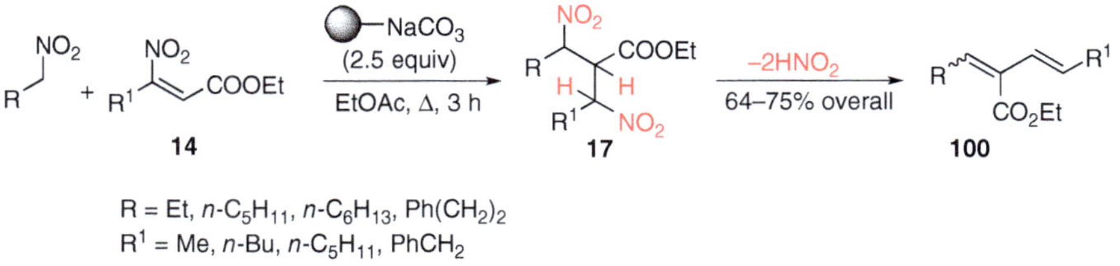

Scheme 8.27 Synthesis of 1,3-butadienes **100**.

Michael addition of primary nitroalkanes to dimethyl α-(bromomethyl)fumarate **101** affords two different products, depending from the reaction conditions, as reported in Scheme 8.28.

Initially, Béji et al. [24a] found that the reaction, using NaOH as base, produces (path a) the dienic system **103**, through spontaneous elimination of nitrous acid from the intermediate-adduct **102**, in good overall yields. Later, a more selective procedure was developed [24b] under aqueous CTAOH (cetyl trimethyl ammonium

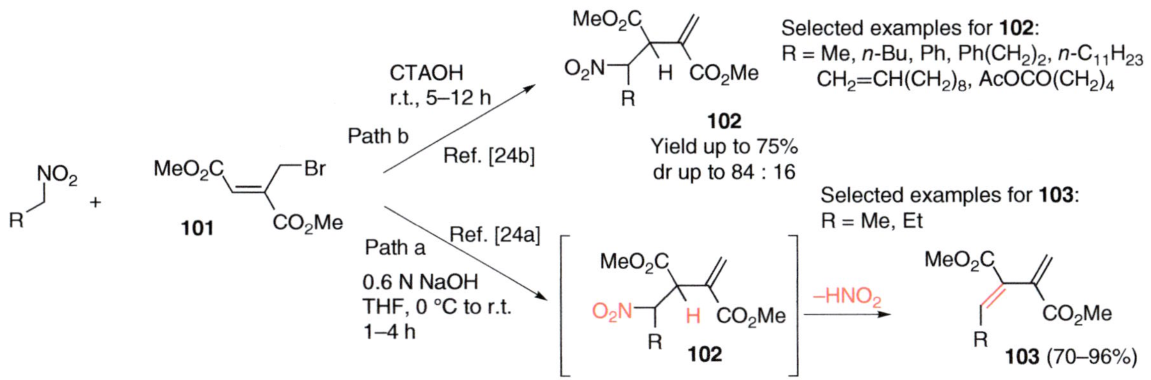

Scheme 8.28 Synthesis of nitrodiesters **102** or 1,3-butadienes **103**.

hydroxide) catalysis (path b). In fact, the elimination of nitrous acid can be avoided, with the exclusive formation of the Michael adduct **102**, without any more transformations.

In addition, both the structures (**102** and **103**) obtained were used, in the same papers, for the preparation of (Eq. (a)) α-alkylidene-γ-lactams **105** [24a] and (Eq. (b)) tetrasubstituted γ-lactams **106** [24b], as reported in Scheme 8.29.

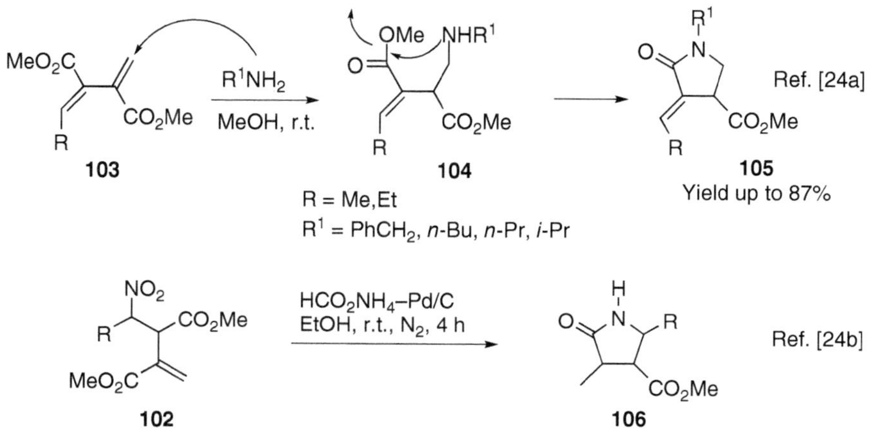

Scheme 8.29 Synthesis of γ-lactams **105** and **106**.

8.4.1 Asymmetric Synthesis of Electron-Poor Alkenes

Few years ago, Palomo and coworkers [25a] reported an important procedure for asymmetric synthesis of **111** bearing γ-stereogenic center, with excellent chemoselectivity (ee >98%, Scheme 8.30).

The method consists in a one-pot process, catalyzed by chiral pyrrolidine **109**, in which conjugate addition of **107** to enals **108** affords the adduct **110**, prone to nitrous acid elimination, with the formation of **111** in excellent overall yields (up to 90%) and enantioselectivity (>93%).

Thus, the method is performed in a one-pot three-step operation without the need of intermediate isolation. Moreover, high chemical efficiency is attained.

In addition, the authors exploited the chemical utility of **111a,b** for the construction of tetrahydrofurans **112** and β^3-homoprolines **113.** The synthesis provides excellent diastereomeric ratios of **112** and a concise stereocontrolled access to **113**.

One year later, the same group, as development of the above method, reported [25b] a new procedure for vinylation of enals **108**, via a one-pot process, starting from conjugate addition of nitroalkane **107a** to enals **108** (Scheme 8.31). The scope of the reaction is to produce the in situ (i) enantioselective Michael addition under chiral pyrrolidine **109**, (ii) elimination of both HNO_2 and $PhSO_2$, strongly helped by the presence of Mg and additives (TMSCl and EDB), and (iii) acetal protection of the formed aldehyde ($CH(OMe)_3$/*p*-TsOH) to **114**.

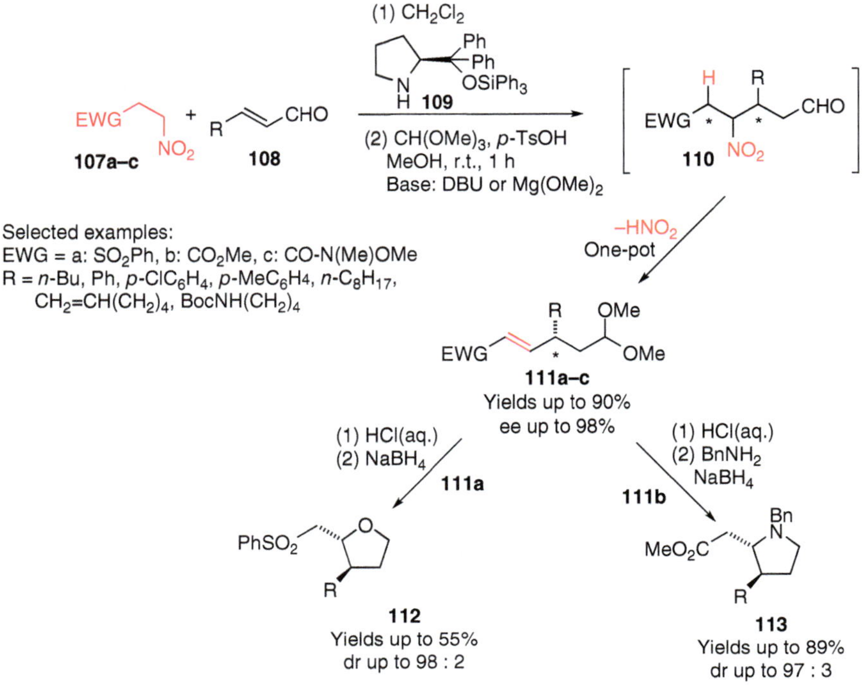

Scheme 8.30 Asymmetric synthesis of conjugate alkenes **111** and their applications to **112** and **113**.

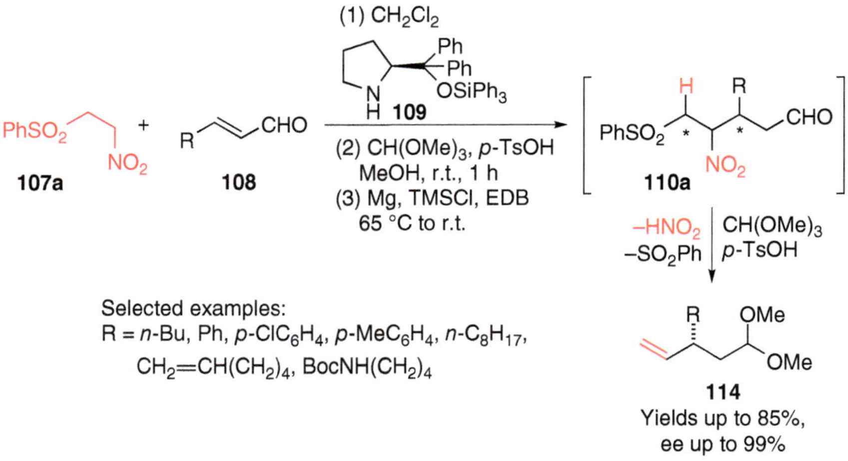

Scheme 8.31 Asymmetric one-pot vinylation of enals **114**.

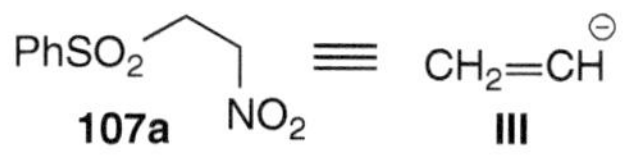

Figure 8.4 Nitroalkanes **107a** as synthetic equivalent of the vinyl anion synthon **III**.

In this procedure, it is important to point out that β-nitroethyl sulfone **107a** acts as a masked vinyl anion **III** (Figure 8.4) in the conjugate additions, and the methodology could help the gap of methods for the direct β-vinylation of enals.

References

1 Ballini, R. and Palmieri, A. (2019). *Adv. Synth. Catal.* 361: 5070–5097.
2 (a) Ballini, R. and Rinaldi, A. (1994). *Tetrahedron Lett.* 35: 9247–9259. (b) Ballini, R. and Bosica, G. (1995). *Tetrahedron* 51: 4213–4222.
3 Ballini, R. and Bosica, G. (1996). *Liebigs Ann.*: 2087–2089.
4 Ballini, R. and Palmieri, A. (2006). *Adv. Synth. Catal.* 348: 1154–1156.
5 Ballini, R., Bosica, G., Fiorini, D. et al. (2004). *Synthesis*: 605–609.
6 Ballini, R., Fiorini, D., and Palmieri, A. (2004). *Tetrahedron Lett.* 45: 7027–7029.
7 Palmieri, A., Gabrielli, S., and Ballini, R. (2010). *Adv. Synth. Catal.* 352: 1485–1492.
8 Ballini, R. and Bosica, G. (1994). *J. Organomet. Chem.* 59: 5466–5467.
9 Nomland, A. and Hills, I.D. (2008). *Tetrahedron Lett.* 49: 5511–5514.
10 Ballini, R., Gabrielli, S., Palmieri, A., and Petrini, M. (2010). *Adv. Synth. Catal.* 352: 2459–2462.
11 Kim, J.N., Ji, Y.I., Gong, J.H., and Lee, K.Y. (2001). *Tetrahedron Lett.* 42: 4195–4197.
12 Lee, M.J., Lee, K.Y., Gowrisankar, S., and Kim, J.N. (2006). *Tetrahedron Lett.* 47: 1355–1358.
13 Ballini, R., Bosica, G., Fiorini, D., and Giarlo, G. (2001). *Synthesis*: 2003–2006.
14 Palmieri, A., Gabrielli, S., and Ballini, R. (2010). *Chem. Commun.* 46: 6165–6167.
15 Ballini, R., Gabrielli, S., Palmieri, A., and Petrini, M. (2014). *RCS Adv.* 4: 43258–43261.
16 Yang, Y., Gao, M., Deng, C. et al. (2012). *Tetrahedron* 68: 6257–6262.
17 Ballini, R., Barboni, L., Fiorini, D., and Palmieri, A. (2004). *Synlett*: 2618–2620.
18 Ballini, R. and Bosica, G. (1996). *Synlett*: 1115–1116.
19 Ballini, R., Marcantoni, E., and Perella, S. (1999). *J. Organomet. Chem.* 64: 2954–2957.
20 Ballini, R., Bosica, G., Masè, A., and Petrini, M. (2000). *Eur. J. Org. Chem.*: 2927–2932.
21 Ballini, R., Bosica, G., Fiorini, D., and Righi, P. (2002). *Synthesis*: 681–685.
22 Ballini, R., Bosica, G., Fiorini, D. et al. (2001). *Org. Lett.* 3: 1265–1267.
23 Yavari, I., Moradi, L., Mokhtarporyani-Sanandej, A., and Mirzaeri, A. (2007). *Helv. Chim. Acta* 90: 392–394.

24 (a) Béji, F., Lebreton, J., Villiéras, J., and Amri, H. (2001). *Tetrahedron* 57: 9959–9962. (b) Ballini, R., Gabrielli, S., and Palmieri, A. (2010). *Tetrahedron Lett.* 51: 1233–1235.

25 (a) López, R., Zalacain, M., and Palomo, C. (2011). *Chem. Eur. J.* 17: 2450–2457. (b) Gianelli, C., Lopez, R., Puente, A. et al. (2012). *Eur. J. Org. Chem.*: 2774–2779.

9

α-Nitrocycloalkanones, Synthesis, and Reactivity

It has been well demonstrated how the aliphatic nitro group is crucial for the formation of C—C bond by reaction with electrophiles under basic conditions (see, for example, Michael and Henry reactions). However, the nitro group of nitroalkanes plays an important role even in the cleavage of C—C bond, and this peculiarity is strongly favored in the presence of a carbonyl in α-position (α-nitro ketones). The juxtaposition on two adjacent positions of nitro group and carbonyl offers a variety of reactivities, peculiar to this class of compounds. In fact, the results that can be performed via C,C cleavage of cyclic α-nitro ketones are of extraordinary importance. Thus, the topic of this chapter is to highlight the main synthetic sources of α-nitrocycloalkanones and their chemical potentiality [1].

9.1 Synthesis of Cyclic α-Nitro Ketones

Some of the simplest cyclic α-nitro ketones **1** are easily available from chemical suppliers, however these compounds can be easily prepared from three main sources: (i) cyclic alkenes **3**, (ii) enol acetates **2**, and (iii) enol silyl ethers **4**.

As reported in Figure 9.1, the enol acetate **2**, prepared from the corresponding ketone, can be nitrated with trifluoroacetyl nitrate, generated through trifluoroacetic anhydride and ammonium nitrate [2]. Cyclic alkenes **3** convert, in good yields, into α-nitro ketones in presence of a mixture of trimethylsilylnitrate and chromium trioxide [3]. α-Nitration of cycloalkanones, using enol silyl ethers **4**, can be performed with $C(NO_2)_4$ (tetranitromethane, TNM) [4], an inert and neutral reagent commercially available.

In Table 9.1 some representative examples of preparation of cyclic α-nitro ketones are reported.

Steroidal ketones are diffused polycyclic structures and their conversion into the corresponding α-nitro ketones could be of great interest for both (i) theoretical studies of their biological activities and (ii) synthesis of new classes of lipids.

A very efficient procedure for the synthesis of the above nitro ketones was reported [5] through formation of the corresponding enol acetates followed by nitration with trifluoroacetyl nitrate (TFAN, Scheme 9.1).

Nitroalkanes: Synthesis, Reactivity, and Applications, First Edition.
Roberto Ballini and Alessandro Palmieri.

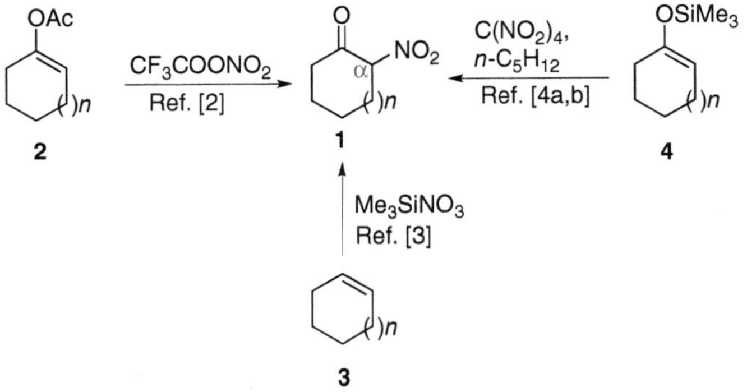

Figure 9.1 Synthesis of α-nitrocycloalkanones **1**.

Table 9.1 Selected examples of α-nitrocycloalkanones **1**.

Entry	Starting material	α-Nitrocycloalkanones 1	Yield (%)
1	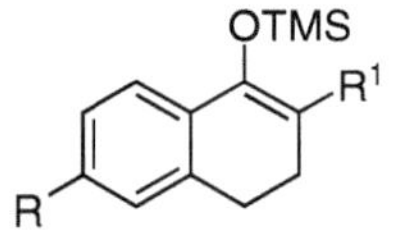 R = H, OMe; R^1 = H, Me	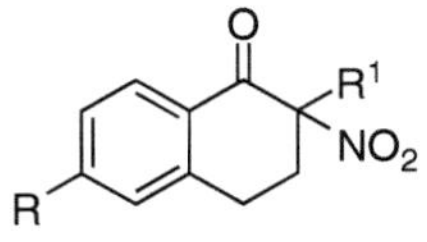	92–95
2	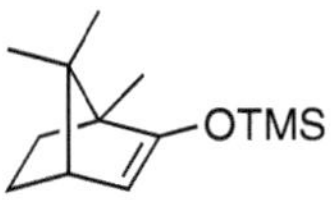	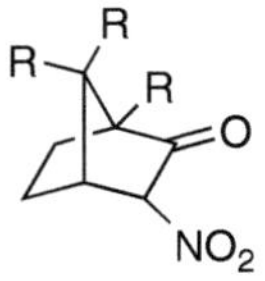	54
3	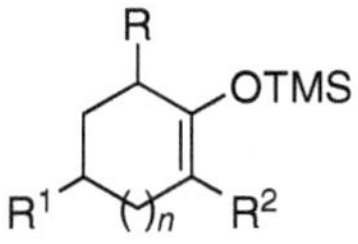 n = 0–2; R = H, Me; R^1 = H, *t*-Bu; R^2 = H, Me	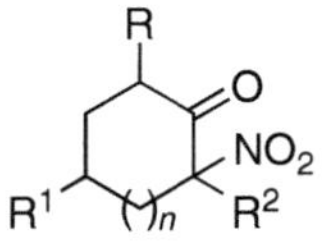	68–95
4	 n = 0–2, 7	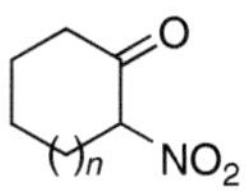	27–72
5	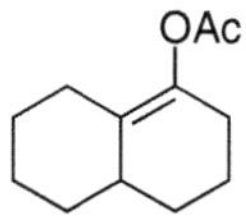	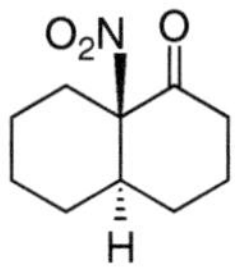	Quantitative

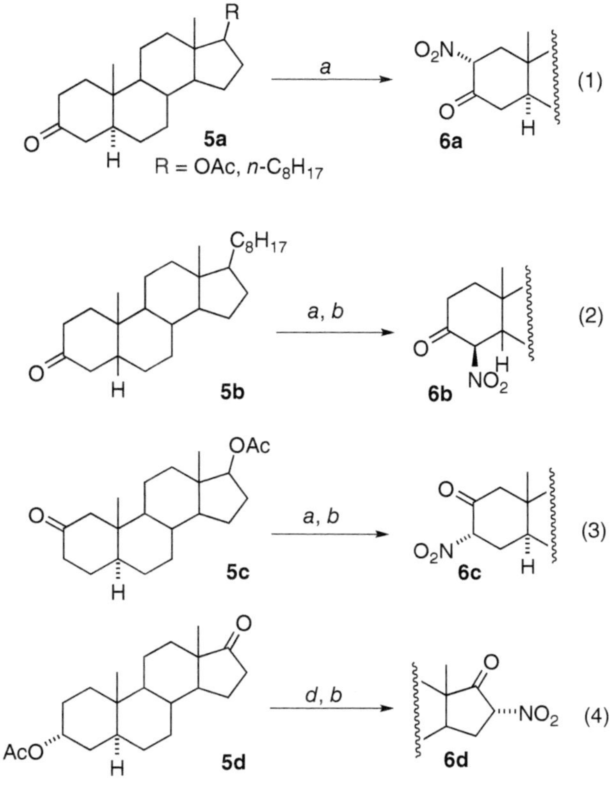

Reaction conditions:
a = $(CH_3CO)_2O$, *p*-TSA, reflux 3–10 h
b = TFAN, CH_2Cl_2, 0 °C, 10–20 min
c = *t*-BuOK, THF, 0 °C, 1 h, AcOH
d = isopropenylacetate, H_2SO_4, reflux

Scheme 9.1 Synthesis of steroid α-nitro ketones **6a–d**.

The nitro ketones **6a–d** are obtained in near quantitative yields in stereocontrolled manner, probably due to the steric control of the relatively large nitrating agent (TFAN).

9.2 Ring Cleavage of Cyclic α-Nitro Ketones

As exhibited in Figure 9.2, the presence of electrophilic center (carbonyl) and an extraordinary EWG (NO_2) in α-position favors a nucleophilic attack at C_1 with simultaneous ring cleavage (C_1–C_2) strongly facilitated in the presence of NO_2.

Thus, α,ω-difunctionalized molecular structures **7** are achieved.

Moreover, it is very important to consider this ring cleavage as a good source of functionalized nitroalkanes.

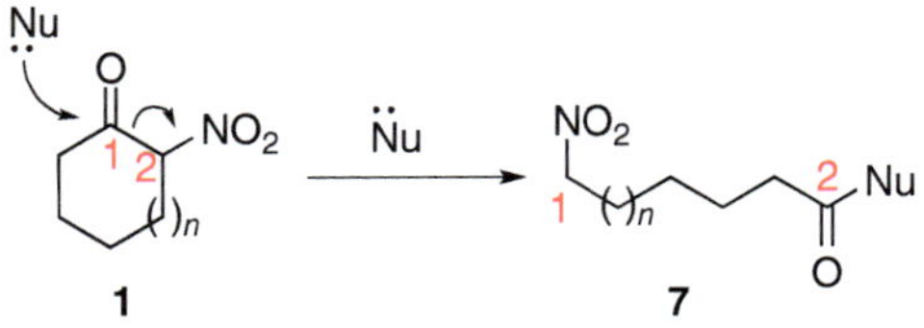

Figure 9.2 Ring cleavage of α-nitrocycloalkanones.

9.2.1 Cleavage to ω-Nitro Acids and ω-Nitro Esters

α-Nitro cycloalkanones can be cleaved by hydroxyl derivatives such as water and alcohols. In fact, as reported in Scheme 9.2, treatment of nitro ketone **8** with refluxing water or refluxing alcohol, achieves good yields of ω-nitro acids **9** or ω-nitro esters **10**, respectively [6].

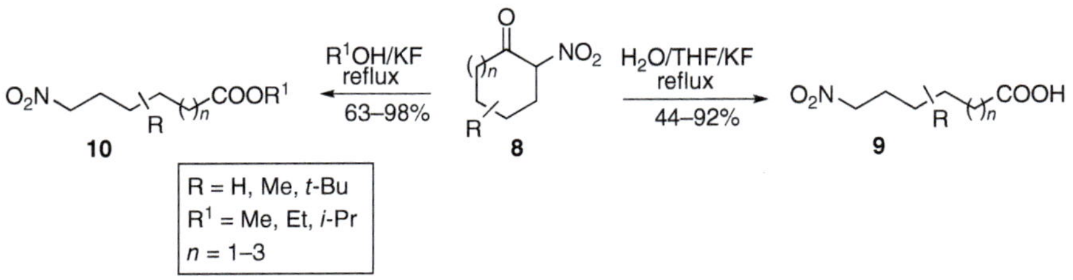

Scheme 9.2 Synthesis of ω-nitro acids **9** and ω-nitro esters **10**.

Few years later, the formation of ω-nitro esters was improved via cleavage of nitro cycloalkanones under heterogeneous conditions (Amberlyst A21) in refluxed MeOH (Scheme 9.3) [7a].

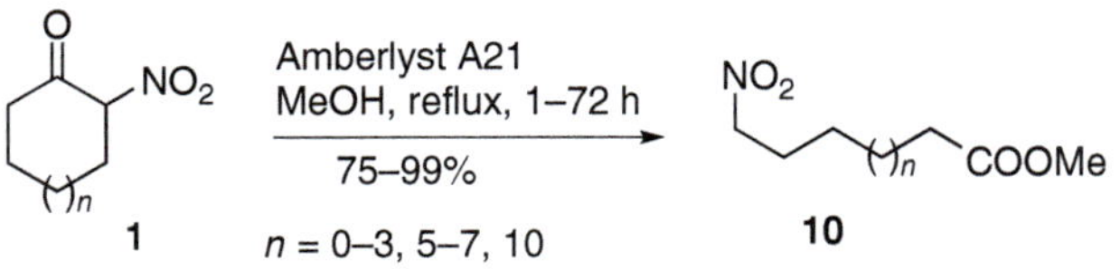

Scheme 9.3 Cleavage of nitro cycloalkanones with MeOH/Amberlyst A21.

The method offers several advantages such as excellent yields (up to 99%) and easy work-up (simple filtration of Amberlyst A21 and evaporation).

Successively, a valuable, low impact, and general method for the preparation of linear and branched ω-nitro esters was reported involving the use 2-*tert*-butylimino-2-diethylamino-1,3-dimethyl-perhydro-1,3,2-diazaphosphorine on polystyrene (PS-BEMP) (a commercially available polymer bound phosphazene base) as catalyst that provides the target compounds in excellent yields and low process mass intensity and *E*-factor values [7b].

As application, ω-nitro ester **10a** was used as starting compound for the synthesis of Exaltolide **13**, an important perfume isolated from angelica root oil. The synthesis starts (Scheme 9.4) with the Nef conversion (**10a** to **11**), followed by reduction

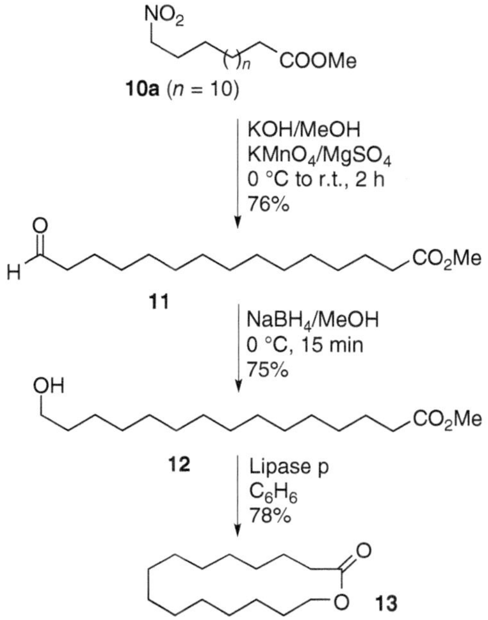

Scheme 9.4 Synthesis of Exaltolide **13**.

to the corresponding hydroxyl ester **12** and successive "Lipase p" cyclization to Exaltolide **13**.

Aqueous NaOH/cethyltrimethylammonium chloride (CTACl) was found [8] an high efficient catalyst for the cleavage to ω-nitro acids **9** (Scheme 9.5) since good yields are obtained in short reaction times (three to eight hours).

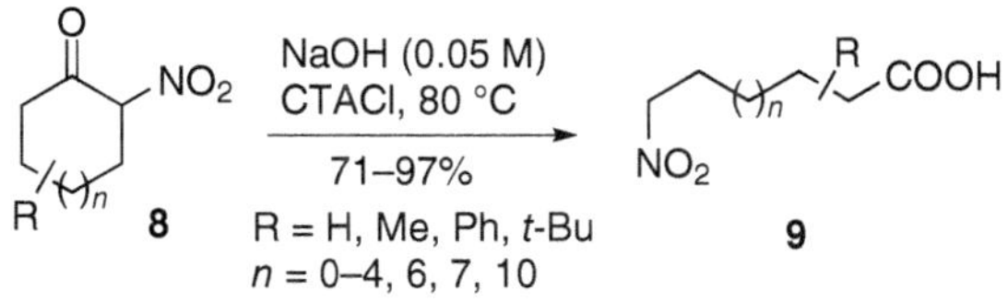

Scheme 9.5 Ring cleavage to ω-nitro acids **9** by CTACl.

Moreover, the aqueous medium, with respect to organic solvents, is less expensive, less dangerous, and more sustainable and grants an easy control of pH. The nitro acids **9** are ideal precursors of the corresponding amino acids **14** (Scheme 9.6), key

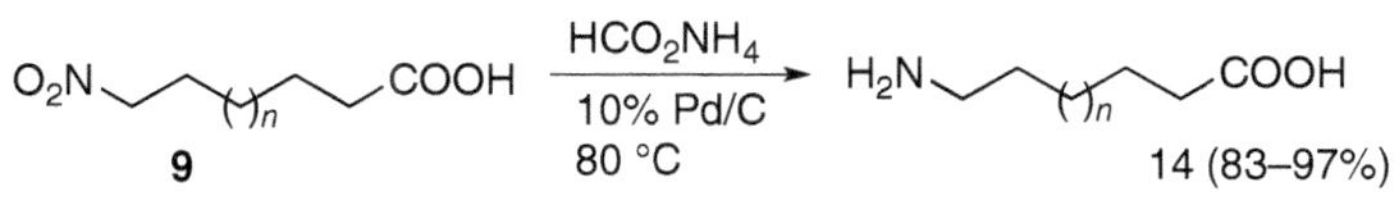

Scheme 9.6 Synthesis of ω-aminoacids **14**.

building block for the preparation of polyamides, elastomers, inhibitor of malignant melanoma, anti-inflammatory, and anti-convulsant.

9-Oxodecanoate **18** is a key prostaglandin intermediate and good starting material for (2*E*)-9-oxo-2-decenoic acid (queen substance) **19**. Both targets have been prepared from α-nitro cyclohexanone (Scheme 9.7).

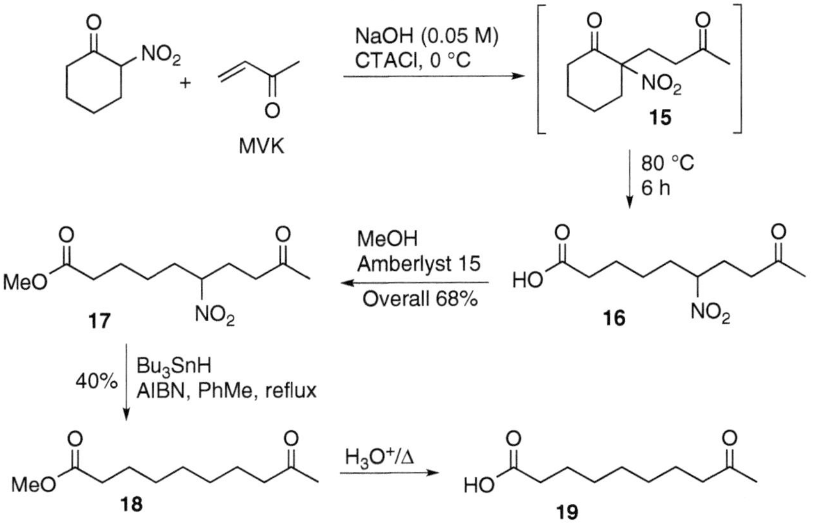

Scheme 9.7 Synthesis of keto ester **18** and keto acid **19**.

The process begins with the Michael addition of nitrocyclohexanone to methyl vinyl ketone (MVK), under aqueous NaOH/CTAC, at 0 °C. The obtained adduct **15**, heated at 80 °C, converts into the open chain structure **16**. Methylation of the latter followed by denitration achieves the target keto ester **18**. Finally, acidic hydrolysis of **18** yields the keto acid **19**.

9.2.2 Cleavage to Methyl ω-Oxoalkanoate

Ring cleavage of **8**, with methanol as nucleophile, under KOH catalysis, produces the formation nitronate ester **20** as intermediate, due to the basic conditions [9]. Thus, prompt addition of $KMnO_4/MgSO_4$ allows the formation of **21** by in situ oxidative-Nef reaction (Scheme 9.8).

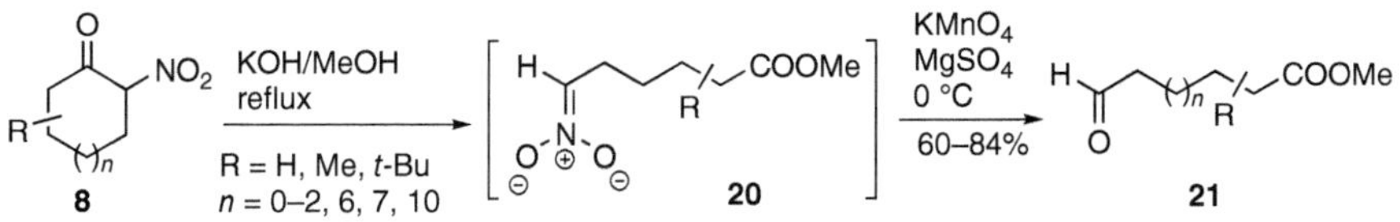

Scheme 9.8 Synthesis of methyl ω-oxoalkanoates **21**.

Good yields of methyl ω-oxoalkanoates are obtained from a variety of α-nitrocycloalkanones **8**.

9.2.3 Reductive Cleavage of α-Nitrocycloalkanones

Reductive cleavage of α-nitrocycloalkanones achieves the chemo- and regioselective formation of ω-nitroalcohols **22** in high overall yields (Scheme 9.9) [10].

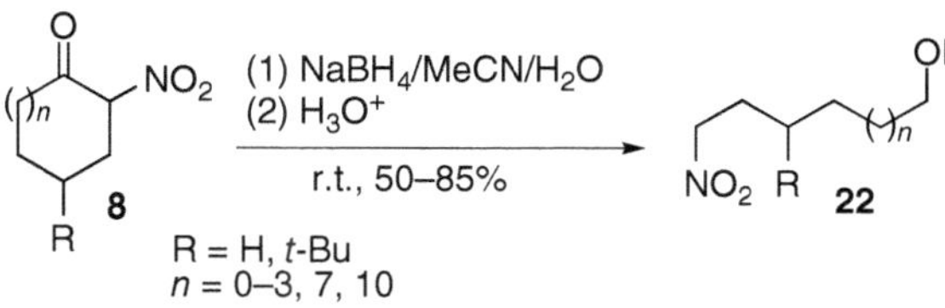

Scheme 9.9 Synthesis of ω-nitroalcohols **22**.

The reaction is realized with a large excess (five times) of $NaBH_4$ in $MeCN/H_2O$, at room temperature.

ω-Nitroalcohols **22** are an important class of functionalized aliphatic nitro compounds having two functional groups in two different oxidation state and they are susceptible to further transformations, owing to the versatility of the nitro and hydroxyl groups.

An important use of this cleavage is the synthesis of *R*-(+)-Patulolide A (**30a**) and *R*-(−)-Patulolide B (**30b**), two important macrolides isolated from *Penicillium urticae* mutant (Scheme 9.10) [11], starting from α-nitrocyclooctanone **23**. In fact, its reductive ring cleavage affords the ω-nitroalkanol **24**, and then converted into the corresponding aldehyde **25** (pyridinium chlorochromate (PCC)) oxidation. Methylation of **25** with CH_3MgI gives the (*R*)-(+)-**26**, successively transformed into the acetyl derivative **27** by goat liver lipase. The Michael addition of **27** to methyl propriolate, depending from the reaction conditions, can be converted into the *E*-adduct **28a** or *Z*-adduct **28b**, respectively. Finally, compounds **28a,b** are simply turned, in few steps, into the target molecules **30a** and **30b**.

9.2.4 Oxidative Cleavage of α-Nitrocycloalkanones

Oxidative cleavage of α-nitrocycloalkanones involves the in situ Nef reaction with the formation of a variety of α,ω-dicarbonyl derivatives difficult to obtain in other ways.

9.2.4.1 Cleavage into α,ω-Dicarboxylic Acids

Treatment of α-nitrocycloalkanones with 30% hydrogen peroxide and potassium carbonate, in MeOH solution, gives α,ω-dicarboxylic acids or keto-acids, depending from the nature (secondary or tertiary) of the nitro group, as reported in Scheme 9.11 [12].

The reaction is performed at room temperature, minimizing any possible oxidative side reactions.

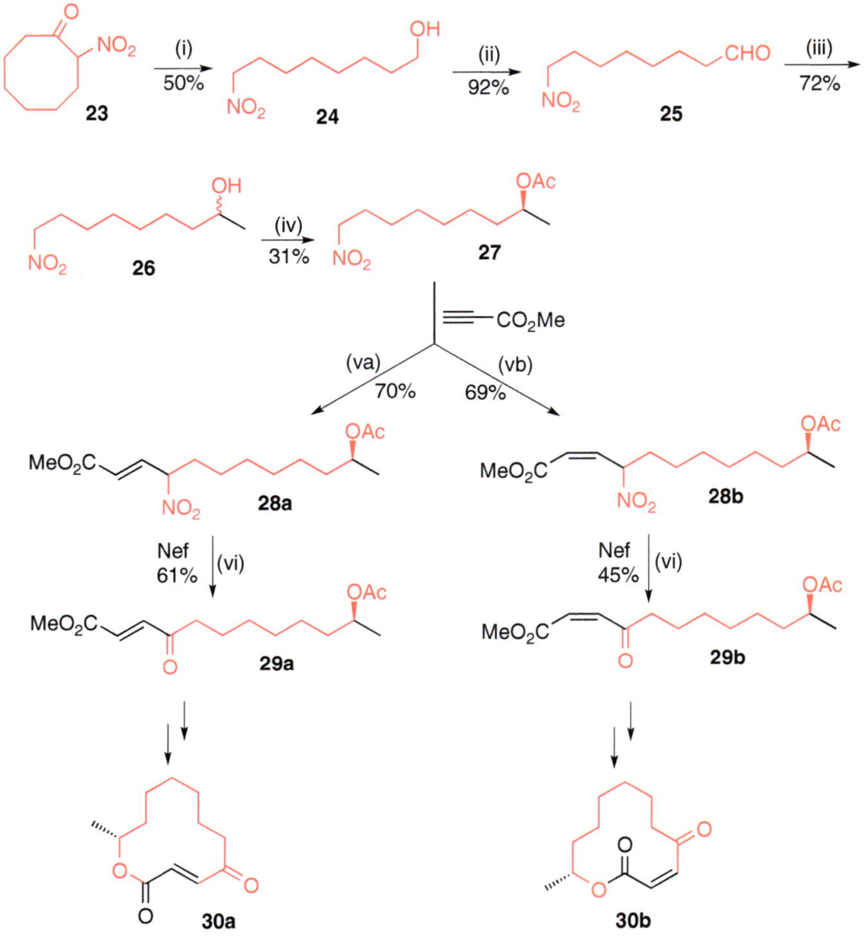

(i) = $NaBH_4/MeCN/H_2O$, (ii) = PCC/CH_2Cl_2, (iii) = $MeMgI/Et_2O$, 0 °C, (iv) = goat liver lipase, (va) = KF/Bu_4NBr/methyl propriolate/DMSO, (vb) = $Et_3N/MeCN$/methyl propriolate, (vi) = $TiCl_3/THF/NH_4OAc$.

Scheme 9.10 Synthesis of (*R*)-Patulolide A and B **30a** and **30b**.

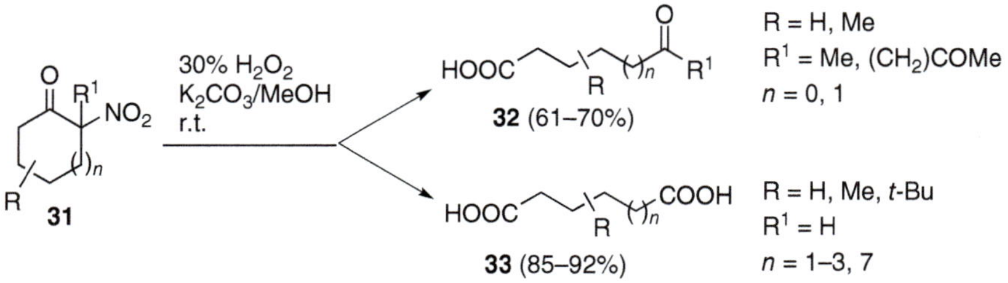

Scheme 9.11 Synthesis of α,ω-dicarboxylic acids **33** and keto acids **32**.

9.2.4.2 Cleavage to α,ω-Dicarboxylic Acids Dialkyl Esters

α,ω-Dicarboxylates are powerful molecules due to their possible transformations; moreover, long chain diesters are found in natural products and used as perfumes, stabilizing for resin and for conservation of food and cosmetics.

These compounds can be easily obtained by oxidative cleavage of α-nitrocycloalkanones **8**, at 80 °C, with potassium persulfate ($K_2S_2O_8$), in R^1OH and in the presence of H_2SO_4 (Scheme 9.12) [13].

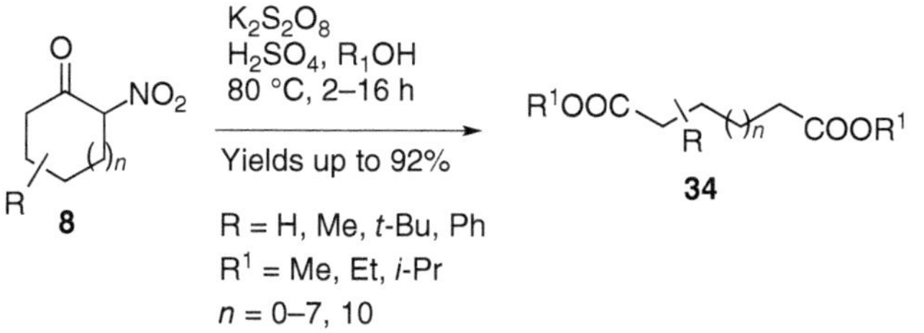

Scheme 9.12 Synthesis of α,ω-dicarboxylic acids dialkyl esters **34**.

This procedure works well with a variety of substrates, in few hours, and can be even applied to an important molecular structure such as 2α-nitro-5α-cholestan-3-one **35**, easily cleaved to dimethylesters **36**, as reported in Scheme 9.13.

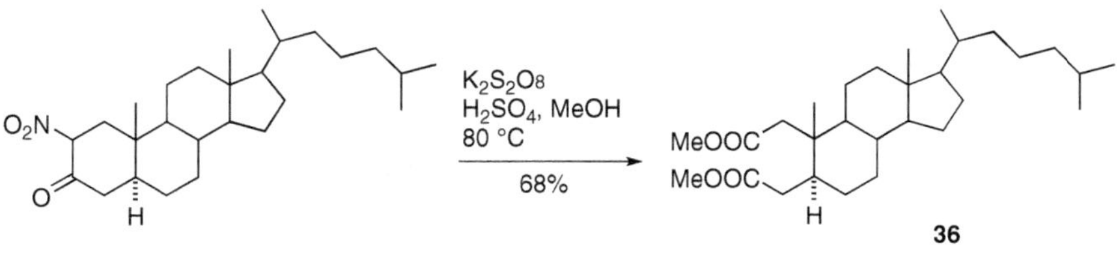

Scheme 9.13 Ring cleavage of steroid nitro ketone **35**.

9.2.4.3 Cleavage to α,ω-Dicarboxylic Acids Monomethyl Esters

Modular oxidative cleavage of 2-nitrocycloalkanones by Oxone® can allow the formation of dicarboxylic acids **33** or dicarboxylic acid monomethyl ester **37**, depending from the reaction conditions, as reported in Scheme 9.14 [14].

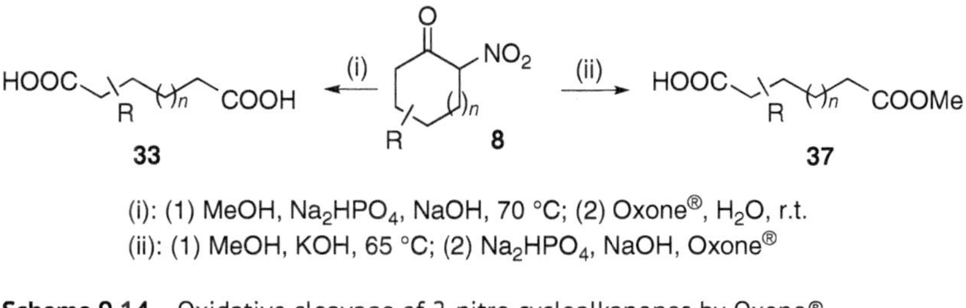

Scheme 9.14 Oxidative cleavage of 2-nitro cycloalkanones by Oxone®.

In fact, treating **8**, in a solution of 0.5 M Na_2HPO_4 and 1 M NaOH, with Oxone gives dicarboxylic acids **33**, while **32** can be obtained heating **8** in a methanolic solution of KOH, followed by addition of Oxone in aqueous 0.5 M Na_2HPO_4 and 1 M NaOH.

9.2.4.4 Cleavage to Methyl ω,ω-Dihalo-ω-nitroalkanoates

Oxidative cleavage of 2-nitro cycloalkanones **1** with sodium hypochlorite, in KOH methanolic solution, leads [15] the corresponding dichloro-ω-nitroalkanoic acids **38** that, in presence of MeOH/Amberlyst 15, convert into the analogous methyl esters **39**, in good overall yields (Scheme 9.15).

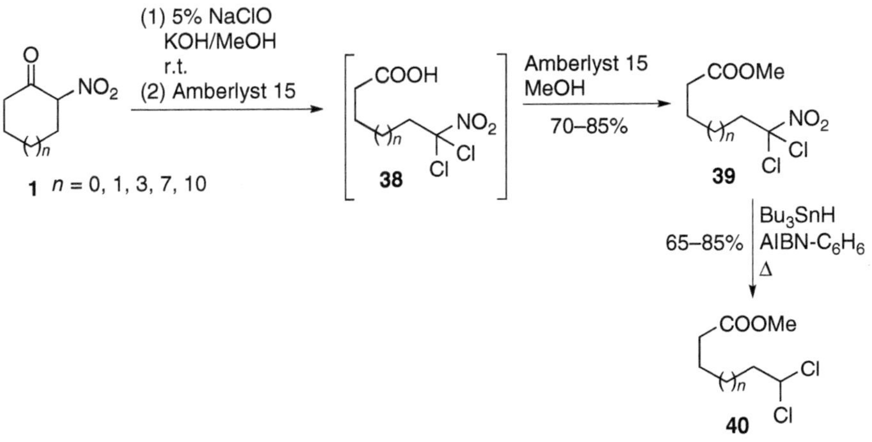

Scheme 9.15 Oxidative cleavage of 2-nitro cycloalkanones by NaClO into **39**.

A possible development is the denitration of **39**, via radical procedure using the Ono method, and formation of ω,ω-dichloroesters **40**.

9.2.5 Reaction of α-Nitrocycloalkanones with Organometallic Reagents

Reactions of α-nitrocyclohexanone with Grignard reagents produce the unexpected nucleophilic addition to the carbonyl moiety and formation of tertiary nitroalcohol adduct **41** instead of the expected ring cleavage (Scheme 9.16) [16]. However, Barua and coworkers found [17] an efficient procedure for the cleavage of cyclic

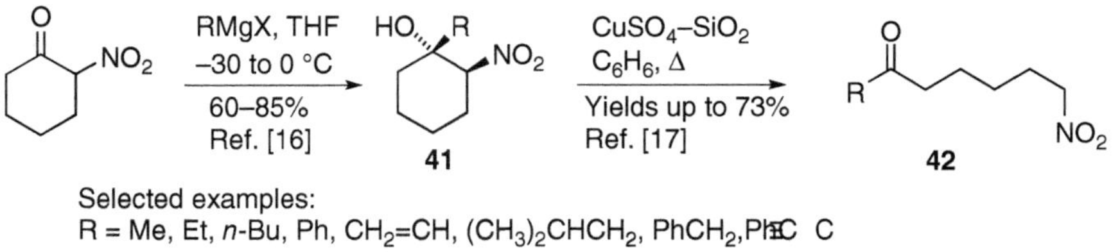

Scheme 9.16 Nucleophilic addition of Grignard reagents to α-nitrocyclohexanone and ring cleavage (**41** to **42**).

nitroalkanols **41** into nitro ketones **42**, in refluxing benzene and in presence of SiO_2–$CuSO_4$.

This pair of results has been strategic for the synthesis of two important targets: (±)-Phoracantholide **45**, a metabolite isolated from the metasternal gland of eucalypt longicorn *Phoracantha synonyma* (Scheme 9.17) [17] and *R*-(+)-α-Lipoic acid **48**, a vital cofactor in the multienzyme complexes, widely distributed in plants and animals (Scheme 9.18) [18].

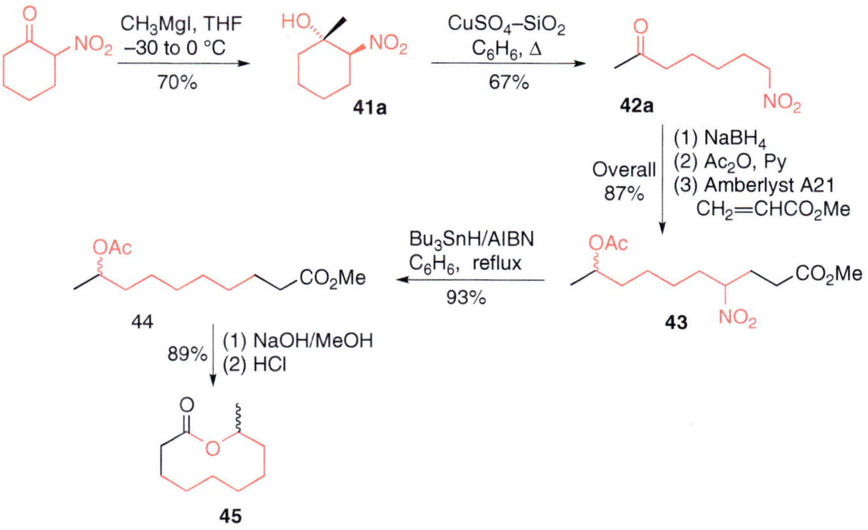

Scheme 9.17 Synthesis of (±)-Phoracantholide **45**.

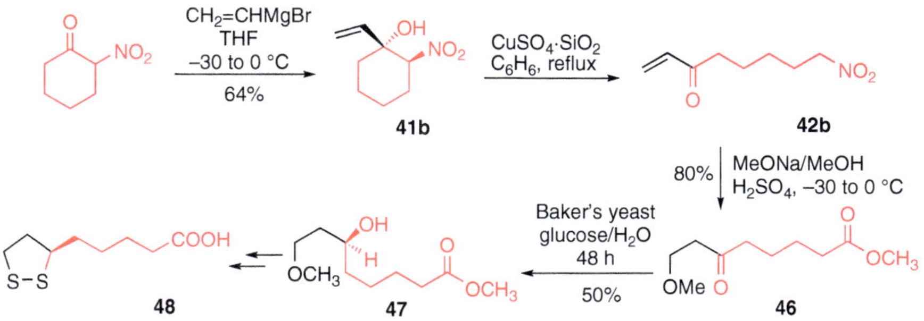

Scheme 9.18 Synthesis of (*R*)-(+)-α-Lipoic acid **48**.

The first step of Scheme 9.17 consists in the nucleophilic addition of a methyl anion to α-nitrocyclohexanone, with the formation of tertiary alcohol **41a**. Successive ring cleavage of the latter ($CuSO_4$–SiO_2) affords the open-chain nitro ketone **42a**.

$NaBH_4$ reduction-acetylation of **42a** followed by Michael addition to methyl acrylate, under Amberlyst A21 catalysis, achieves the adduct **43**. Denitration of the latter and successive cyclization of the obtained **44** affords the target compound **45** in 34% overall yield.

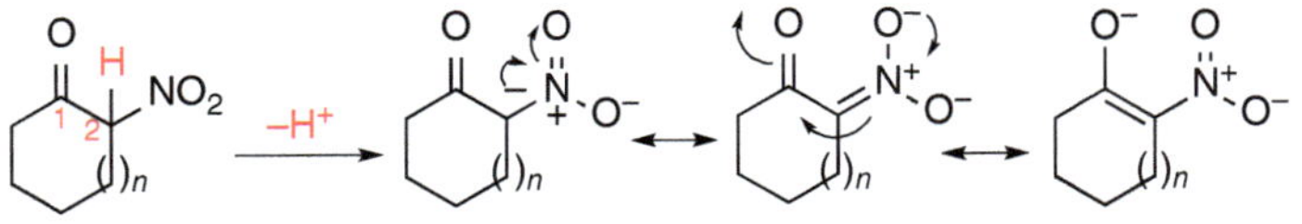

Figure 9.3 Acidity of hydrogen in 2-position.

The synthetic process of Scheme 9.18 opens with the vinylation of 2-nitrocyclohexanone with CH_2=CHMgBr and formation of the adduct **41b**.

Ring cleavage of the latest to nitro enone **42b** and in situ Michael addition (MeO^-)-Nef reaction afford the keto ester **46** that, employing baker's yeast enantioselective reduction produces the structure **47**, easily converted into (*R*)-α-Lipoic acid **48** in few simple steps.

9.3 α-Nitrocycloalkanones and Michael Reaction

The Michael addition of 2-nitrocycloalkanones to electron-poor alkenes is strongly favored by the acidity of hydrogen in 2-position because it is, simultaneously, germinal to the nitro-group and in α-position to the carbonyl (Figure 9.3).

Thus, it is frequently possible to conduct the Michael reaction without the need of any base, mainly in water [19].

Anyway, the Michael reaction of 2-nitrocycloalkanones is often the key starting step for the preparation of important targets and here it will be reported some recent examples on this topic.

(i) An interesting anionic domino process was reported, via a double Michael reaction. In fact, starting from 2-nitrocycloalkanones **8** and two molecules of acrolein in presence of methanolic solution of K_2CO_3, the method affords functionalized nitrocyclohexenes **53** in a one-pot way (Scheme 9.19) [20].

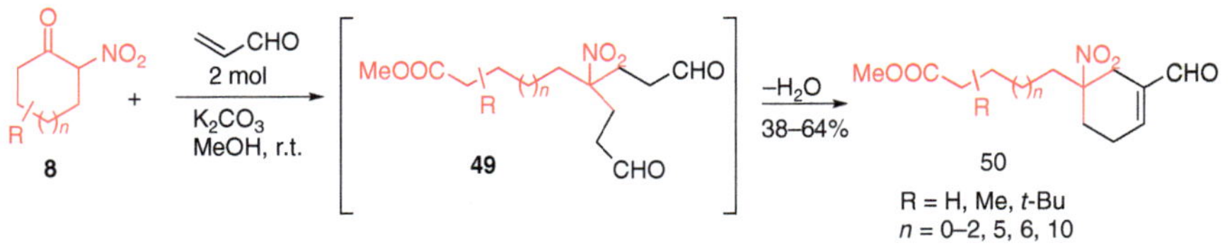

Scheme 9.19 Synthesis of nitrocyclohexenes **50**.

In fact, K_2CO_3 is able to promote the conjugate addition of one molecule of acrolein and in situ ring cleavage (promoted by MeOH) of nitrocycloalkanone moiety, and successive addition of one more molecule of acrolein to **49**. Finally, internal aldol condensation of **49** allows **50** in good overall yields.

(ii) Neat organocatalytic asymmetric Michael addition of α-nitro-cyclohexanone to conjugated nitroalkenes, using thiourea derivative **52** as chiral catalyst, for providing the chiral adduct **53** (Scheme 9.20) [21].

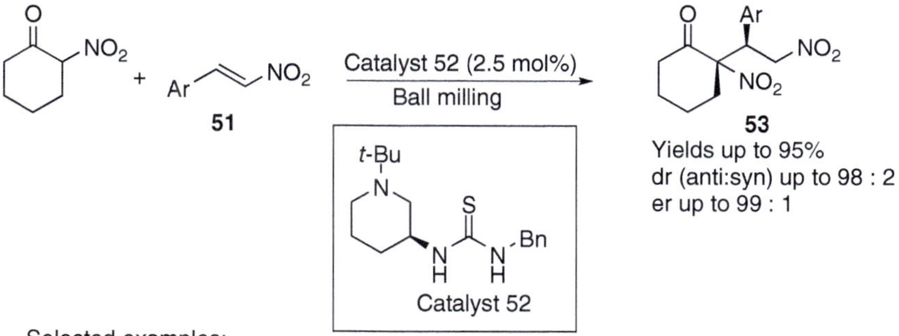

Selected examples:
Ar = 2-furyl, Ph, *p*-FC_6H_4, *p*-MeC_6H_4, *p*-$MeOC_6H_4$

Scheme 9.20 Asymmetric Michael reaction of 2-nitrocyclohexanone to nitroalkenes **51**.

High yields and excellent stereoselectivities are performed, thus the adducts **53** became a powerful class of functionalized and chiral 1,3-dinitroalkanes.

(iii) Organocatalytic enantioselective Michael addition of 2-nitrocyclohexanone to acrolein was developed and applied to the preparation of 1-azaspiro[4.5]decan-6-one **55** [22], a chiral building block for the synthesis of a variety of natural alkaloids (Scheme 9.21).

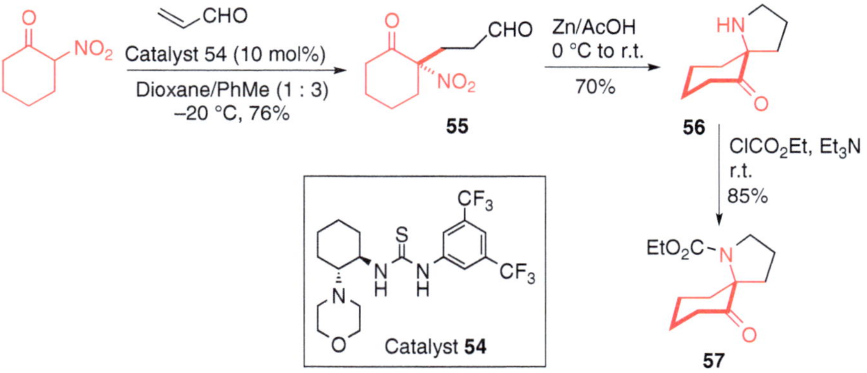

Scheme 9.21 Asymmetric synthesis of 1-azaspiro[4.5]decan-6-one **57**.

(iv) Organocatalytic (catalyst **59**) cascade reaction of 2-nitrocyclohexanone with enals **58** achieves the production of the unexpected bicyclic adducts **60**, in good yields and high enantioselectivities (Scheme 9.22) [23]. The formation of **60** can be explained with the regioselective nucleophilic attack of C_6 (due to the less acidic methylene group) and intramolecular Henry reaction (C_1–C_2). In presence of 1,4-diazabicyclo[2.2.2]octane (DABCO) as base.

If necessary, the bicyclic structure **60** can be expanded to the macrocyclic molecule **61** by action of MeOH. Finally, **61** can be converted into the more powerful eight-membered cyclic ketone **63**, found in many important natural products, applying the Nef reaction, in excellent enantioselectivity and good overall yield.

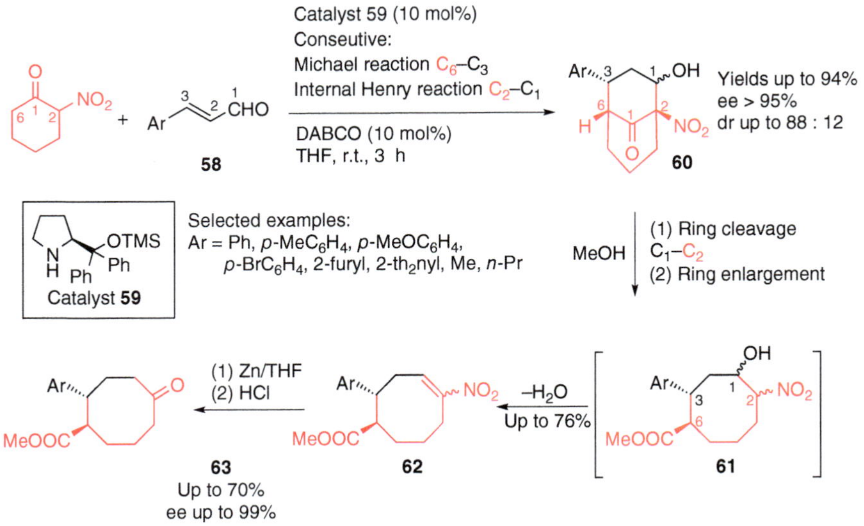

Scheme 9.22 Asymmetric synthesis of bicyclic **60** and macrocyclic **63**.

The process depicted in Scheme 9.22, where a six-membered ring can be expanded to eight-membered ring, is named "Zip Reaction" and 2-nitrocycloalkanones are often strategic in this field. Thus, some representative examples will be reported at the end of this chapter.

9.4 α-Nitrocycloalkanones and Henry Reaction

Nitroaldol (Henry) reaction of 2-nitrocycloalkanones with aldehydes has been the starting procedure for the formation of C—C bond, accomplished by the ring cleavage of the cyclic nitro ketone.

(ia) The Henry reaction of commercially available 2-nitrocyclododecanone **64** with hexanal **65** affords nitroalcohol **66**, prone to open into the corresponding open chain nitroalkanol **67** (Scheme 9.23) [24].

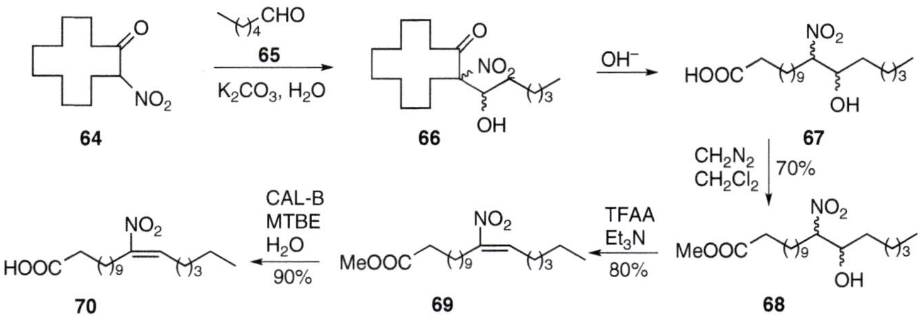

Scheme 9.23 Synthesis of (*E*)-12-nitrooctadecadec-12-enoic acid methyl ester **67**.

Methylation (CH_2N_2) of carboxylic functionality **67** affords **68** that dehydrated allows into the corresponding (*E*)-nitroalkene **69**. Successive hydrolysis of ester-functionality with Lipase B from *Candida antarctica* (CAL-B), in methyl *t*-butyl ether (MTBE), affords the nitro-fatty acid (NO_2-FAs) **70**, very important molecule because in several bioactivities including vasodilatation, inhibition of inflammation and of platelet activation and function is involved.

(iia) The reaction of nitrocycloalkanones with aromatic dialdehydes **71**, in DBU/THF/H_2O, affords the one-pot synthesis of nitroindane-1,3-diol **72** via a double Henry reaction (Scheme 9.24) [25].

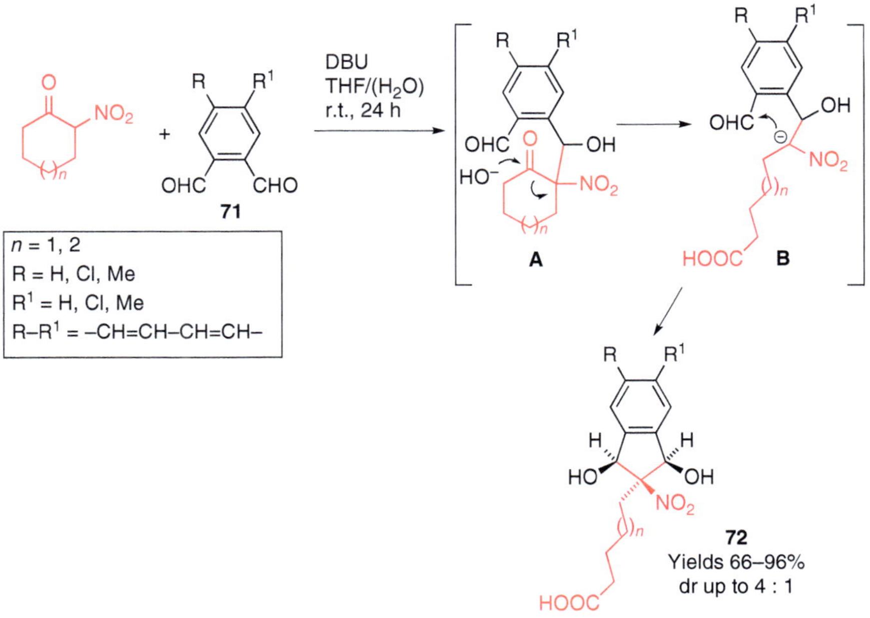

Scheme 9.24 Synthesis of Indane derivatives **72**.

An initial nitroaldol reaction affords the intermediate **A** that undergoes ring cleavage of nitrocyclohexanone structure to **B**. One more internal nitroaldol reaction of **B** achieves *cis*-Indane derivatives **72** in good overall yields (up to 96%) and dr (up to 4 : 1).

9.5 "Zip Reaction"

A variety of reactions of 2-nitrocycloalkanones with external nucleophiles have been described in this chapter, together with some of the synthetic applications of the open chain structures.

On the other hand, reaction with internal nucleophiles can afford ring enlarged products (Figure 9.4) and this why the latter reactions are named "Zip Reactions."

Thus, here some of the more recent, representative examples are reported.

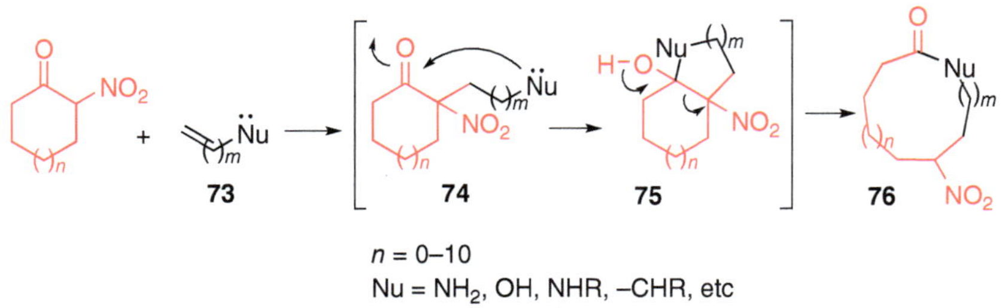

Figure 9.4 Zip reactions.

9.5.1 Synthesis of Bicyclic Macrolactones

Macrolactones (≥8-membered ring) are important structures for their wide range of biological activities. The most synthetic procedures for their preparation consist in the esterification of a suitable ω-hydroxy acid or derivatives; however, a general drawback is the low yields obtained. A novel one-pot, efficient, and stereoselective procedure for the preparation of functionalized, bridget, bicyclic medium-sized lactones **80** was developed starting from Michael addition of 2-nitrocycloalkanones to α-alkyl-α,β-unsaturated aldehydes **77** in aqueous K_2CO_3, via domino reaction (Scheme 9.25) [26].

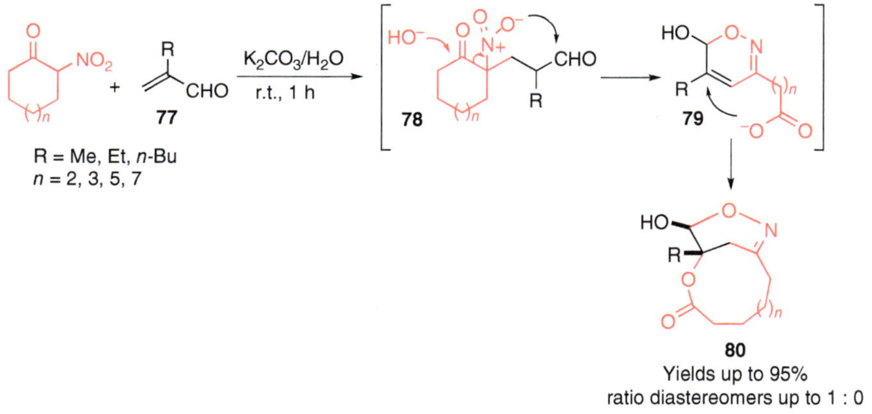

Scheme 9.25 Synthesis of functionalized macrolactones **80**.

Good yields and high stereocontrol (depending from the ring size) are achieved under sustainable and mild conditions (aqueous K_2CO_3/r.t.) in short reaction time (one hour).

9.5.2 Synthesis of 12-Oxotetradecan-14-lactam

Macrolactams are an important family of natural products that display a variety of biological activities such as antitumor, antifungal, cytotoxicity, antimicrobial, and inhibitor of superoxide generation. An interesting procedure reported a pilot synthesis of 15-membered lactam (12-oxotetradecan-14-lactam) **85**, through ring enlargement of 2-nitrocyclododecanone **64** (Scheme 9.26) [27]. The first step involves an unusual Michael addition to acrylamide **81**, under refluxing Et_2O, and formation of the adduct **82**. Conversion of **82** into **83** (Hofmann rearrangement) and successive simultaneous ring cleavage of **83** provide lactam **84**. Deprotection of the latter (TFA/CH_2Cl_2, r.t., 0.5 hour), and successive Nef reaction (MeONa, HCl, 0 °C), gives the target 12-oxotetradecan-14-lactam **85** in 81% overall yield from **83**.

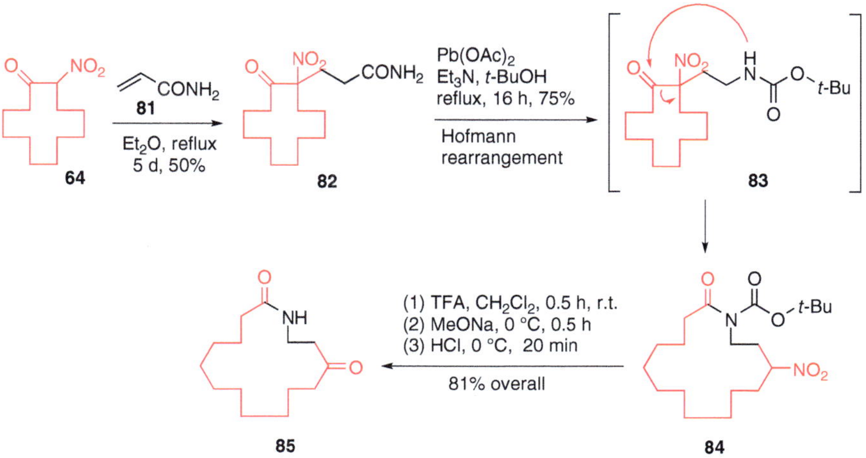

Scheme 9.26 Synthesis of functionalized macrolactam **85**.

9.5.3 Asymmetric Synthesis of Bicyclic Hemiketals

Bifunctional organocatalysis (catalyst **88**), in a mixture of CH_2Cl_2/MTBE (methyl-tertbutylether) as solvents, achieves highly enantioselective inverse-electron-demand hetero-Diels–Alder reaction of α-nitrocyclohexanone with the enol-form (of **86**) **87** with the one-pot formation of densely functionalized chiral bicyclic skeletons **89** that possess three stereogenic centers (up to 83% yield, 10 : 1 dr and 97% ee, Scheme 9.27) [28].

Moreover, compounds **89** are precursors of chiral 10-membered lactones **90** in a retro-Henry-type cleavage performed in a single step, under catalytic amount of n-Bu_4NF.

9.6 Arylation of Nitrocycloalkanones

Arylation of cyclic α-nitroketones improves the chemical potential of these compounds and their synthetic applicability.

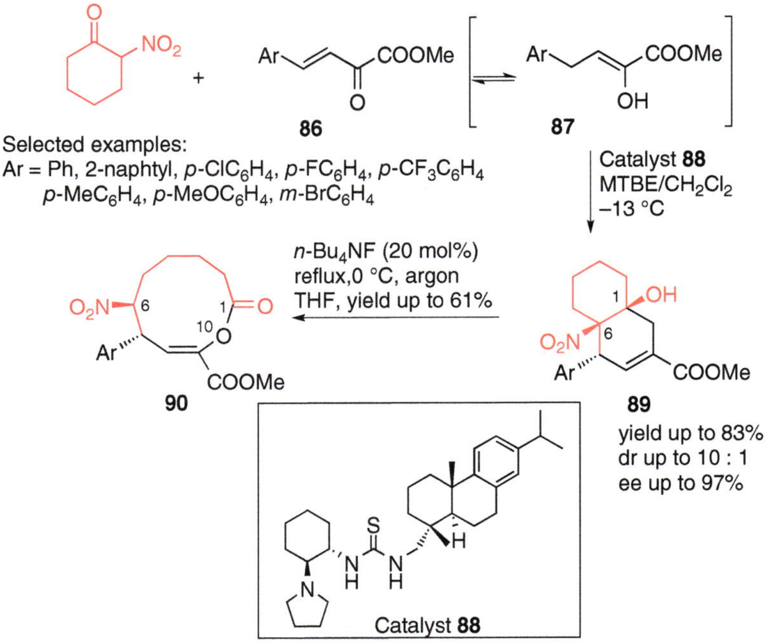

Scheme 9.27 Asymmetric synthesis of bicyclic skeleton **89** and ring cleavage to ten-member lactone **90**.

9.6.1 Synthesis of Benzo- and Naphtho-fused Bicyclo[*n*.3.1]structures

Areno-fused bicycle[*n*.3.1]alkane derivatives **93** are the key structure of many natural products sowing important biological activities (see, for example: dezocine, obtusanal B, huperzine, *N*-methylwelwistatin, etc.).

A decade ago it was reported that these classes of compounds can be easily obtained from 2-nitrocycloalkanones and α-halobenzyl hadides **91** (Scheme 9.28) [29]. In fact, selective benzylation of nitroketone, in presence of DBU/THF, provides the adduct **92** in excellent yield. The latter, without the need of any purification step, allows Pd-catalyzed ([Pd$(Ph_3P)_2$/Cs_2CO_3/toluene, reflux]) intramolecular C-arylation reactions, affording bicyclic compounds **93** in very good yields.

The mild reaction conditions required tolerate the presence of a variety of functional groups in the aromatic region, with either electron-releasing or electron-withdrawing character such as halogen, nitro group, and oxygen (Scheme 9.29).

However, in some of the reactions from α-nitrocycloheptanone or α-nitrocyclo-octanone ($n = 1, 2$), the cyclization step affords indane or indene derivatives, arising from nitrous acid elimination, instead of the expected bicyclic compounds.

The same authors reported the ring cleavage of nitrobicyclic compounds **94** to 2,8-dioxabicyclo[3.2.1]octanes **95**, under reductive conditions ($LiAlH_4$ or $NaBH_4$) [29b].

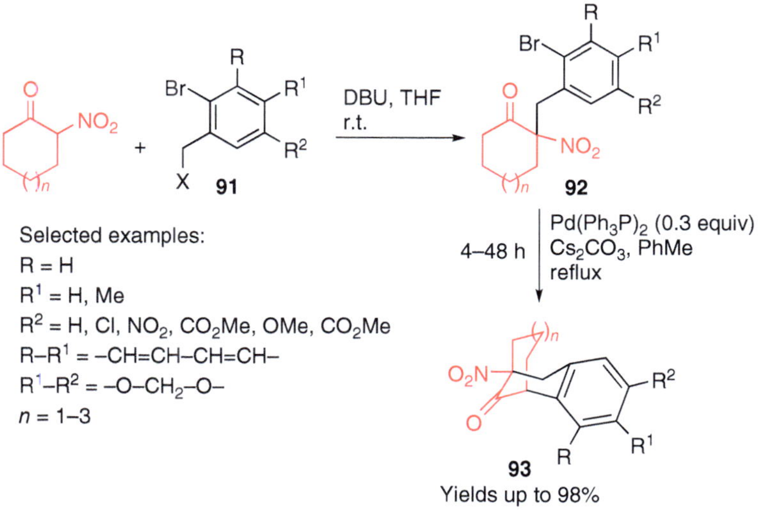

Scheme 9.28 Synthesis of benzo- and naphtho-fused bicyclo[*n*.3.1]structure **93**.

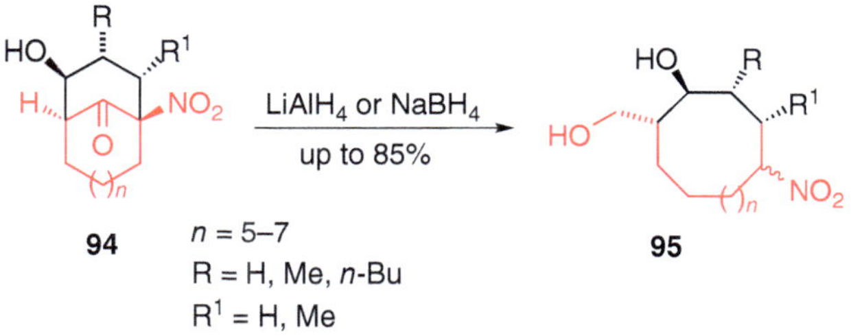

Scheme 9.29 Synthesis of bridget 2,8-dioxabicylclo[3.2.1]octanes **95**.

9.6.2 α-Arylation of 2-Nitrocycloalkanones

Tertiary α-aryl-α-nitrocycloalkanones **95** are important structures present in some bioactive molecule such as herbicides. Moreover, the simultaneous presence of carbonyl, aliphatic nitro group, and aromatic ring offers a large variety of chemical reactivity potentiality, very useful in organic synthesis.

Very recently, these molecules have been efficiently obtained by transition-metal-free α-arylation of α-nitro cycloalkanones with diaryliodonium triflate ($ArIAr^1$)OTf **96** (Scheme 9.30) [30].

The diaryliodonium salts **96** can be symmetric ($Ar = Ar^1$) **96a** or unsymmetric (Ar^1 = mesityl) **96b**, depending from the substrates. However, satisfactory to excellent yields of **97** (up to 90%) are obtained, and a variety of functionalities, such as ester, ether, and aromatic nitro group, can be tolerated.

The procedure has been applied to the synthesis of Tiletamine **100**, a clinical employed as dissociative anesthetic compound for treatment of various animal diseases (Scheme 9.31).

Selected examples:
Ar = Ph, o-FC$_6$H$_4$, m-MeC$_6$H$_4$, p-BrC$_6$H$_4$, p-i-PrC$_6$H$_4$,
p-OMeC$_6$H$_4$, p-CO$_2$MeC$_6$H$_4$, p-PhC$_6$H$_4$, p-NO$_2$C$_6$H$_4$
n = 0–2, 7

Scheme 9.30 Synthesis of tertiary α-aryl-α-nitrocycloalkanone **97**.

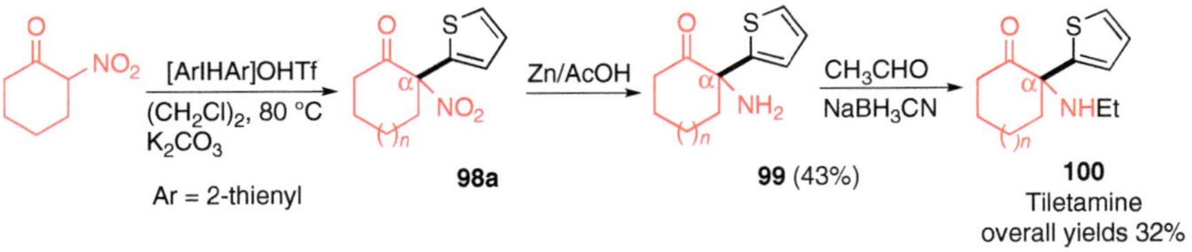

Scheme 9.31 Synthesis of Tiletamine **100**.

The synthesis needs of three steps, as follow: (i) α-arylation of 2-nitrocyclohexanone as 2-thienyl **98a**, (ii) nitro to amine (Zn/AcOH) reduction to **99**, and (iii) imine-formation–reduction, achieving tiletamine **100** in 32% overall yields.

References

1 Ballini, R. (1999). *Synlett* 7: 1009–1018.
2 Dampawan, P. and Zajac, W.W. Jr., (1983). *Synthesis*: 545–546.
3 Venkat Ram Reddy, M., Kumareswaran, R., and Vankar, Y.D. (1995). *Tetrahedron Lett.* 36: 7149–7152.
4 (a) Rathore, R., Lin, Z., and Kochi, J.K. (1993). *Tetrahedron Lett.* 34: 1859–1862. (b) Rathore, R. and Kochi, J.K. (1996). *J. Org. Chem.* 61: 627–639.
5 Rank, W. (1991). *Tetrahedron Lett.* 32: 5353–5356.
6 Ballini, R. and Petrini, M. (1986). *Synth. Commun.* 16: 1781–1788.
7 (a) Ballini, R., Petrini, M., and Polzonetti, V. (1992). *Synthesis*: 355–357; (b) Chiurchiù, E., Gabrielli, S., Ballini, R., and Palmieri, A. (2017). *Green Chem.* 19: 4956–4960.
8 Ballini, R., Papa, F., and Abate, C. (1999). *Eur. J. Org. Chem.*: 87–90.
9 Ballini, R., Bosica, G., and Gigli, F. (1998). *Tetrahedron* 54: 7573–7580.
10 Ballini, R., Petrini, M., and Rosini, G. (1990). *Tetrahedron* 46: 7531–7538.
11 Kalita, D., Khan, A.T., Barua, N.C., and Bez, G. (1999). *Tetrahedron* 55: 5177–5184.
12 Ballini, R., Marcantoni, E., Petrini, M., and Rosini, G. (1988). *Synthesis*: 915–918.
13 Ballini, R. and Bosica, G. (1997). *Tetrahedron* 53: 16131–16138.

14 Ballini, R., Curini, M., Epifano, F. et al. (1998). *Synlett*: 1049–1050.

15 Ballini, R., Petrini, M., and Polimanti, O. (1996). *J. Org. Chem.* 61: 5652–5655.

16 Ballini, R., Bartoli, G., Gariboldi, P.V. et al. (1993). *J. Org. Chem.* 58: 3368–3372.

17 Saikia, A.K., Hazarika, M.J., Barua, N.C. et al. (1996). *Synthesis*: 981–985.

18 Bezbarua, M.S., Saikia, A.K., Barua, N.C., and Kalita, D. (1996). *Synthesis*: 1289–1290.

19 Giorgi, G., López-Alvado, P., Miranda, S. et al. (2013). *Eur. J. Org. Chem.*: 1327–1336.

20 Ballini, R., Barboni, L., Bosica, G. et al. (2001). *Tetrahedron* 57: 6079–6081.

21 Jörres, M., Mersmann, S., Raabe, G., and Bolm, C. (2013). *Green Chem.* 15: 612–616.

22 Ding, X.-H., Cui, W.-C., Li, X. et al. (2013). *Tetrahedron Lett.* 54: 1956–1959.

23 Kuan, Y.-N., Chen, Z.-Y., and Yan, M. (2014). *Chem. Commun.*: 10471–10473.

24 Zanoni, G., Bendjeddou, M.V.L., Porta, A. et al. (2010). *J. Org. Chem.* 75: 8311–8314.

25 Giorgi, G., Arroyo, F.J., López-Alvarado, P., and Menéndez, J.C. (2010). *Synlett*: 2465–2467.

26 Giorgi, G., Miranda, S., López-Alvarado, P. et al. (2005). *Org. Lett.* 7: 2197–2200.

27 Jia, Y.-M., Liang, X.-M., Chang, L., and Wang, D.-Q. (2007). *Synthesis* 5: 744–748.

28 Jiang, X., Wang, L., Kai, M. et al. (2012). *Chem. Eur. J.* 18: 11465–11473.

29 (a) Giorgi, G., Maiti, S., López-Alvarado, P., and Menéndez, J.C. (2011). *Org. Biomol. Chem.* 9: 2722–2730; (b) Giorgi, G., López-Alvarado, P., and Menéndez, J.C. (2012). *Org. Biomol. Chem.* 10: 5131–5136.

30 An, Y., Zhang, X.-M., Li, Z.-Y. et al. (2019). *Chem. Commun.* 55: 119–122.

10

Acyclic α-Nitro Ketones: Synthesis and Reactivity

Acyclic α-nitro ketones are a strategic class of functionalized nitroalkanes due to the presence of a nitro group and a carbonyl in two adjacent positions that provide a variety of synthetic potential. However, although α-nitrocycloalkanones and acyclic α-nitro ketones possess the same functionalities, they have unlike reactivities and are often used differently and the chemical power of acyclic α-nitro ketones was exhaustively reported in two important reviews [1, 2].

10.1 Synthesis of α-Nitro Ketones

α-Nitro ketones are synthetically available from a variety of sources; thus a huge number of these molecules can be easily accessible.

10.1.1 Synthesis of α-Nitro Ketones from Henry Reaction

The main spring involves (Scheme 10.1) the formation of C—C bond by (i) Henry reaction of the appropriate nitroalkane **2** with the right aldehydes **1** and (ii) oxidation of the formed β-nitroalcohol **3** into nitro ketone **4a** using a variety of oxidative methods (i–iii).

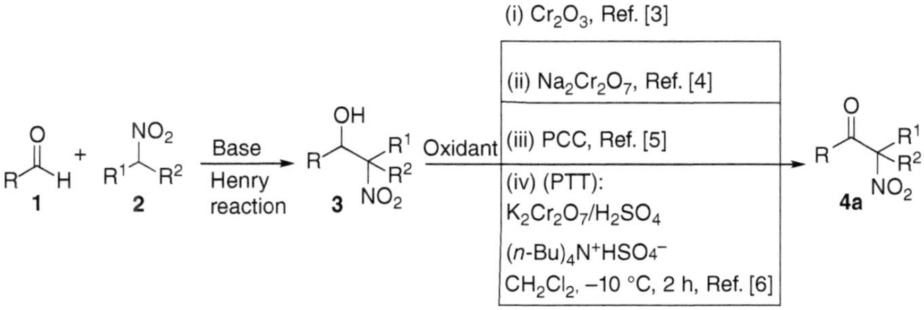

Scheme 10.1 Synthesis of α-nitro ketones **4a** via the Henry reaction.

Nitroalkanes: Synthesis, Reactivity, and Applications, First Edition.
Roberto Ballini and Alessandro Palmieri.

(i) The pioneering oxidative process of nitroalkanols **3** was performed reacting the nitroalcohols **2** with chromium trioxide[3] or sodium dichromate [4] in strong acids. However, these severe conditions often achieve low yields and acid-labile functionalities cannot be preserved.

(ii) An efficient and chemoselective oxidation with PCC (pyridinium chlorochromate) in CH_2Cl_2 and good yields (up to 87%) are achieved, preserving acid-labile functionalities such as ester, ether, ketal, tetrahydropyranyl, etc. [5].

(iii) Afterwards, an improved method was elaborated under the phase-transfer technique (PTT). Thus, addition of $K_2Cr_2O_7/H_2SO_4$ to a solution of nitroalcohol **3** and $(n\text{-Bu})_4N^+HSO_4^-$, CH_2Cl_2, at −10 °C affords α-nitro ketones **4a**, after two hours, with high chemoselectivities (functionalities preserved: ester, acetal, ketal, ether, C=C double bond) and excellent yields (up to 95%) [6].

(iv) With particular attention to the sustainability, a one-pot procedure has been realized from aldehyde **1** and nitroalkane **2** to nitro ketone **4a** (Scheme 10.2), through a solvent-free procedure, via nitroaldol reaction on activated neutral alumina (20 hours), completed by in situ oxidation (20 hours) to **4a** with wet alumina supported chromium(VI) oxide [7].

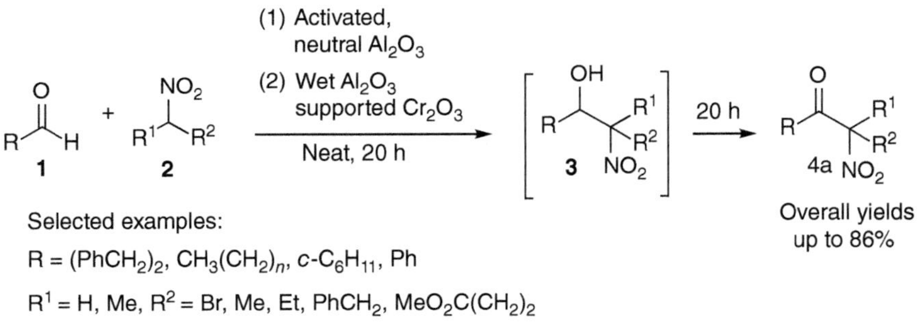

Scheme 10.2 One-pot synthesis of α-nitro ketones **4a** via the Henry reaction.

Good to excellent overall yields (up to 86%) are obtained with an easy work-up consisting in extraction with Et_2O and evaporation of the solvent, the crude product is directly submitted to flash chromatography.

(v) An interesting procedure, employing the Henry reaction, was recently reported starting from primary nitroalkanes and phenylselenyl acetaldehyde **5** catalyzed by basic alumina in 2-methyltetrahydrofuran (2-MeTHF) as green solvent (Scheme 10.3) [8]. Acetylation of the formed nitroalcohol **6** with Ac_2O/Amberlyst 15 under neat conditions produces nitro ester **7**.

The latter, submitted to oxidation with *m*-chloroperbenzoic acid (*m*-CPBA), under MW and flow conditions in order to avoid any side reactions usually observed in batch process, allows elimination of phenylselenyl group and formation of the enol-acetate **8** in satisfactory to good yields (up to 81%). Hydrolysis of **8** (H_2SO_4/EtOH/60 °C) achieves the easy formation of α-nitro ketones **4b**.

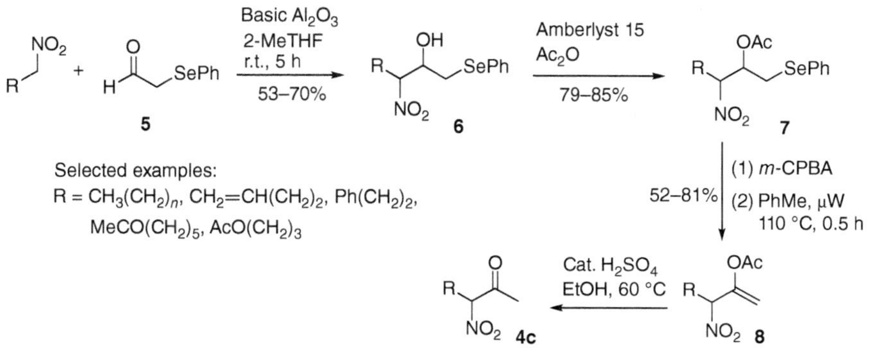

Scheme 10.3 Synthesis of α-nitro ketones **4c** from phenylselenyl acetaldehyde **5**.

10.1.2 Synthesis of α-Nitro Ketones from Carboxylic Acid Derivatives

C-Acylation of nitroalkanes is an alternative procedure to prepare 2-nitro ketones from acyl halides, anhydrides, or activated esters; however, the first report achieves modest results.

More promising results was obtained (Seebach and Leher) [9] via reaction of doubly metalated complex $(R\text{-}C{=}NO_2)^{2-}2Li^+$ with acyl chlorides or esters.

(i) A direct acylation of sodium or potassium methanenitronate with *N*-acylimidazole **9** was next published (Scheme 10.4) [10]. The procedure works well with a large variety of *N*-acylimidazoles achieving primary α-nitro ketones **4c** difficult to prepare in other ways.

(ii) Acylation of **9** (Scheme 10.4) with primary and secondary nitroalkane–lithium salts (stored as dry form) in a solution of dimethylsulfoxide furnishes nitro ketones **4a** in satisfactory to excellent yields (up to 87%) [11].

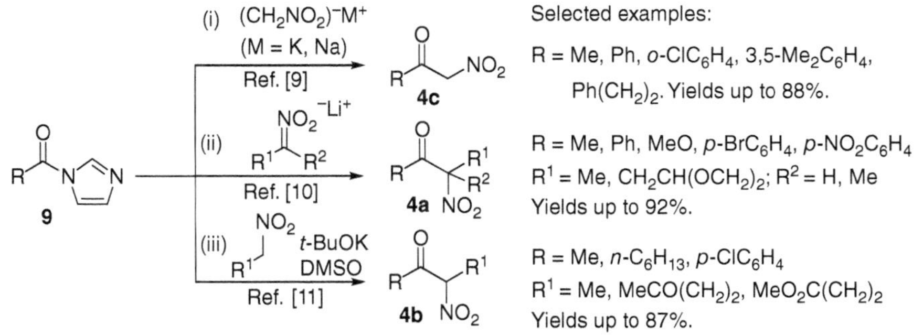

Scheme 10.4 Synthesis of α-nitro ketones **4a–c** from acylimidazoles **9**.

(iii) An alternative and more practical method, in which is possible to overcome the tedious use of dry forms, has been proposed reacting **9** with primary nitroalkanes, in a solution of potassium *t*-butoxide as base (Scheme 10.4) [12].

(iv) An alternative method employs acyl chloride **10** and trichloronitromethane **11** in presence of tin(II)chloride, as reported in Scheme 10.5 [13]. Good yields

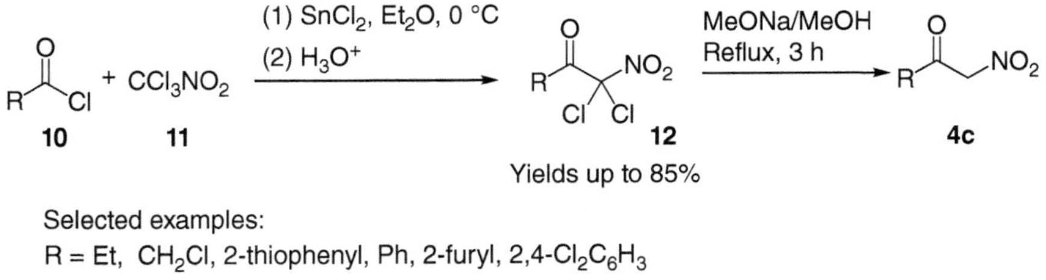

Scheme 10.5 Synthesis of α-nitro ketones **4c** from acylchloride **10**.

(up to 85%) of dichloronitro ketones **12** are obtained. Successive treatment of **12** with refluxing MeONa/MeOH affords primary nitroketones **4c**.

10.1.3 Synthesis of α-Nitro Ketones from Alkenes

Possible precursors of α-nitro ketones are the corresponding alkenes and nitroalkenes, easily accessible with a variety of synthetic procedures or commercially available.

(i) Olefins (acyclic and cyclic) **13** can conveniently converted into the correlates α-nitro ketones **4b**, in good yields, by treatment with trimethylsylil nitrate and chromium trioxide (Scheme 10.6) [14].

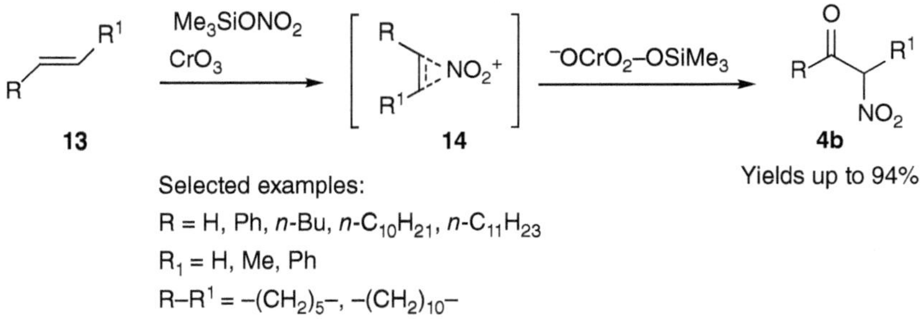

Scheme 10.6 Synthesis of α-nitro ketones **4b** from alkenes **13**.

Probably the formation of NO_2^+ is due to a species such as NO_2-O-CrO_2-($SiMe_3$). The formed intermediates **14**, under $^-CrO_2(OSiMe_3)$ oxidation, provide nitro ketones **4b** in good overall yields (up to 94%).

(ii) β-Nitrostyrenes **15** are the key precursors of α-nitro acetophenones **16** when treated with *t*-butylhydroperoxide (*t*-BuOOH) and butyllithium (Scheme 10.7) [15].

10.1.4 Synthesis of α-Nitro Ketones from Silyl Enol Ethers

A powerful procedure has been reported by nitration of silyl enol ethers **16** (Scheme 10.8) [16].

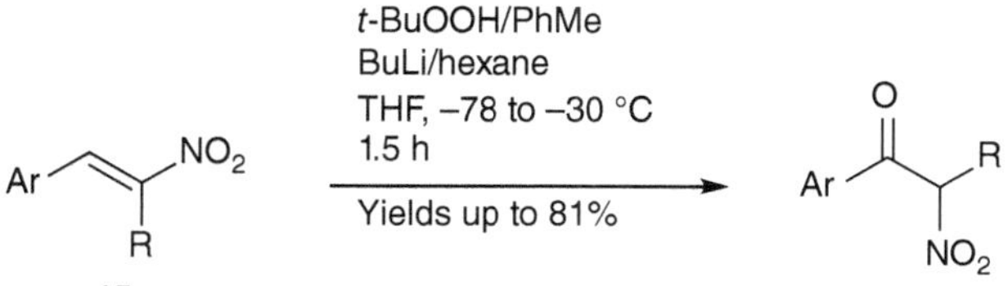

15 **4d**

Selected examples:
Ar = Ph, *p*-$MeOC_6H_4$, *p*-MeC_6H_4, *o*-$NO_2C_6H_4$, 3,4-$MeOC_6H_3$, 3,4-$(OCH_2O)C_6H_3$
R = H, Me

Scheme 10.7 Synthesis of α-nitro ketones **4d** form nitroalkenes **15**.

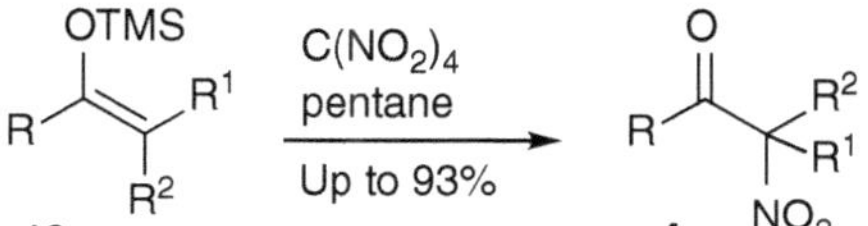

16 **4a**

Selected examples:
R = Aryl, *n*-Pr, t-Bu
R^1 = H
R^2 = H, Et, *n*-Bu, Bn
R^1–R^2= Cycloalkyl

Scheme 10.8 Synthesis of α-nitro ketones **4a** from silyl enol ether **16**.

Compounds **16** are easily prepared from the corresponding ketones and their oxidation is performed with tetranitromethane, as mild nitrating agent. Thus, the method allows good yields of a variety of nitro ketones **4a** in satisfactory to excellent yields (up to 93%).

10.2 Reactivity of Acyclic α-Nitro Ketones

Acyclic α-nitro ketones are a very important class of functionalized nitroalkanes due to the presence of two highly versatile functional groups in two vicinal positions. In addition, these compounds contain an acidic α-hydrogen atom with pK_a values [17] close to carboxylic acids and in aqueous solution high enol content was observed due to the strong effect of the nitro group [18].

10.2.1 Replacement of the Nitro Group of α-Nitro Ketones

Replacement of the nitro group, belonging to α-nitro ketones, with another atom or group can be a great opportunity for synthetic applications of these compounds.

10.2.1.1 Replacement of the Nitro Group with Hydrogen

Replacement of the nitro group with hydrogen is usually called "denitration" and this substitution can be performed in few methods.

(i) The most used procedure for this substitution was developed, some decades ago ([19], Scheme 10.9) via a direct procedure that involves a free-radical process, using tributyl tin hydride (Bu_3SnH) in refluxing benzene and in presence of a catalytic amount of azobisisobutyronitrile (AIBN) as radical activator. Good yields are achieved with a variety of substrates.

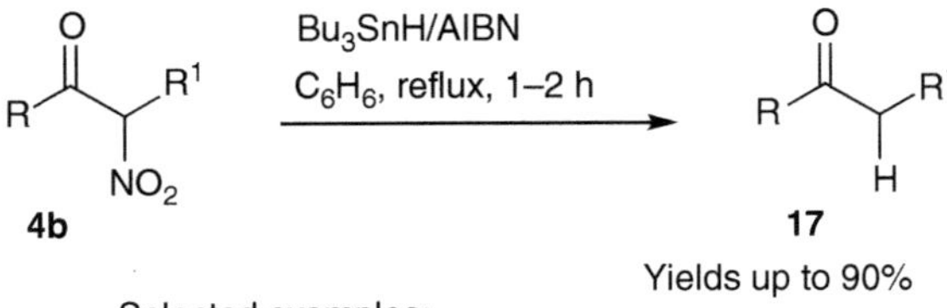

Selected examples:
R = Alkyl, aryl
R^1 = Me, $MeOCO(CH_2)_n$, $MeCO(CH_2)_2$

Scheme 10.9 Denitration of α-nitro ketones **4b** by Bu_3SnH.

(ii) The nitro group can be also substituted with hydrogen on treatment with sodium dithionite, as soft reducing agent [$Na_2S_2O_4/Et_3SiH$ in hexamethylphosphoramide (HMPA)/H_2O]. The procedure works well with a variety of nitro ketones preserving other functionalities (CN, COOMe, Br), thus giving access to functionalized ketones **18** (Scheme 10.10) [20].

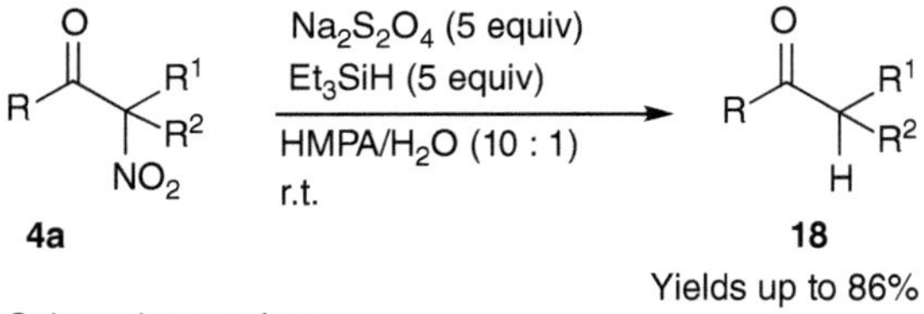

Selected examples:
R = Alkyl, aryl
R^1 = H, Me
R^2 = $MeCO(CH_2)_2$, $NCCH_2(Me)CH$, $MeOCO(CH_2)_2$
$MeCHBr(CH_2)_2$

Scheme 10.10 Denitration of α-nitro ketones **4a** by $Na_2S_2O_4$.

(iii) Nitro group of primary nitro ketones **4c** can be replaced with hydrogen through an indirect procedure and treatment with ethanethiol in presence of $AlCl_3$, as the Lewis acid (Scheme 10.11) [21].
A plausible mechanism includes the formation of protected nitro ketone **19** that, through the simultaneous elimination of $(EtS)_2$ and nitrous acid, converts into the enol form **20** that quickly converts into protected ketone **21**, easily precursor of free methyl ketones **22**.

(iv) Nitro ketones **4a** can be denitrated from their tosylhydrazones, with lithium aluminum hydride, in dry THF (Scheme 10.12) [22].

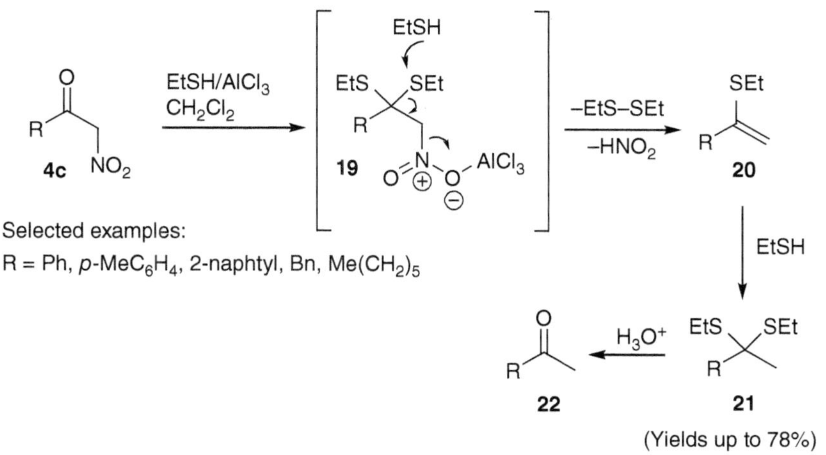

Scheme 10.11 Denitration of α-nitro ketones **4c** into **21**.

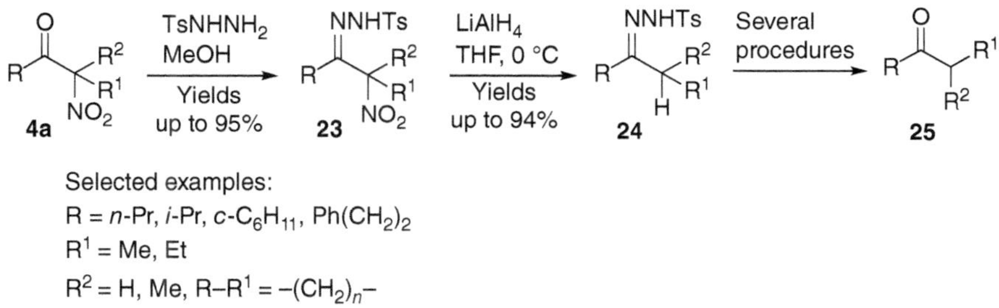

Scheme 10.12 Denitration of α-nitro ketones **4a** into **25**.

The method produces high yields of denitrated tosylhydrazones **24** with both secondary and tertiary nitroderivatives. Moreover, **24** can be easily cleaved, using several methods, into denitrated ketones **25**. It is important to observe that this method is an indirect procedure for denitration of α-nitro ketones and offers important advantages from purification point of view since all the tosylhydrazone derivatives involved (**23** and **24**) can be simply purified by crystallization.

Thus, the easy availability of acyclic α-nitro ketones from a good number of precursors, associated with the possibility to replace the nitro group with a hydrogen, renders these molecules "strategic building blocks" for the preparation of important targets. Here are reported some representative examples.

(*Z*)-5-Undecen-2-one **30**, the principal volatile component of the pedal gland exudates of the bontebox *Damaliscus dorcas dorcas*, has been prepared starting from nitroalkene **26** and ethanal with in situ nitroaldol/oxidation reactions to nitro ketone **27** (Scheme 10.13) [23]. Conversion of the latter into the corresponding tosylhydrazone, followed by reductive ($LiAlH_4$/THF) denitration and removal of the hydrazine-protection, affords the target compound **30** in 55% overall yield from **26** and 70% from nitro ketone **27**.

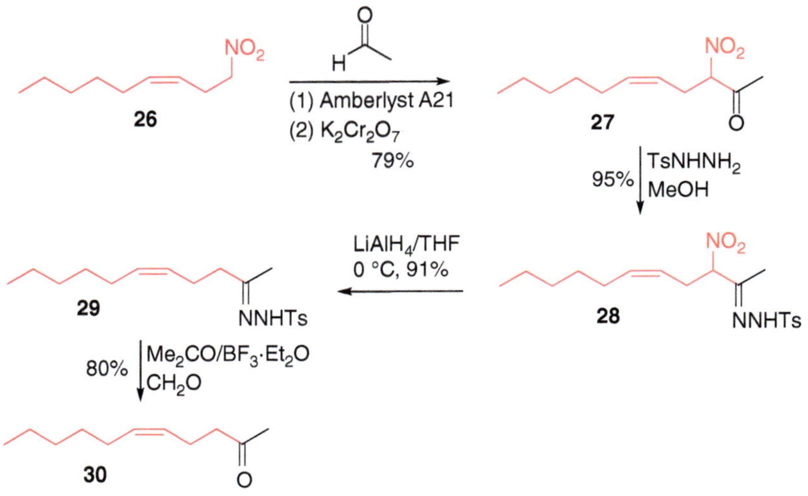

Scheme 10.13 Synthesis of (*Z*)-5-undecen-2-one **30**.

It is important to point out that both the hydrazones **28** and **29** are purified by crystallization with ethanol/H_2O.

Two important prostanoid building blocks, 9-keto esters **35a,b**, have been prepared from ω-nitro ester **31** (easily obtained via ring methanolic cleavage of 2-nitrocyclooctanone) [24].

Thus (Scheme 10.14), the Henry reaction of **31** with **32** gives nitroalcohols **33** that, by PCC oxidation, converts into the corresponding nitro ketone **34**. Denitration of the latter with Bu_3SnH/C_6H_6 affords methyl 9-oxodecanoate **35a** and methyl 9-oxo-tridecenoate **35b** in 46% and 45% overall yields, respectively.

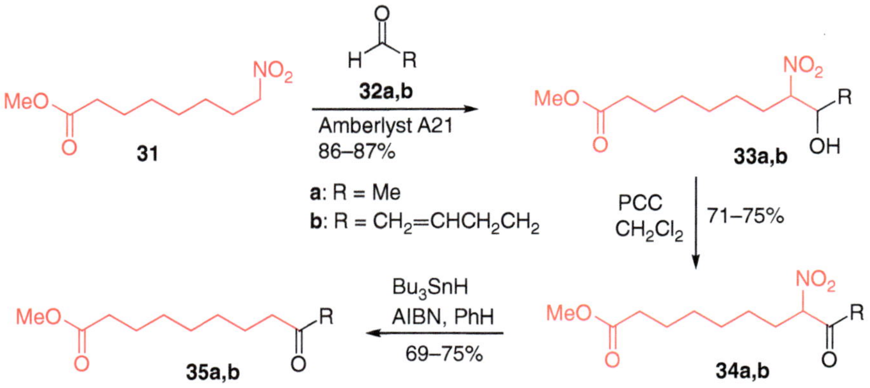

Scheme 10.14 Synthesis of methyl 9-oxodecanoate **35a** and methyl 9-oxo-12-tridecenoate **35b**.

(*Z*)-Heneicos-6-en-11-one **41**, sex pheromone of Douglas Fir Tussock Moth (*Orgyia pseudotsugata*), a severe defoliator of fir forest, can be synthesized via nitro ketones as reported in Scheme 10.15 [25]. Nitroaldol reaction of nitroalkene **36**

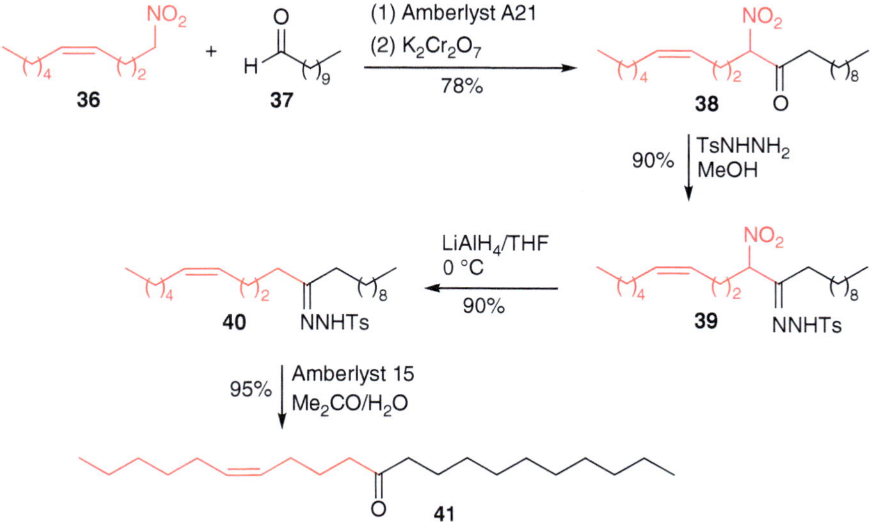

Scheme 10.15 Synthesis of (*Z*)-heneicos-6-en-11-one **41**.

with undecanal **37** on Amberlyst A21, under neat conditions and in situ oxidation ($K_2Cr_2O_7$ under phase transfer procedure) of the formed nitroalcohol allows one-pot formation of nitro ketone **38**.

Successive conversion of the latter into the corresponding tosylhydrazone **39** followed its denitration with $LiAlH_4$, in THF, furnishes **40**. Deprotection of **40** with acetone/water, under acidic conditions (Amberlyst 15), achieves the formation of the target compound **41** in 60% overall yield and without any isomerization of the (*Z*)-C=C double bond.

Observing the role of starting nitroalkanes employed (**26**, **31**, and **36**) in the last three Schemes (10.13–10.15), it is possible to note that they act as synthetic equivalents of the carbanions **Ia–c**, as reported in Figure 10.1.

10.2.1.2 Tandem Denitration–Deoxygenation

Denitration–deoxygenation consists in the removal of both nitro- and carbonyl-groups under reductive conditions.

(i) The first tandem denitration–deoxygenation procedure of α-nitro ketones was realized reducing the corresponding tosylhydrazones with $LiAlH_4$, in THF at 60 °C (Scheme 10.16) [26].

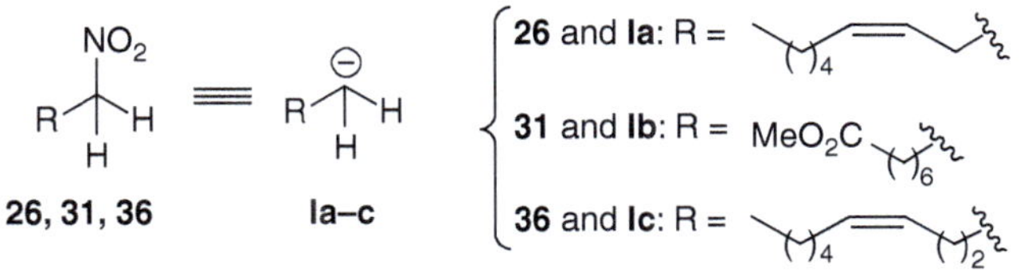

Figure 10.1 Nitroalkenes **26**, **31**, and **36** as synthetic equivalents of carbanion synthons **Ia–c**.

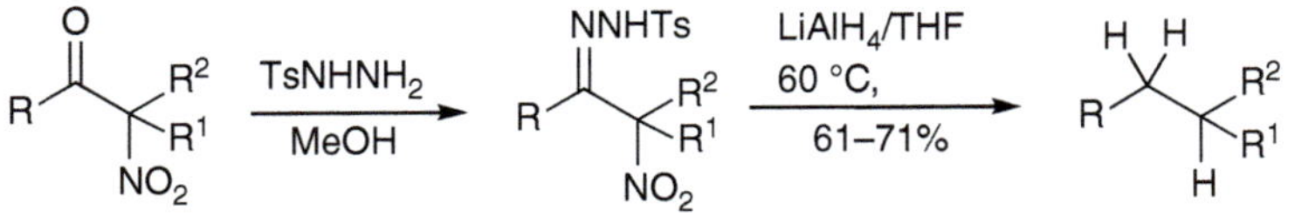

Selected examples:
R = $Ph(CH_2)_2$, $n\text{-}C_{11}H_{23}$
R^1 = Me, $n\text{-}C_6H_{11}$, $n\text{-}C_8H_{17}$
R^2 = H, Me
R–R^1 = $-(CH_2)_{13}-$

Scheme 10.16 Denitration–deoxygenation of α-nitro ketones using $LiAlH_4$.

The same authors applied this method to the synthesis (Scheme 10.17) of (*Z*)-9-tricosene **47**, sex pheromone component of mature female housefly (*Musca domestica*). Thus, the Henry reaction of 1-nitropentane **44** with oleic aldehyde **43**, on basic alumina under neat conditions, followed by in situ oxidation of the obtained nitroalcohol with $K_2Cr_2O_7$, provides the direct formation of nitro ketone **45** in 85% yield. Conversion of the latter into tosylhydrazone **46** is performed in 92% yield after recrystallization. Finally, reduction of **46** with $LiAlH_4$ at 60 °C gives **47** in 66% yield (52% overall yield).

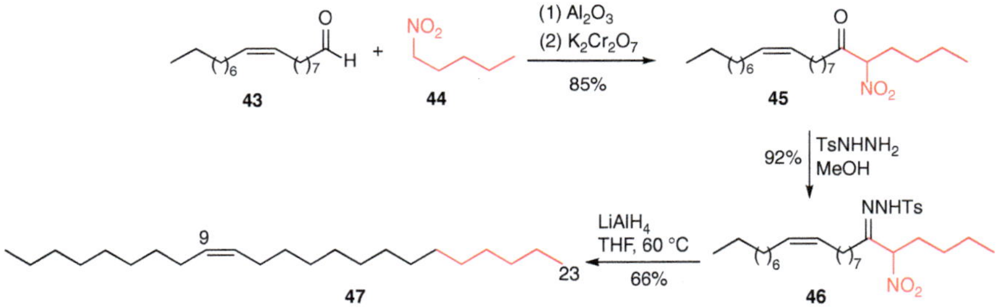

Scheme 10.17 Synthesis of (*Z*)-9-tricosene **47**.

(ii) An efficient, one-pot, chemoselective "tandem denitration–deoxygenation" process has been realized reducing the nitro ketone tosylhydrazones with $NaBH_4$/MeOH, at 80 °C (Scheme 10.18) [27]. Thus, reaction of nitro ketones **4b** and tosylhydrazine in methanol (10 hours) followed by $NaBH_4$ reduction of the formed tosylhydrazone **23a** at 80 °C (10 hours), furnishes good overall yields (up to 71% from **4b**) of denitrated–deoxygenated compounds **48**.

Following this procedure important functionalities (such as aromatic-nitro-group and ester) are preserved.

The applicability of this method was tested for the synthesis of 2-methylheptadecane **52** (Scheme 10.19), a very important natural component of (i) sex attractant pheromone of a number of species of the *Artidae* family (tiger moth), (ii) flavor in

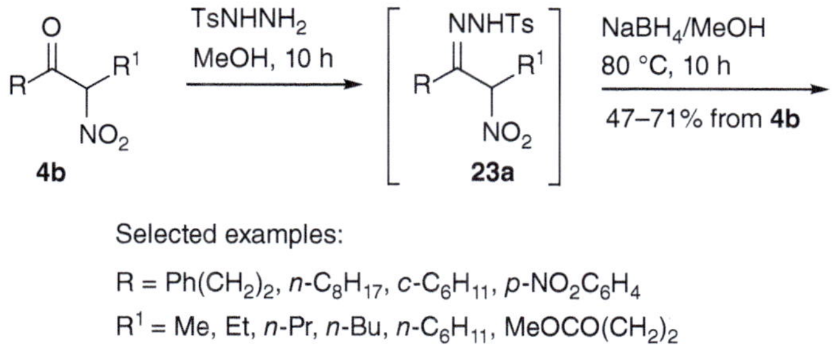

Scheme 10.18 Denitration–deoxygenation of α-nitro ketones using $NaBH_4$.

four mango varieties and in black soya beans, (iii) flavor of the blue–green algae, and (iv) essential oil from *Clematis hexapetala* Pall and *Inula nervosa* Wall [28].

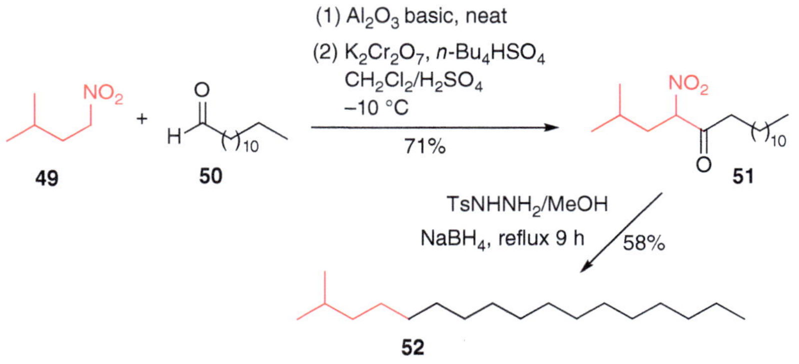

Scheme 10.19 Synthesis of 2-methylheptadecane **52**.

Nitroaldol reaction of 3-methylnitrobutane **49** with tridecanal **50**, under neat conditions under basic alumina, and successive in situ oxidation ($K_2Cr_2O_7$ under PTT) of the nitroalkanol prepared, directly affords the nitro ketone **51** in 71% yield. Conversion of **51** into the corresponding tosylhydrazone and in situ treatment with $NaBH_4$, at 80 °C, gives the target compound **52** in just two steps and in 41% overall yield.

In consideration of the role of starting nitroalkanes (**44** and **49**) and aldehydes (**43** and **50**) employed in Schemes 10.17 and 10.19, it is possible to assert (Figure 10.2)

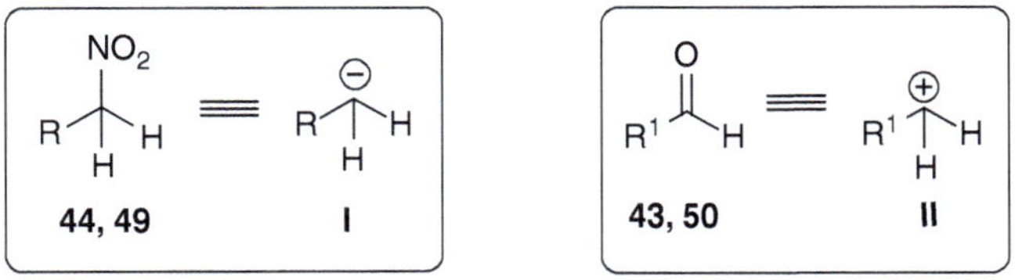

Figure 10.2 Nitroalkanes **44** and **49** as synthetic equivalents of carbanion synthon **I**. Aldehydes **43** and **50** as synthetic equivalents of carbanions **II**.

that nitroalkanes act as synthetic equivalents of the carbanions **I**, while aldehydes act as synthetic equivalents of carbocations **II**.

10.2.1.3 Replacement of the Nitro Group with Deuterium

The possibility to introduce of deuterium atoms in molecules is very important, mainly for studies of mechanisms. In this context, α-nitro ketones are versatile compounds to produce deuterated compounds.

Firstly, Ono et al. reported (Scheme 10.20) [29] the regioselective C-α mono- and di-deuteration of alkyl ketones by H/D exchange of the α-hydrogen, from α-nitro ketones **4b**.

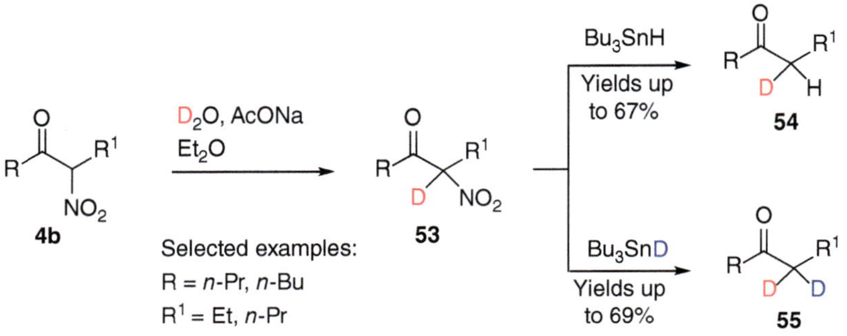

Scheme 10.20 Synthesis of α-deuterated ketones **53–55**.

Thus, treatment of **4b** with D_2O/AcONa in Et_2O results in a clean H/D exchange yielding monodeuterated nitro ketone **53**. Successive treatment of the latter with Bu_3SnH yields α-monodeuterated ketones **54**, while treatment with Bu_3SnD gives α,α-dideuterated ketones **55**, both in good yields.

Deuteration of α-nitro ketones **4a** can be carried out through the corresponding tosylhydrazones **56**, affording a variety of deuterated molecules, depending from the reactants and the reaction conditions (Scheme 10.21) [26, 30].

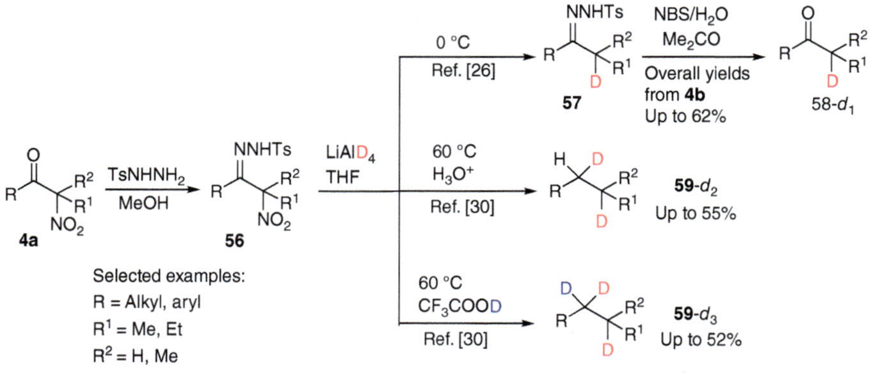

Scheme 10.21 Deuteration of α-nitro ketones via tosylhydrazones.

Thus, treatment of tosylhydrazones **56** with $LiAlD_4$ at 0 °C affords the corresponding mono-deuterated derivatives **57** that, after deprotection, achieves the α-d_1 ketones **58** [30].

If the reduction of **56** is completed at 60 °C instead of 0 °C, followed by H_3O^+ acidification, it allows di-deuterated molecules **59**-d_2. Finally, **56** is transformed into tri-deuterated structures **59**-d_3 reducing with $LiAlD_4$ at 60 °C, and successive work-up with CF_3COOD.

Thus, α-nitro ketones have demonstrated to be powerful precursors of mono-, di-, and tri-deuterated molecules.

10.2.1.4 Replacement of the Nitro Group with Phenylthio Group

The α-position of α-nitro ketones can be replaced with phenylthio group via reaction with phenylthiol or its potassium salt (Scheme 10.22)[31].

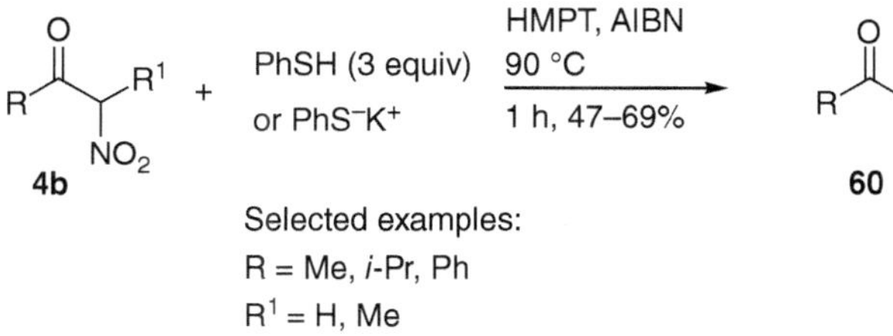

Scheme 10.22 Replacement of nitro group with "SPh."

The reaction moves on following an electron-transfer mechanism and works well with primary ($R^1 = H$) and secondary nitro derivatives. The procedure is simple since a mixture of **4b**, phenylthiol and AIBN, in hexamethylphosphoric triamide (HMPT), at 90 °C for one hour, allows the thio-substituted compound **60** in satisfactory yields (up to 69%).

10.2.2 α-Nitro Ketones to Conjugated Enones

Nitroaldol reaction of α-nitro ketones with formaldehyde constitutes the first step of an efficient method for regioselective preparation of α-methylene ketones **63** (Scheme 10.23) [32].

In fact, reaction of **4b** with 37% CH_2O, in alcoholic (*i*-PrOH) solution of Ph_3P, and successive acetylation of the formed nitroalkanol, gives **61** in excellent yields. Successive denitration (Bu_3SnH/AIBN) converts **61** into **62**, prone to elimination of acetic acid under basic (1,8-diazabicyclo[5.4.0]undec-7-ene [DBU]) conditions, achieving **63** in excellent overall yields.

An interesting procedure, in order to convert α-nitro ketones into conjugate enones, was elaborated via their tosylhydrazones **64** under basic conditions (DBU), as reported in Scheme 10.24 [33].

In fact, 1,4-elimination of nitrous acid provides tosylazoalkenes **66a** that, under DBU, tautomerize into the more stable structure **66b**. The azo–diazo conversion

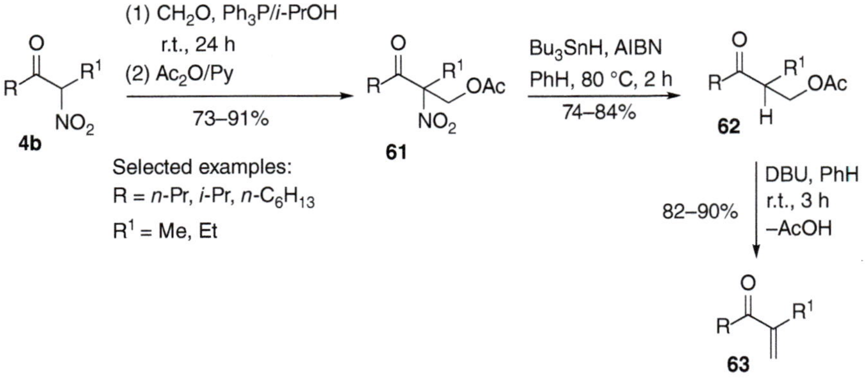

Scheme 10.23 Synthesis of conjugate enones **63**.

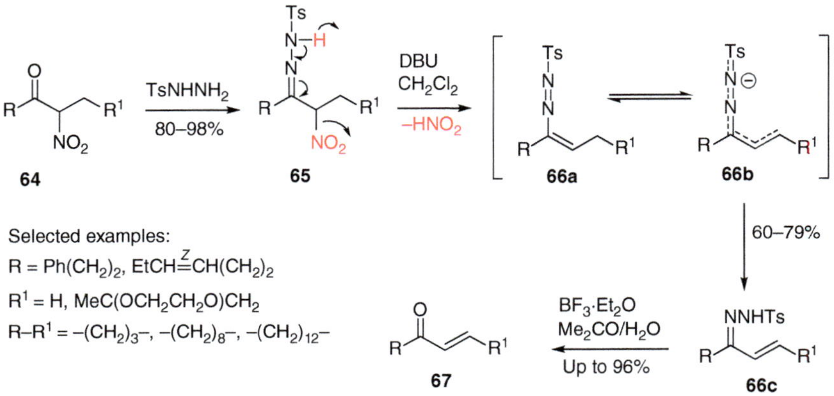

Scheme 10.24 α-Nitro ketones into enones **67** via tosylhydrazones.

of **66b–c** can be explained by the electron withdrawing effect of tosyl group that strongly polarizes the allylic hydrogen of the alkene. Tosylhydrazones are in general crystalline and can be easily purified by crystallization; moreover they are highly versatile structures in organic synthesis due to their versatility for further chemical transformations. However, if necessary, tosylhydrazones **66c** can be readily cleaved to enones **67** using a variety of procedures, including the use of $BF_3 \cdot Et_2O/Me_2CO/H_2O$, as reported in Scheme 10.24.

In addition, because it is possible to prepare nitro ketones with the nitro group in either α- (**69**) or α^1-position (**74**) with the right choice of the starting molecules, the aforementioned procedure can be employed as regioselective approach for the preparation of α,β- or α^1,β^1-unsaturated ketones **72** and **77**, respectively (Scheme 10.25).

In fact, solvent-free Henry reaction (1-nitrohexane **68** with propanal or heptanal **73** with nitroethane) on basic alumina followed by in situ oxidation ($K_2Cr_2O_7$) under phase transfer reaction allows, directly, nitro ketones **69** and **74**. Successive formation of tosylhydrazones **70** and **75**, and treatment with DBU, achieves **71** and **76**

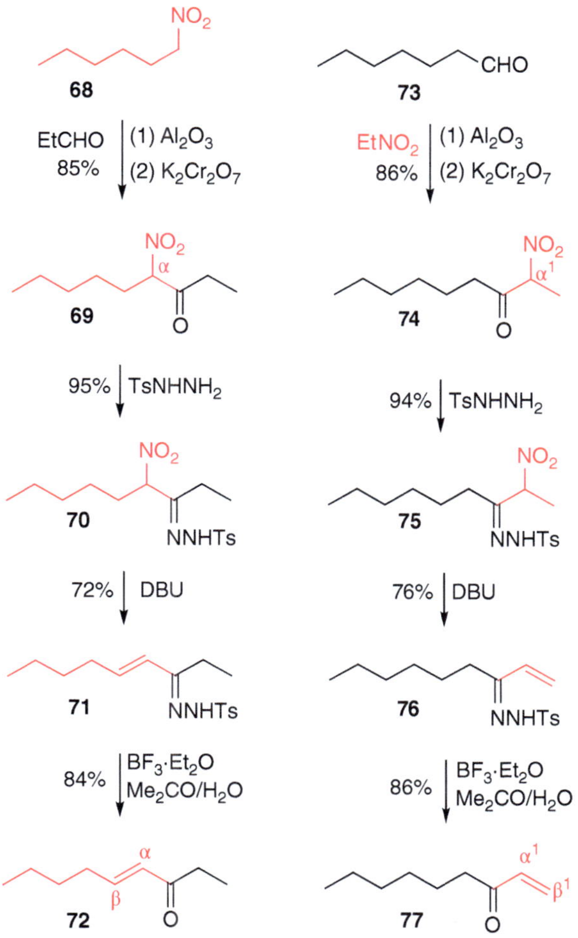

Scheme 10.25 Regioselective formation of α,β-unsaturated ketones **72** and **77**.

via elimination of nitrous acid. Finally, regeneration of carbonyl affords, regioselectively, the non-4-en-3-one **72** (49% overall yield) and non-1-en-3-one **77** (53% overall yield), respectively.

One more synthetic potential of α-nitro ketones has been demonstrated in the preparation of (*E*)-conjugated enediones **81** (Scheme 10.26) [34], a powerful class of intermediates for the preparation of valuable targets such as prostaglandins, rethrolones, perfumes, pheromones, macrocycles, etc.

The process starts with the Henry reaction of γ-nitroalkanol **78** with an aldehyde, promoted by Amberlyst A21, under neat conditions. The obtained nitrodiols **79**, after purification, are oxidated with $K_2Cr_2O_7$ under PTT and the formed crude nitro diones **80** are, in situ, treated with silica gel/cyclohexane/EtOAc (7 : 3), under vigorous stirring in order to help the elimination of nitrous acid, with consequent formation of (*E*)-enediones **81** in good overall yields. As note, for R = *n*-Bu the

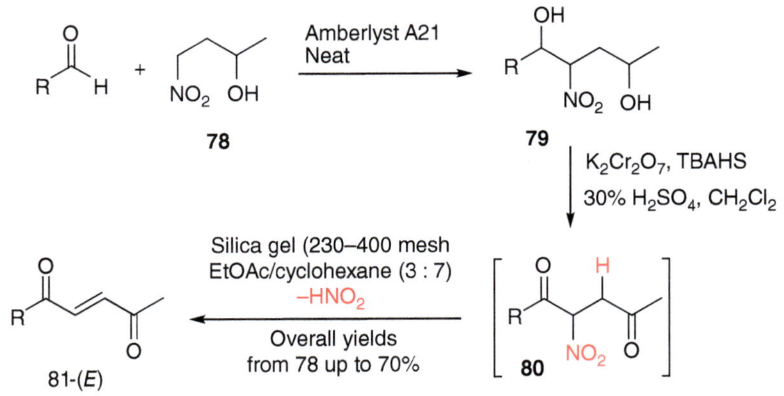

Selected examples:
R = *i*-Pr, *n*-Bu, *n*-C_5H_{11}, *c*-C_6H_{11}, *n*-$C_{10}H_{21}$, $Ph(CH_2)_2$, $Me(CH_2)_2CH\overset{Z}{=}CH(CH_2)_7$

Scheme 10.26 Synthesis of (*E*)-2-ene-1,4-diones **81**.

corresponding (*E*)-enedione **81** is the main component of volatile compounds derived from cephalic secretion of workers of *Trigona tataira*.

10.2.3 α-Nitro Ketones into Nitroalkanols

α-Nitroalcohols are usually prepared by Henry reaction; however, often the level of stereoselectivity is low because the basicity needed for this reaction favors the epimerization of the formed nitroalkanols. In this context, alternative asymmetric ways have been proposed from α-nitro ketones.

The first reduction of α-nitro ketones **4b** into *anti*-β-nitro alcohols **82** was reported two decades ago (Scheme 10.27) [35].

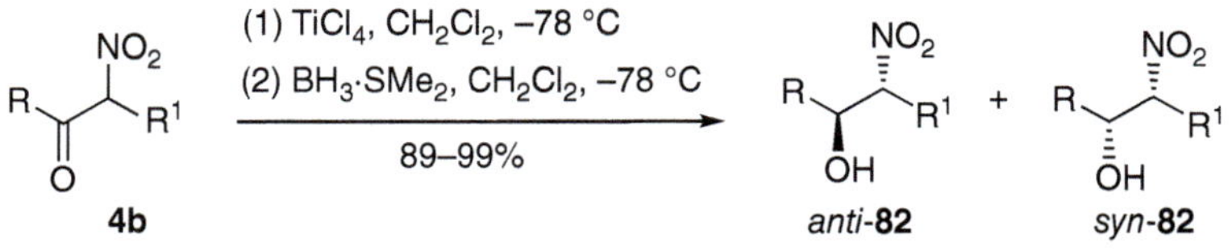

Selected examples:
R = Ph, *o*-$NO_2C_6H_4$, *n*-C_8H_{17}, *i*-Pr
R^1 = Me, *n*-Pr, *i*-Bu

anti/*syn* up to 99/1

Scheme 10.27 Asymmetric reduction of α-nitro ketones to nitroalkanols **82**.

The reaction is performed at −78 °C with borane-dimethyl sulfide ($BH_3·SMe_2$) in the presence of $TiCl_4$. Excellent yields (89–99%) and *anti*-selectivity (up to 99 : 1) were obtained. It is obvious that the increase in bulkiness of R^1 shifts the conformational equilibrium toward the *anti/syn* ratio.

Few years later, Crich et al. [36] developed the asymmetric reduction of α-nitro ketones **4a** to trisubstituted β-nitro alcohols **84** (Scheme 10.28) by borane dimethyl sulfide, under chiral 4,4,5-trisubstituted-(*S*)-oxazaborolidine **83** catalysis.

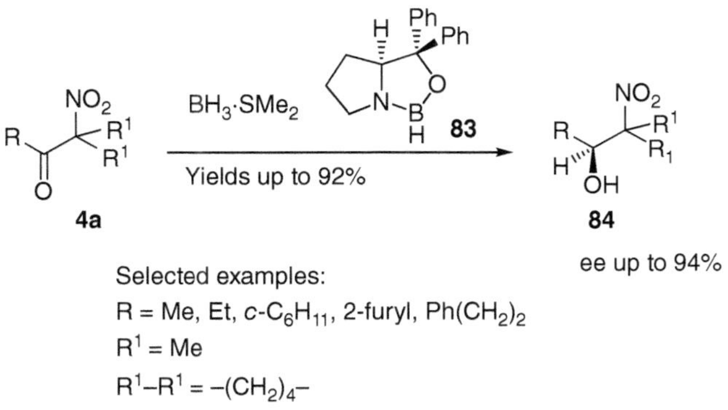

Scheme 10.28 Symmetric reduction of α-nitro ketones **4a** to nitroalkanols **84**.

Good to excellent yields of **84** (up to 92%) and enantioselectivities (up to 94%) are achieved.

Asymmetric reduction of 2-nitro acetophenones **85** (Scheme 10.29) [37] by HCOOH/NEt_3 containing chiral Ru(II) catalyst {RuCl[(*S,S*)-*N*-(*p*-toluenesulfonyl)-1, 2-diphenylethylene diamine]η^6-*p*-cymene} **86** affords nitroalkanols **87** in good yields and *ee*. The latter can be easily converted into aminoalcohols **88**, strategic chiral building blocks for the synthesis of molecules with biological activities.

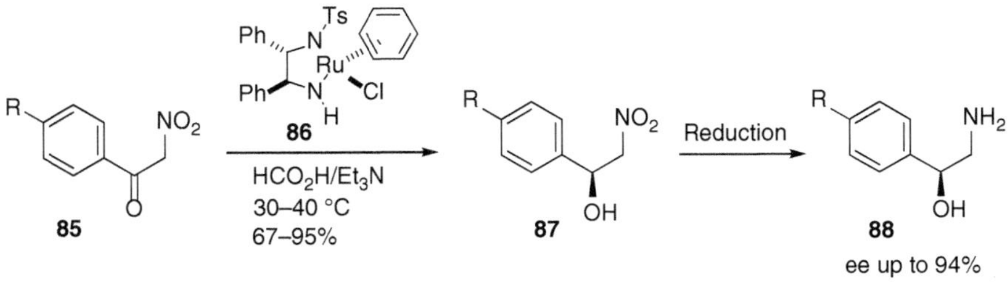

Scheme 10.29 Asymmetric reduction of α-nitro acetophenones **85** into nitroalkanols **87**.

Because α-nitro ketones can be employed as electrophilic species due to the presence of a carbonyl, they can be transformed into tertiary β-nitro alcohols, thanks to nucleophilic addition of organometallic species. This conversion can be achieved by reaction of α-nitro ketones with two equivalents of organomagnesium reagents that are not able to deprotonate the α-acidic proton of α-nitrocycloalkanones that strongly coordinate with the carbonyl and the nitro oxygen. Thus, a second equivalent of reagent is needed to carry out the addition. Grignard reagents are unable to attack monoanion **a** of Figure 10.3, while organolithium compounds are stronger nucleophiles and can attack the deprotonated substrates.

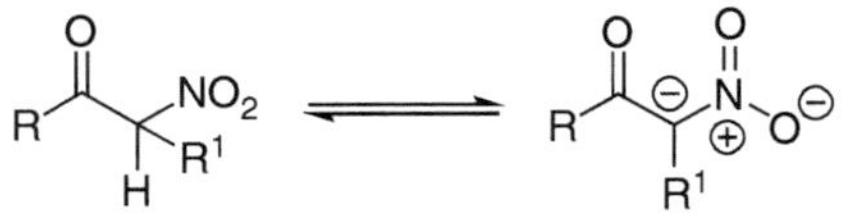

Figure 10.3 α-Acidic proton of α-nitro ketones **4b,c**.

The reactions of Grignard reagents with α-nitrocyclohexanone produce exclusively *trans*-nitroalcohols; whereas organolithium derivatives show little or no selectivity. Contrary, lithium reagents exhibit excellent stereoselectivity with open-chain substrates, allowing the *anti*-diastereomers **89** (Scheme 10.30) [38].

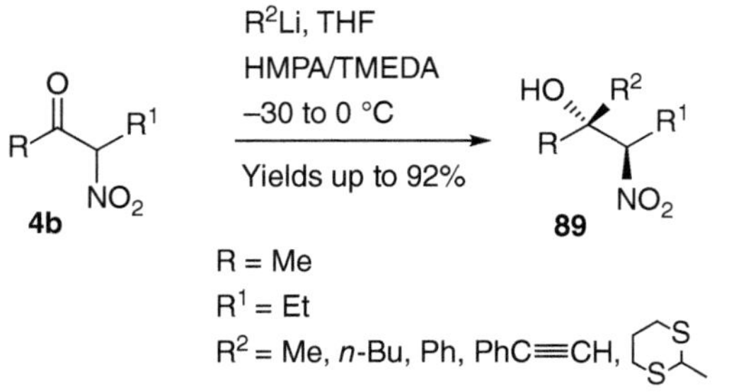

Scheme 10.30 Asymmetric nucleophilic addition of R^2Li to α-nitro ketones **4b** yielding nitroalkanols **89**.

10.2.4 Chemoselective Reduction of α-Nitro Ketones to Amino Ketones

The reduction of α-nitro ketones usually affords the chemoselective conversion of carbonyl into hydroxyl (see Schemes 10.27–10.29) with consequent formation of nitroalkanols.

However, chemoselective reduction of nitro ketones into α-amino ketones can be performed by hydrogenation, using 5% Pt-sulfide-on-carbon (Pt-S-C) as catalyst (Scheme 10.31) [39].

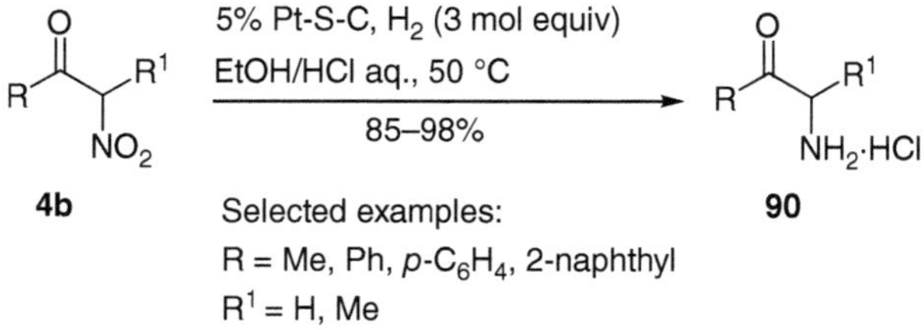

Scheme 10.31 Chemoselective reduction of the nitro group of α-nitro ketones **4b–90**.

Deactivation of catalyst is essential in order to suppress the formation of further reduced derivatives such as nitroalkanols or nitroalkanes (due to the possible reduction of nitroalkanols).

As application, this procedure was employed for the synthesis of terephthaloyl-bis (methylamine) dihydrochloride **94**, a key building block for the preparation of aromatic oligomers (Scheme 10.32).

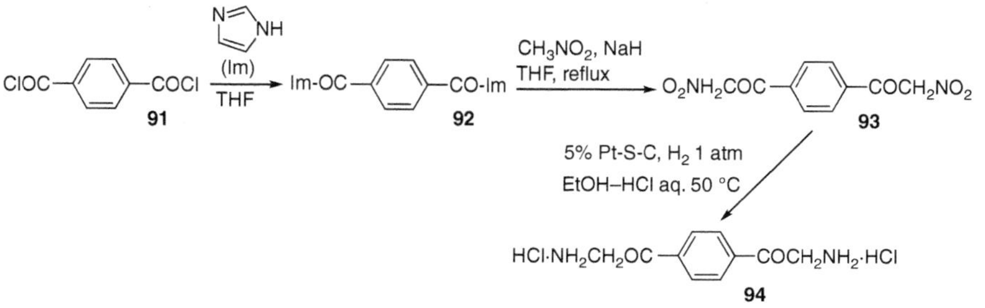

Scheme 10.32 Synthesis of **94**.

Commercially available terephthaloyl chloride **91** is converted into the corresponding diimidazole **92** that, by reaction with CH_3NO_2/NaH, affords dinitrocompound **93**. Successive hydrogenation with 5% Pt-S-C yields the target diamine **94**.

Selective reduction of the carbonyl vs. nitro group can be even performed by H_2/Pd/C and the method was tested in a new synthesis of α-aminolevulinic acid (5-ALA) **98**, important molecule involved in heme biosynthesis (Scheme 10.33) [40].

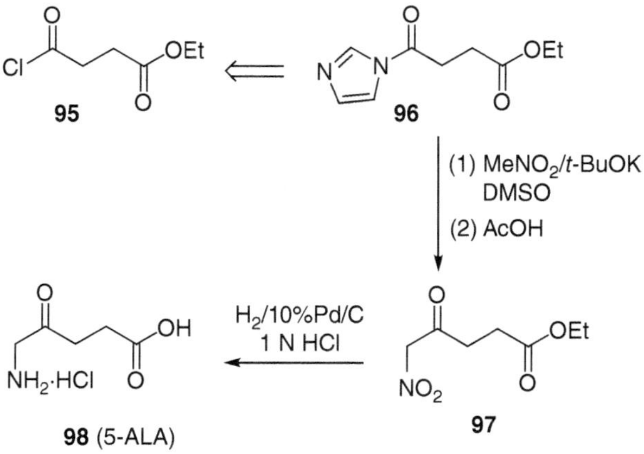

Scheme 10.33 Synthesis of 5-ALA **88**.

Imidazole **96**, obtained from the corresponding acyl chloride **95**, is transformed into nitro ketone **97** with $MeNO_2$/*t*-BuOK. Treatment with H_2/10%Pd/C and acidic (HCl) work-up converts the latter into 5-ALA **98** through simultaneous reduction of the nitro group and ester hydrolysis.

Chemoselective reduction of the nitro group of α-nitro ketones can be also performed by octylviologen (an electron transfer reagent) [41]. This opportunity was

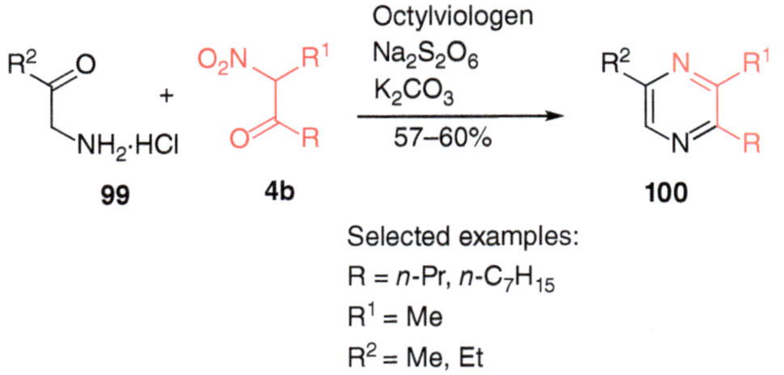

Scheme 10.34 Synthesis of trialkyl-substituted pyrazines **100**.

applied to the one-pot synthesis of alkylpyrazines **100**, a flavor component in food and pheromones in various insect species, as reported in Scheme 10.34 [42].

Reaction of α-amino ketone **99** and α-nitro ketone **4b** in the presence of octylviologen/$Na_2S_2O_6$/K_2CO_3 yields one-pot synthesis of trialkyl pyrazines **100**. However, the role of octylviologen (an electron-transfer agent) is strategic because provides the in situ reduction of the nitro group of **4b** to amino group.

10.2.5 Alkylation of α-Nitro Ketones

α-Nitro ketones possess three different, main points susceptible to the formation of C—C bond with both nucleophiles or electrophiles (Figure 10.4).

Thus, depending on the reaction conditions, the alkylation can be performed on the α- (carbon 1) or α^1-position (carbon 3) or on the carbon of carbonyl (carbon 2).

10.2.5.1 α^1-Alkylation of α-Nitro Ketones

α-Nitro ketones can be alkylated in α^1-position by alkyl halides **101**, through a doubly deprotonated form (Scheme 10.35) [43], yielding R^2-alkylated α-nitro ketones **102**.

Doubly deprotonation is carried out at −70 °C using lithium diisopropylamide (LDA)/HMPA (5 : 1) as solvent, while the alkylation is performed at 0 °C (method A).

In order to enhance the nucleophilicity of the dianion, method B uses an LDA solution of tetramethylethylenediamine (TMEDA), a lithium-chelating compound. This procedure involves the alkylation step at −50 °C with a considerable increase of yields (up to 85% vs. 68%).

An interesting application of this method was devoted to the synthesis of (±)-Muscone **107** (Scheme 10.36) [44], a rare and valuable perfumery ingredient isolated from the male musk deer *Moschus moschiferus*.

The synthesis starts from 2-nitrocyclopentadecanone **103**, easily prepared as reported in Chapter 9, α^1-methylated with CH_3I in LDA/TMEDA/DMPU/THF

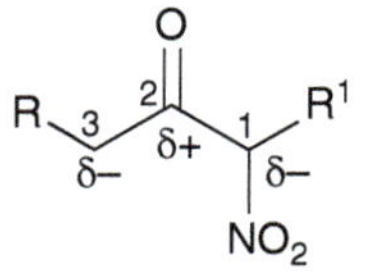

Figure 10.4 α-Nitro ketones: main three centers prone to form C–C bond.

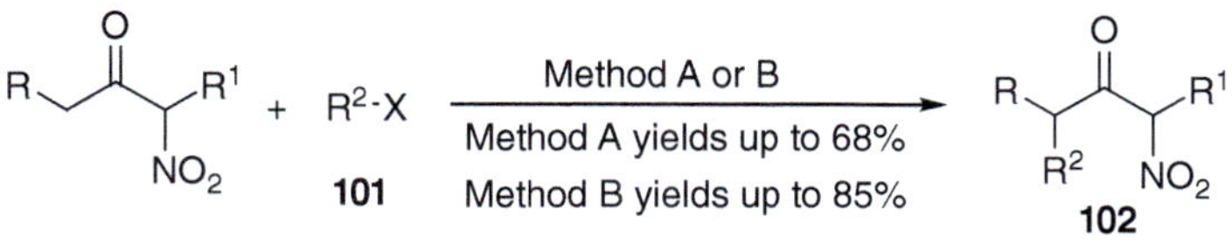

R = H, Me, Et. R^1 = Me, Et

R^2 = Bn, *n*-Bu, *n*-C_7H_{15}, MeCH=CHCH$_2$, *o*-BrC$_6$H$_4$CH$_2$Br

X = Br, I

Method A: (1) LDA, THF/heptane, −35 °C, HMPA, 15 min
(2) α-Nitro ketone (5 mmol), R^2–X (10 mmol), 60 min
(3) AcOH

Method B: (1) LDA, THF, −35 °C, TMEDA, 15 min
(2) HMPA, α-nitro ketone (5 mmol), −70 °C
(3) R^2–X (10 mmol), −50 °C, 3–5 h
(4) AcOH, −50 °C

Scheme 10.35 α^1-Alkylation of α-nitro ketones.

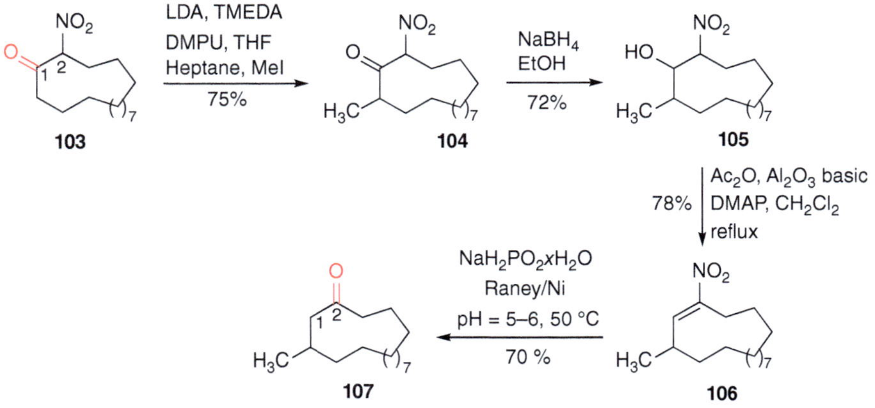

Scheme 10.36 Synthesis of (±)-Muscone **107**.

(method B). The alkylated compound **104** is submitted to reduction with $NaBH_4$ and formation of β-nitroakanol **105**. Dehydration of the latter and successive reduction of the formed nitroalkene **106** (Nef reaction) furnish (±)-Muscone **107** in 30% overall yield from **103**.

An important point of this synthetic strategy is the possibility to shift the carbonyl from 1-position to 2-position from α-nitro ketones, via the Nef reaction.

10.2.5.2 α-Allylation of α-Nitro Ketones

A pioneering α-allylation of α-nitro ketones **4b** was easily promoted by Pd(0) catalysis under stirring a mixture of **4b**, allylic carbonates **108**, and Pd(PPh_3)$_4$ (5 mol%), in THF (Scheme 10.37) [45].

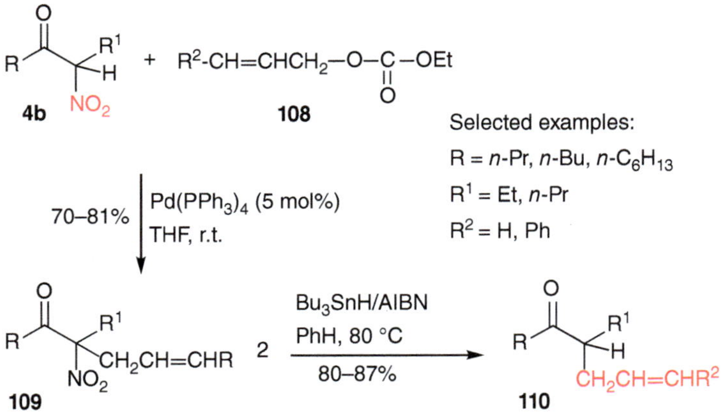

Scheme 10.37 α-Alkylation of α-nitro ketones **4b** into **109** and **110**.

Allylated nitro ketones **109** are achieved in good yields (up to 81%). Moreover, the nitro group of compounds **109** can be replaced with hydrogen (Bu_3SnH) yielding ketones **110** in excellent yields (up to 87%). Thus, this procedure can be considered as formal substitution of the nitro group with an allylic group and, on the other hand, constitutes a new approach for the regioselective α-allylation of ketones.

More recently, Tunge and Grenning [46] developed an asymmetric bisallylation of α-nitro acetone derivatives (as nucleophiles) through a "deacylative" process, taking advantages from previous results (Scheme 10.38) by which nitro ketones **4b** can be mono-allylated in two different ways under Pd(0) catalysis. In fact, reaction of **4b** with allylic alcohol **111a,b** acetate (R^1 = Ac) or carbonate (R^2 = Boc) gives mono-allylated structure **112**, while treatment of **4b** with allylic alcohols allows the allylated nitroalkane **115**. In fact, the hypothesis is that allylic alcohol **111c** could deacylate the intermediate α-nitro ketone **4b** with simultaneous formation of nitronate anion **113** (nucleophile) and allylic acetate **114** (electrophile). Thus, because allylic alcohols are relatively nonreactive toward palladium catalysis, the nitroacetone is expected to undergo rapid, selective Tsuji–Trost allylation [47] with

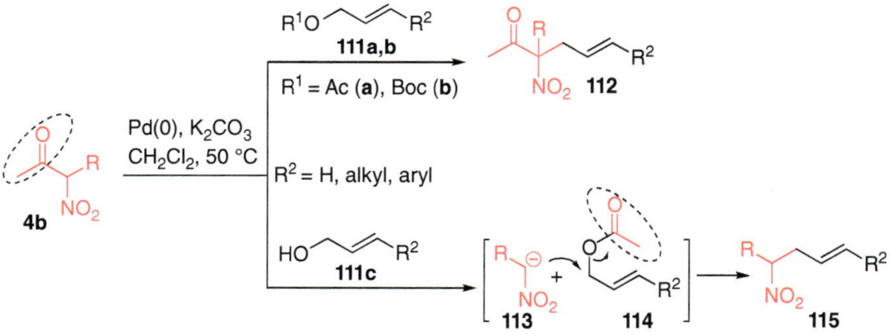

Scheme 10.38 Mono-allylation of nitroacetone derivatives via deacylation of **4b**.

allyl acetate **114**. The resulting nitro ketone would be able to give further allylation only though a deacilative–allylation process.

Based on the previous observations, authors developed a new, three component procedure for deacylative bis-allylation of nitroalkanes, from α-nitro ketones **4b** (Scheme 10.39).

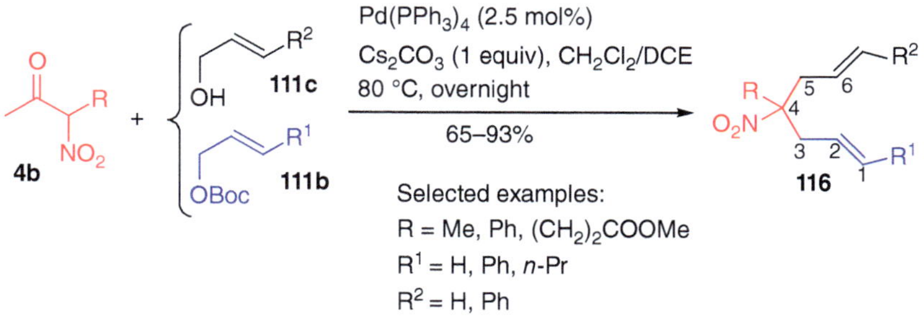

Scheme 10.39 Unsymmetric bis-allylation of nitroalkanes **116** from α-nitro ketones **4b** by a three components reaction.

This three-component reaction can be considered an efficient procedure for the synthesis of unsymmetrically 1,6-dienes **116**.

As expansion of the potential of the method, a triallylation of primary nitroacetone **4c** (Scheme 10.40) was developed. This four-component reaction is carried out with 2 equiv of an allylic carbonate such as **111b** and 1.3 equiv of an allylic alcohol such as **111c**.

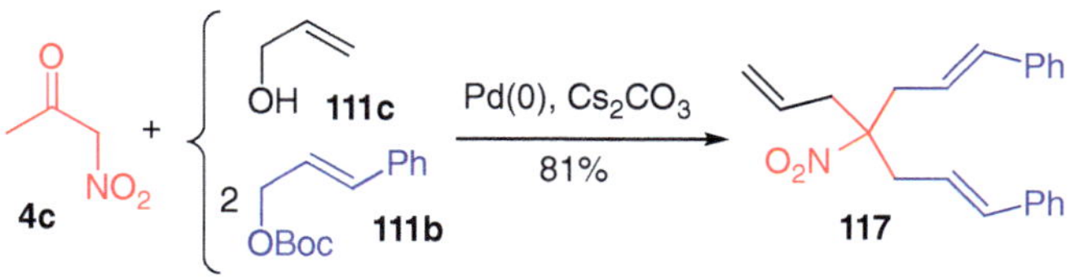

Scheme 10.40 Tris-allylation of nitro ketone **4c**.

During studies devoted to the allylation of 1,4-dienes by C–H activation, Trost et al. [48] found that α-nitro ester **118** can be allylated into compounds **120** from 1,4-dienes **119** under palladium catalysis (Scheme 10.41).

The process needs of 1.1 equiv of E_3N, 1 equiv of 2,6-dimethylbenzoquinone (2,6-DMBQ), 5 mol% of $Pd(OAc)_2$, and 10 mol% of PPh_3. The products **120** are prepared in satisfactory to good yields (up to 80%) with consequent conjugation of the dienic system.

More recently it has been reported that Pd(II)/bis-sulfoxide catalyses allylation reactions employing α-branched nitro ketones [49]. A variety of allyl group can be introduced in moderate to good yields (up to 91%) and stereoselectivity ($E/Z > 20:1$).

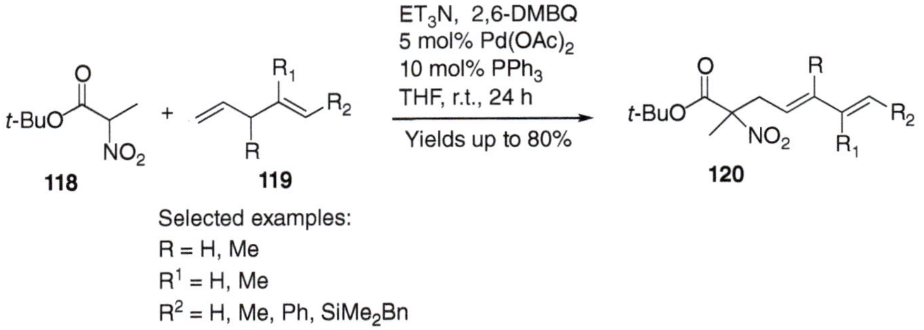

Scheme 10.41 Allylation of 1,4-dienes by C–H activation of α-nitro esters **118**.

10.2.5.3 α-Alkylation of α-Nitro Ketones by Michael Reaction Followed by Nitrous Acid Elimination

The Michael reaction is a fundamental way to perform C—C bond from nitroalkanes and electron-poor alkenes (Chapter 5), while the elimination of nitrous acid is the key step for generation of C=C double bond (Chapter 8). Thus, the combination of the (i) Michael reaction and (ii) nitrous acid elimination offers a good number of synthetic applications for preparation of a variety of important targets.

(a) Synthesis of α,β-Unsaturated Hydrazones

Reaction of α-nitro ketones **4b** with conjugated aza-alkenes **121** produces the one-pot formation of α,β-unsaturated hydrazones **123** in high yields (Scheme 10.42) [50]. Firstly, conjugate Michael addition of **4b–121**, under basic conditions (MeONa), gives the adduct **122** prone to eliminate nitrous acid and in situ formation of **123**.

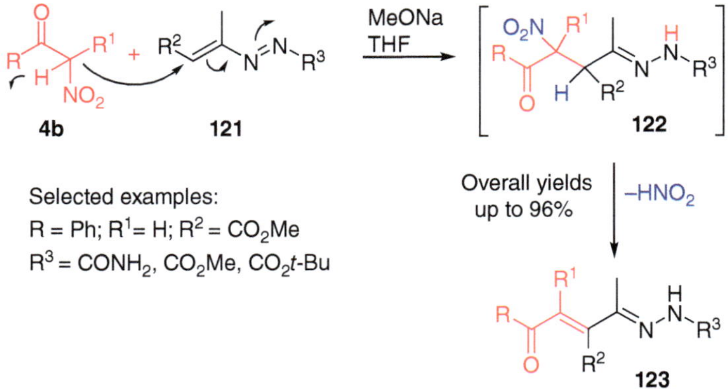

Scheme 10.42 One-pot synthesis of α,β-unsaturated hydrazones **123**.

With this procedure a variety of conjugated hydrazones can be easily achieved by the right choice of the alkyl groups (R, R^1, R^2, and R^3).

(b) Synthesis of Furans

The Michael addition of α-nitro carbonyl compounds **4c** to enones **124**, under basic conditions, followed by in situ denitration represents a convenient protocol for the one-pot preparation of 2,3-dihydrofurans **126** via the Michael adduct **125** (Scheme 10.43) [51].

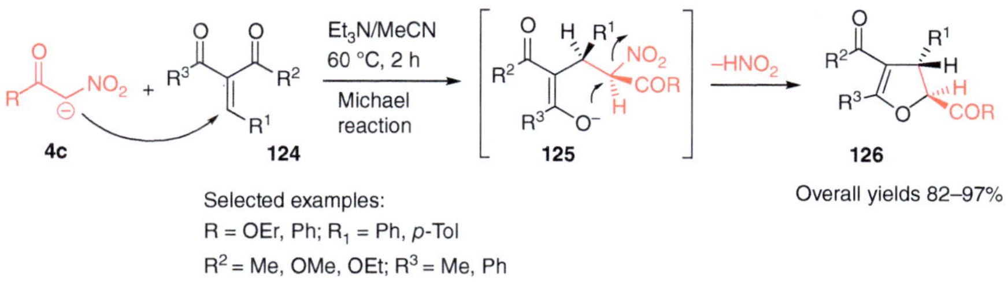

Scheme 10.43 Synthesis of 2,3-dihydrofurans **126**.

The reaction is applicable to a range of enones and α-nitro compounds with a variety of functional groups.

Very recently [52], a new synthetic strategy for dihydrofurans, from α-nitro ketones **4c** and α-cyano-enones **127**, has been reported (Scheme 10.44). The reaction proceeds through conjugate addition of a nitronate anion to enone **127**, and successive cyclization–denitration. Good overall yields of **130** (up to 92%) and diastereoselectivity (up to 20 : 1) were acquired.

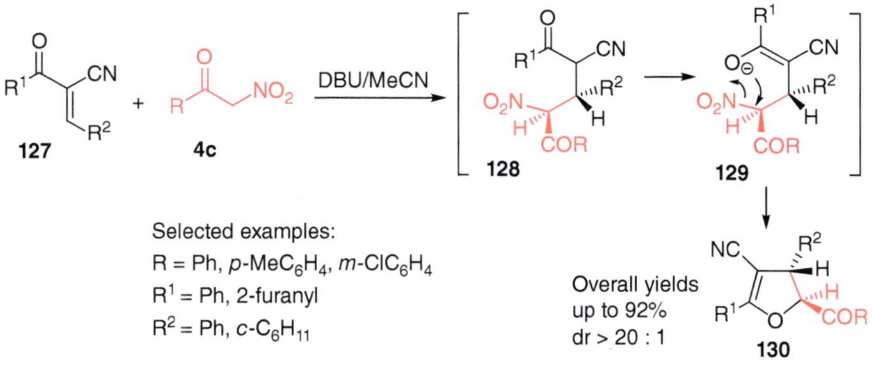

Scheme 10.44 Synthesis of dihydrofurans **130**.

A regioselective synthesis of polysubstituted furans was reported [53] starting from the Michael reaction of α-nitro acetophenone **4c** and Morita–Baylis–Hillman (MBH) acetate **131**, under 1,4-diazabicyclo[2.2.2] octane (DABCO) catalysis in THF (Scheme 10.45).

The method proceeds via a one-pot S_N2' nucleophilic attack and successive intramolecular oxa-Michael reaction followed by HNO_2 elimination. Satisfactory overall yields of a good number of tetrasubstituted furans **134** are achieved.

(c) Synthesis of Trisubstituted Isoxazoles

A highly efficient domino method to prepare tri-substituted isoxazoles **138** has been reported reacting 2 mol of α-nitro carbonyl compounds **4c** with 1 mol of 1,2-dicarbonyls **135** (prepared in situ from methyl ketones from terminal aryl alkenes) [54].

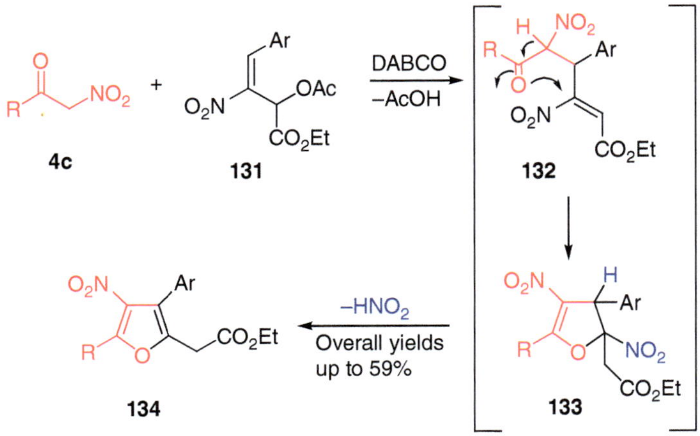

Selected examples:

R = Ph; Ar = Ph, 1-naphthyl, o-$NO_2C_6H_4$, 2-furyl, p-MeC_6H_4

Scheme 10.45 Synthesis of furans **134**.

The first step consists (Scheme 10.46) in a nitroaldol condensation with formation of the electrophilic intermediate **136**, predisposed to the Michael reaction with **4c** allowing the adduct **137**. Nitrous acid and water elimination yield a large variety of isoxazoles **138** in excellent overall yields.

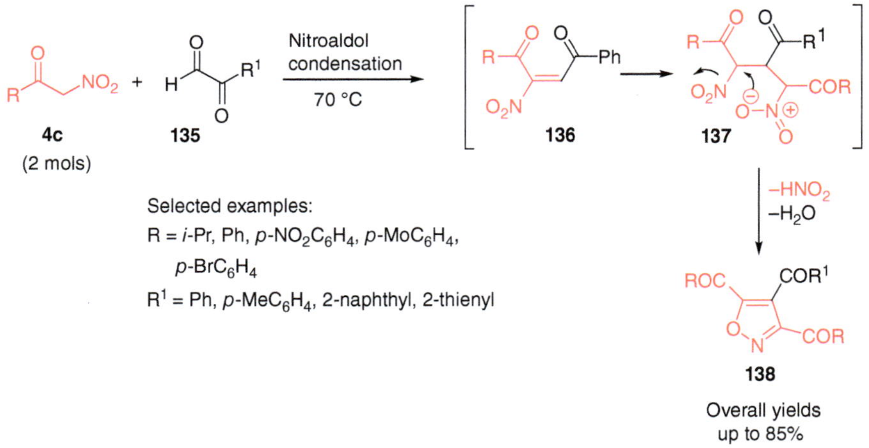

Selected examples:

R = i-Pr, Ph, p-$NO_2C_6H_4$, p-MoC_6H_4, p-BrC_6H_4

R^1 = Ph, p-MeC_6H_4, 2-naphthyl, 2-thienyl

Scheme 10.46 Synthesis of trisubstituted isoxazoles **138**.

(d) Synthesis of α-Pyrones

Oxidative reaction of α-nitro ketones **4c** and enals **139**, in presence of a catalyst and an oxidant (Scheme 10.47) and in presence of Cs_2CO_3 in THF, affords, via intermediates **140–142**, the one-pot synthesis of α-pyrones **143** in moderate to good overall yields (up to 74%) [55]. However, the synthesis proceeds through a Michael addition to acyl azolium ions followed by HNO_2 elimination and cyclization.

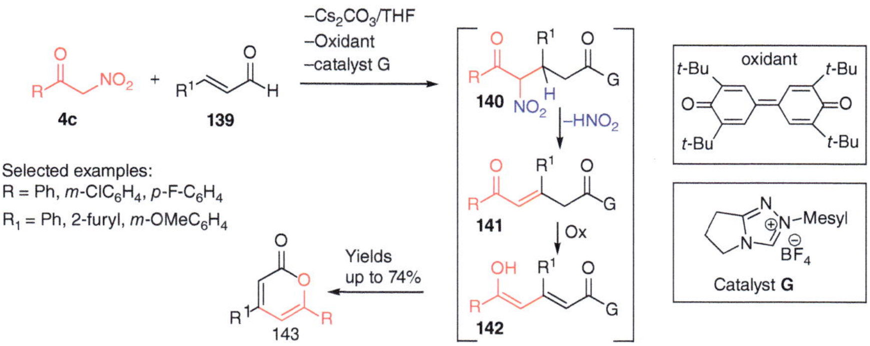

Scheme 10.47 Synthesis of α-pyrones **143**.

10.2.5.4 α-Alkylation of α-Nitro Ketones by the Mannich (or Aza-Henry) Reaction

The first study of the reactivity of trifluoromethyl aldimines **144** with α-nitro ketones **4b,c** has been reported few years ago [56]. Interestingly, organocatalyzed Mannich-type reactions occurred with a complete control of the absolute and relative stereochemistry, giving highly functionalized β-amino-α-nitro-β-(trifluoromethyl) **146** as diastereomerically pure compounds. Inversely, under nitro-Mannich conditions only the facial stereoselectivity can be controlled with the formation of compound **145** (Scheme 10.48).

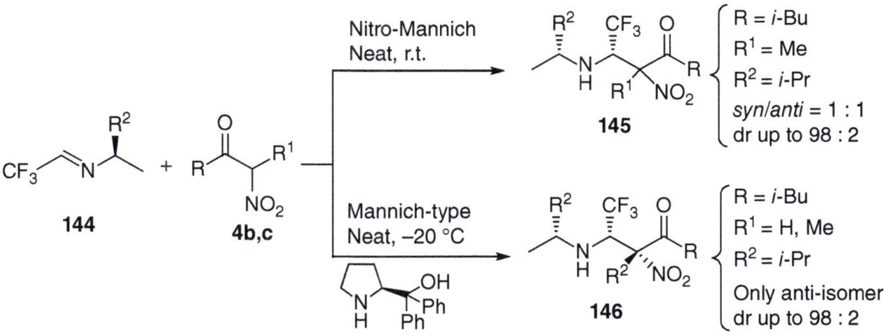

Scheme 10.48 Alkylation of **4b,c** by nitro-Mannich and Mannich-type reactions.

10.3 Other Reactions

The chemical potential of α-nitro ketones is really extensive and a huge number of other reactivities could be reported. However, we'll detail just some of these.

10.3.1 Synthesis of Furoxans

Furoxans **148** can be easily prepared in two steps from α-nitro ketones **4b**. Thus, conversion of **4b** into the corresponding ketoximes **147** can be easily realized under Amberlyst A21 catalysis [57], successive treatment of oximes with acidic alumina in MeCN at 60 °C achieves good yields of furoxans (Scheme 10.49) [58].

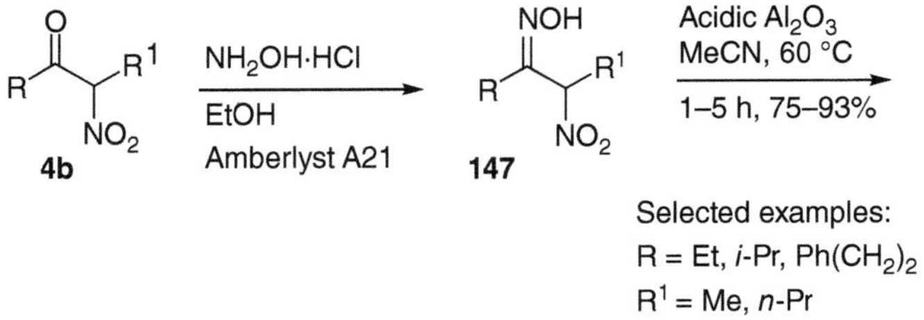

Scheme 10.49 Synthesis of furoxans **148**.

10.3.2 Synthesis of α-Nitro-α-Diazocarbonyl Derivatives

Diazocarbonyls are ideal precursors for several transition and α-nitro derivatives are particularly useful due to the presence of one more important functionality, such as the nitro group.

The conversion **4c** into **149** is performed (Scheme 10.50) [59] at r.t., in a hexane/MeCN solution of trifluoromethanesulfonyl(triflyl)azide, upon addition of pyridine. The yields achieved from α-nitro ketones are generally lower than those realized with α-nitro acetate (R = alkyl-O).

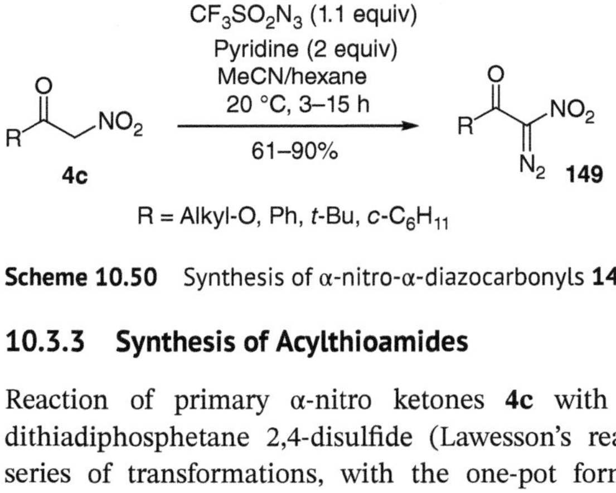

Scheme 10.50 Synthesis of α-nitro-α-diazocarbonyls **149**.

10.3.3 Synthesis of Acylthioamides

Reaction of primary α-nitro ketones **4c** with 2,4-bis(4-methoxyphenyl)-1,3-dithiadiphosphetane 2,4-disulfide (Lawesson's reagent [LR]) proceeds, after a series of transformations, with the one-pot formation of acylthioamides **150** (Scheme 10.51) [60]. However, treatment of the latter with an alcohol converts **150** into acylthioureas **151**.

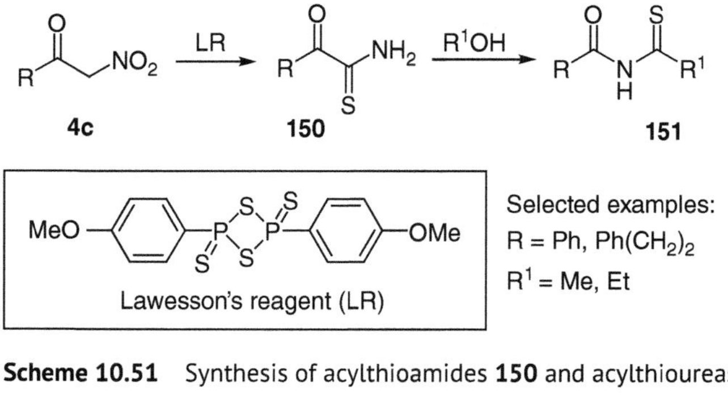

Scheme 10.51 Synthesis of acylthioamides **150** and acylthioureas **151**.

References

1 Ballini, R., Bosica, G., Fiorini, D., and Palmieri, A. (2005). *Tetrahedron* 61: 8971–8993.
2 Gharui, C. and Pan, S.C. (2019). *Org. Biomol. Chem.* 17: 5190–5211.
3 Seler, K. (1966). *Isr. J. Chem.* 4: 7–22.
4 Hurd, C.D. and Nikson, M.E. (1955). *J. Org. Chem.* 20: 927–936.
5 (a) Rosini, G. and Ballini, R. (1983). *Synthesis*: 543–544. (b) Adams, L.L. and Luzzio, F.A. (1989). *J. Org. Chem.* 54: 5387–5390.
6 Rosini, G., Ballini, R., Sorrenti, P., and Petrini, M. (1984). *Synthesis*: 607–608.
7 Ballini, R., Bosica, G., and Parrini, M. (1998). *Tetrahedron Lett.* 39: 7963–7964.
8 Gabrielli, S., Mariotti, N., Chiurchiù, E. et al. (2018). *J. Org. Chem.* 83: 12855–12862.
9 Seebach, D. and Leher, F. (1976). *Angew. Chem. Int. Ed.* 15: 505–506.
10 Baker, D.C. and Putt, S.R. (1978). *Synthesis*: 478–479.
11 Crumbie, R.L., Nimitz, J.S., and Mosher, H.S. (1982). *J. Org. Chem.* 47: 4040–4045.
12 Ono, N., Fujii, M., and Kaji, A. (1987). *Synthesis*: 532–535.
13 Demir, A.S., Tanyeli, C., Aksoy, H. et al. (1995). *Synthesis*: 1071–1073.
14 Shahi, S.P., Gupta, A., Pitre, S.V. et al. (1999). *J. Org. Chem.* 64: 4509–4511.
15 Ashwell, M.A. and Jackson, F.W. (1988). *Synthesis*: 229–231.
16 (a) Rathore, R., Lin, Z., and Kochi, J.K. (1993). *Tetrahedron Lett.* 34: 1858–1862. (b) Rathore, R., Lin, Z., and Kochi, J.K. (1996). *J. Org. Chem.* 61: 627–639.
17 (a) Fontana, A., De Maria, P., Siani, G. et al. (2000). *Eur. J. Org. Chem.*: 1641–1646. (b) Goumont, R., Magnier, E., Kizilian, E., and Terrier, F. (2003). *J. Org. Chem.* 68: 6566–6570.
18 Gasparrini, F., Pierini, M., Villani, C. et al. (2003). *J. Org. Chem.* 68: 3173–3177.
19 (a) Ono, N., Miyake, R., Tamura, R., and Kaji, A. (1981). *Tetrahedron Lett.* 22: 1705–1708. (b) Ono, N., Tamura, R., and Kaji, A. (1983). *J. Am. Chem. Soc.* 105: 4017–4022. (c) Ono, N., Miyake, H., and Kamimura, A. (1985). *Tetrahedron* 41: 4013–4023.
20 Kamimura, A., Kurata, K., and Ono, N. (1989). *Tetrahedron Lett.* 30: 4819–4820.
21 Node, M., Kawabata, T., Ueda, M. et al. (1982). *Tetrahedron Lett.* 23: 4047–4050.
22 Rosini, G. and Ballini, R. (1983). *Synthesis*: 137–138.
23 Rosini, G., Ballini, R., and Petrini, M. (1985). *Synthesis*: 46–48.
24 Rosini, G., Ballini, R., and Petrini, M. (1985). *Synthesis*: 269–271.
25 Ballini, R. (1991). *J. Chem. Soc., Perkin Trans. 1*: 1419–1421.
26 Ballini, R., Petrini, M., and Rosini, G. (1990). *J. Org. Chem.* 55: 5159–5161.
27 Ballini, R., Castagnani, R., and Marcantoni, E. (1992). *J. Chem. Soc., Perkin Trans. 1*: 3161–3162.
28 Ballini, R. and Bosica, G. (1993). *J. Chem. Res. (S)*: 371.
29 Ono, N., Hamamoto, I., Miyake, H., and Kaji, A. (1982). *Chem. Lett.*: 1079–1080.
30 Rosini, G. and Ballini, R. (1983). *Synthesis*: 228–230.
31 Miyake, H. and Yamamura, K. (1986). *Bull. Chem. Soc. Jpn.* 59: 89–91.

32 Ono, N., Miyake, H., Fuji, M., and Kaji, A. (1983). *Tetrahedron Lett.* 24: 3477–3480.

33 Ballini, R. and Giantomassi, G. (1995). *Tetrahedron* 51: 4173–4182.

34 Ballini, R. and Bosica, G. (1994). *J. Org. Chem.* 59: 5466–5467.

35 Ballini, R., Bosica, G., Marcantoni, E. et al. (2000). *J. Org. Chem.* 65: 5854–5857.

36 Crich, D., Ranganathan, K., Rumthao, S., and Shirai, M. (2003). *J. Org. Chem.* 68: 2034–2037.

37 Watanabe, M., Murata, K., and Ikariya, T. (2002). *J. Org. Chem.* 67: 1712–1715.

38 Ballini, R., Bartoli, G., Gariboldi, P.V. et al. (1993). *J. Org. Chem.* 58: 3368–3372.

39 Tamura, R., Oda, D., and Kurokawa, H. (1986). *Tetrahedron Lett.* 27: 5759–5762.

40 Nudelman, A. and Nudelman, A. (1999). *Synthesis*: 568–570.

41 Yu, C., Liu, B., and Hu, L. (2001). *J. Org. Chem.* 66: 919–924.

42 Elmaaty, T.A. and Castle, L.W. (2005). *Org. Lett.* 7: 5529–5530.

43 Ballini, R., Bartoli, G., Castagnani, R. et al. (1992). *Synlett*: 64–66.

44 Ballini, R., Marcantoni, E., and Petrini, M. (1995). *Liebigs Ann.*: 1381–1383.

45 Ono, N., Hamamoto, I., and Kaji, A. (1986). *J. Org. Chem.* 51: 2832–2833.

46 (a) Grenning, A.J. and Tunge, J.A. (2011). *Angew. Chem. Int. Ed.* 50: 1688–1691. (b) Grenning, A.J. and Tunge, J.A. (2011). *J. Am. Chem. Soc.* 133: 14785–14794.

47 Trost, B.M., Toste, F.D., and Pinkerton, A.B. (2001). *Chem. Rev.* 101: 2067–2096.

48 Trost, B.M., Hasmann, M.M., and Thaisrivongs, D.A. (2012). *Angew. Chem. Int. Ed.* 51: 4950–4953.

49 Howell, J.M., Liu, W., Young, A.J., and White, M.C. (2014). *J. Am. Chem. Soc.* 136: 5750–5754.

50 Attanasi, O., Ballini, R., Liao, Z. et al. (1993). *Tetrahedron* 49: 7027–7036.

51 Chuang, C.-P., Chen, K.-P., Hsu, Y.-L. et al. (2008). *Tetrahedron* 64: 7511–7516.

52 Sahoo, S.C., Maity, R., and Pan, S.C. (2019). *ACS Omega* 4: 2792–2803.

53 Mane, V., Kumar, T., Pradhan, S. et al. (2015). *RSC Adv.* 5: 69990–69999.

54 Yang, Y., Gao, M., Deng, C. et al. (2012). *Tetrahedron* 68: 6257–6262.

55 Bera, S. and Studer, A. (2017). *Synthesis* 49: 121–126.

56 Pelagalli, A., Pellacani, L., and Fioravanti, S. (2017). *Eur. J. Org. Chem.*: 3373–3380.

57 Ballini, R., Barboni, L., and Filippone, P. (1997). *Chem. Lett.*: 475–476.

58 Curini, M., Epifano, F., Marcotulli, M.C. et al. (2000). *Tetrahedron Lett.* 41: 8817–8820.

59 Charette, A.B., Wurz, R.P., and Ollevier, T. (2000). *J. Org. Chem.* 65: 9252–9254.

60 (a) Harris, P.A., Jackson, A., and Joule, J.A. (1989). *Sulfur Lett.* 10: 117–122. (b) Harris, P.A., Jackson, A., and Joule, J.A. (1989). *Tetrahedron Lett.* 30: 3189–3192.

11

Nitro Cyclopropanes: Synthesis and Applications

Nitrocyclopropanes are a fascinating class of small nitro-functionalized cycloalkanes that possess conformational rigidity associated with the great reactivity due the presence of a nitro group. The unique structure of the cyclopropane subunit is common to many compounds of biological and pharmacological relevance. As such, there have been significant efforts toward the synthesis of nitro-functionalized cyclopropanes as well as their application in synthesis. In addition, the nitrocyclopropane moiety is the key structure of important natural products such as Hormaomycin **A** (Figure 11.1), produced by the strain W384 *Streptomyces griseoflavus*, that has a direct impact on the secondary metabolite proliferation of certain microorganism [1].

It is clear that nitro cyclopropanes could be the strategic starting chemical structures for many other targets, as depicted in Figure 11.2.

In addition, these strained-ring nitro compounds are of interest as energetic materials playing an important role in aeronautics, weapons industry, and other high-tech fields [1–4].

Thus, in this chapter will be reported both synthetic methods to prepare nitro cyclopropanes and their synthetic applications.

11.1 Synthesis of Nitro Cyclopropanes

Nitro cyclopropanes can be prepared by cyclopropanation of a number of molecules such as bromoderivatives, alkenes, nitroalkenes, and γ-nitroalcohols. Here listed are the main procedures and some important applications of the obtained nitro cyclopropanes that are reported.

11.1.1 Synthesis of Nitro Cyclopropanes from Bromine Derivatives

The presence of bromine is often strategic for nitrocyclopropanation of the appropriate chemical structures, due to its behavior as good leaving atom.

(i) A pioneering paper [5] reported (Scheme 11.1) the synthesis of nitro cyclopropanes starting from γ-nitro ketones **1**, easily obtained by the Michael reaction of primary nitroalkanes with enones.

Nitroalkanes: Synthesis, Reactivity, and Applications, First Edition.
Roberto Ballini and Alessandro Palmieri.

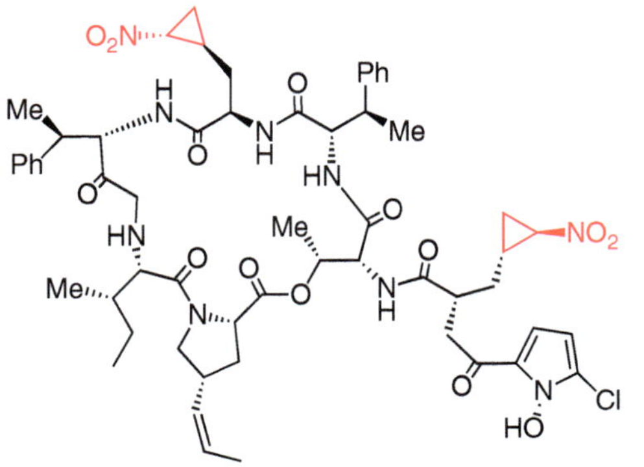

Figure 11.1 Hormaomycin (**A**).

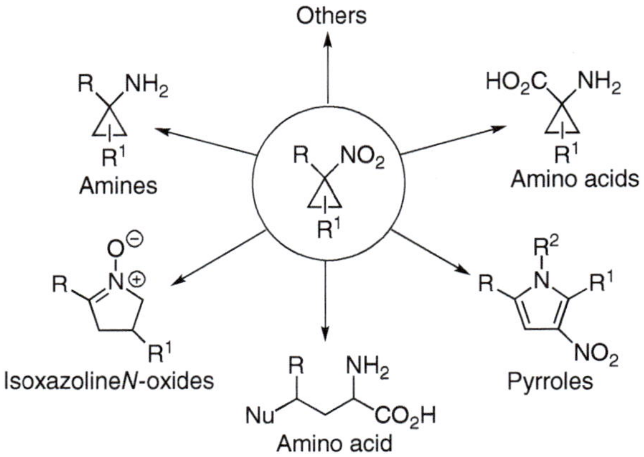

Figure 11.2 Synthetic applications of nitrocyclopropanes.

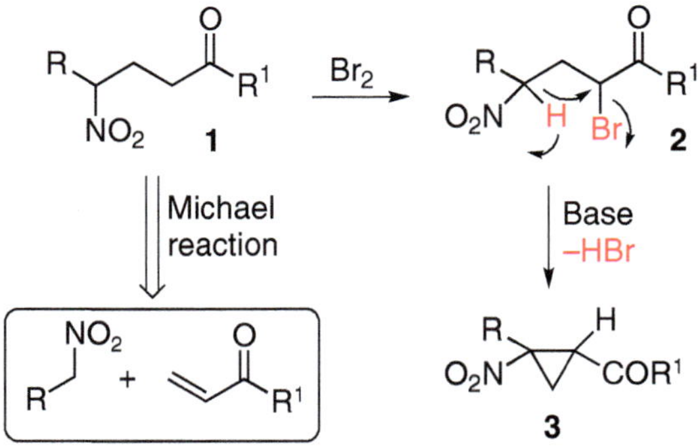

Scheme 11.1 Nitrocyclopropanation of γ-nitro ketones **1**.

Thus, bromination of **1** affords **2** that, under basic conditions, provides cyclopropanation via elimination of HBr, with formation of **3**.

(ii) An interesting procedure was reported for cyclopropanation of steroids (Scheme 11.2) [6], by the Michael reaction of nitromethane with 2-bromo-3-oxo-1,4,6-triene steroids **4**, under basic conditions (*t*-BuOK/*t*-BuOH), achieving the one-pot formation of 1α,2α-(nitromethylene)-3-oxo-4,6-dienes **6** in moderate yields (25–27%), via HBr-elimination from the adduct **5**.

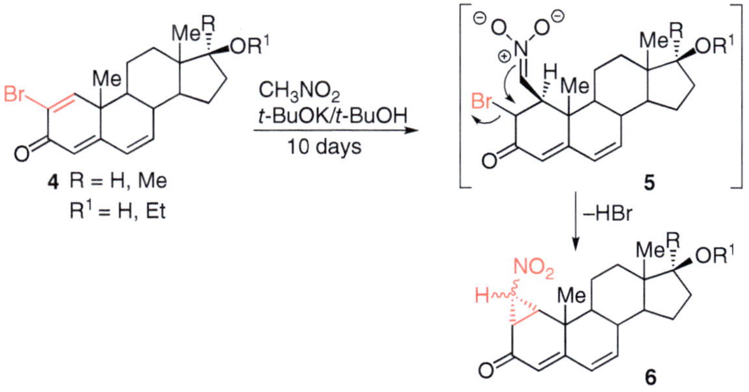

Scheme 11.2 Nitrocyclopropanation of steroids **4**.

(iii) Arai et al. [7] reported an asymmetric cyclopropanation reaction, using quaternary ammonium salt **8** as catalyst for phase-transfer reaction (PTR), via Michael addition of $MeNO_2$ to α-bromocyclopentanone **7** and in situ HBr elimination from **9** and consequent one-pot formation of **10** (Scheme 11.3) in 50% yield and 62% ee.

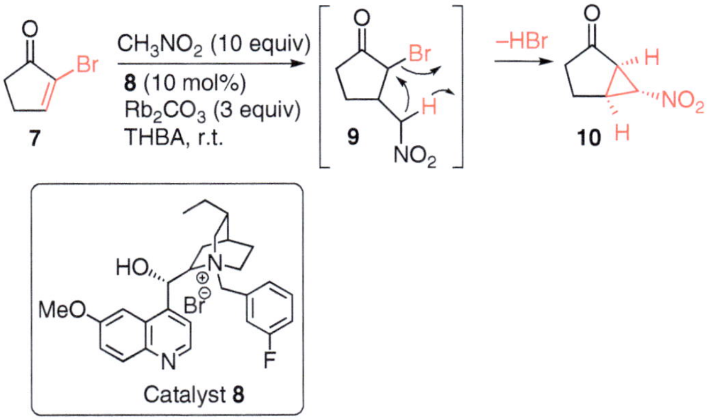

Scheme 11.3 Nitrocyclopropanation of **7** into **10**.

The authors tested several catalysts and found **8** the more efficient in terms of yield (50%) and ee (62%). Moreover, in order to improve the performance of the reaction, they suggested to add tetrahexylammonium bromine (THBA) as an achiral co-PTR and the use of Rb_2CO_3 as base instead of the classic K_2CO_3.

(iv) *t*-Butyl 2,3-diobromopropanoate **11a** has been recognized as possible key building block for one-pot synthesis of functionalized nitrocyclopropanes

(Scheme 11.4) prone to be converted into β-aminoacids or other important targets with biological activity [8].

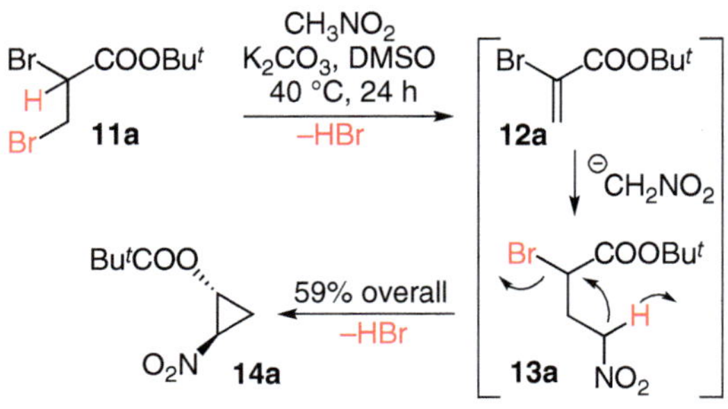

Scheme 11.4 Nitrocyclopropanation of **11a** into **14a**.

The same authors reported the application of compound **14a** as key precursor for the synthesis of 3-(*trans*-2-aminocyclopropyl)alanine ((AcP)Ala) **18** (Scheme 11.5), the key constituent of belactosin A **19** that possesses antitumor activity [9].

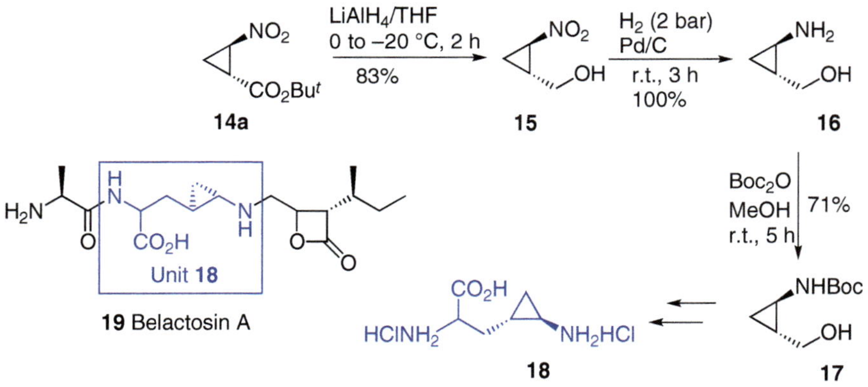

Scheme 11.5 Synthesis of unit **18**, key constituent of Belactosin A **19**.

$LiAlH_4$ reduction of **14a** provides primary alcohol **15** that, submitted to Pd/C hydrogenation, converts into aminoalcohol **16**. Boc-protection of the latter allows **17** ideal precursor of unit **18**, a key structure of Belactosin **19**.

A multigram (20 g) synthesis of *trans*-2-(trifluoromethyl)cyclopropanamine **22** has been reported as one more application, a very interesting molecule from many points of view that (i) possesses the cyclopropyl ring that often has beneficial effects upon biological activity and metabolic stability. (ii) Cyclopropanamine is a small building block in pharmaceuticals and also frequently found in drugs. (iii) Trifluoromethyl group is a privileged substructure within medicinal chemistry. For these reasons, a multigram synthesis of **22** was reported form dibromo structure **11b** (Scheme 11.6) [10].

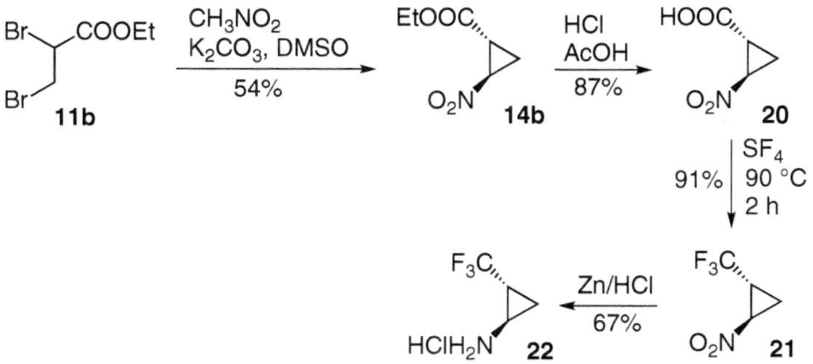

Scheme 11.6 Synthesis of 22.

The synthesis starts with the reaction of nitromethane with dibromines **11b** under basic conditions (K_2CO_3 in DMSO) and the one-pot formation of cyclopropane **14b** in 54% yield. Successive hydrolysis of ester with HCl furnishes the nitrocarboxylic acid **20** in 87% yield. However, the crucial step of the process is the transformation of the carboxy group into trifluoromethyl group, successfully performed with sulfur tetrafluoride in presence of catalytic amount of water in order to produce HF, at 90 °C for two hours. Finally, reduction of **21** (Zn/HCl) completed the synthesis to the target molecule **22**-HCl in 67% yield.

(v) 1,2-Dibromoethylphosphonate **23** was found strategic precursors of a new class of nitrocyclopropane **24** by its reaction with nitromethane under basic conditions. Moreover, compound **24** was found useful precursors of other important molecular structures such as amine **25**, β-nitrocyclopropylphosphonic acid **26** and β-nitrocyclopropylphosphonic acid **27** (Scheme 11.7) [11a].

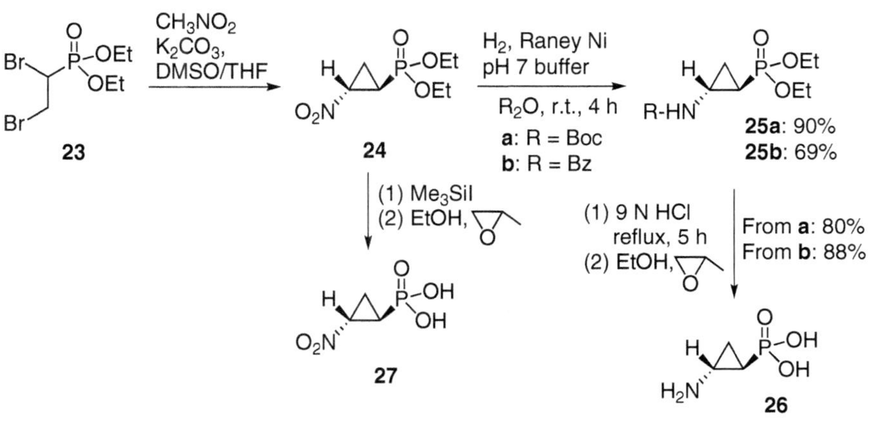

Scheme 11.7 Cyclopropanation of 23.

The cyclopropanation of **23** can be even performed with primary nitro alkanes giving access to unusual trisubstituted cyclopropanes **28** (Scheme 11.8) [11a, b].

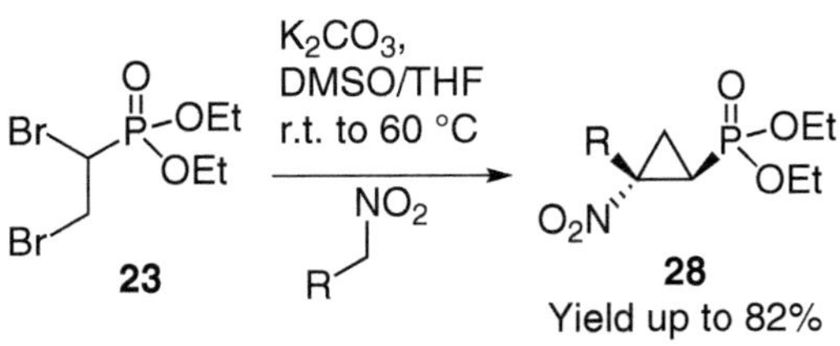

R = Me, Et, n-Pr, Ph, CH_2OBn, CH_2OTBS, CO_2Et, $(CH_2)_2CO_2Et$, SO_2Ph

Scheme 11.8 Cyclopropanation of **23** into **28**.

(vi) Bromonitromethane ($BrCH_2NO_2$) **29**, used as nucleophile in the Michael reaction with the appropriate electrophiles, furnishes one-pot cyclopropanation and formation of a variety of nitrocyclopropanes (Scheme 11.9). Preliminary results [12] reported the cyclopropanation of *N*-benzylmaleimide **30a** by reaction with **29**, under dimethyl-1,3,4,5-tetrahydropyrimidine (DMTHP) catalysis, achieving **31a** in 36% yield.

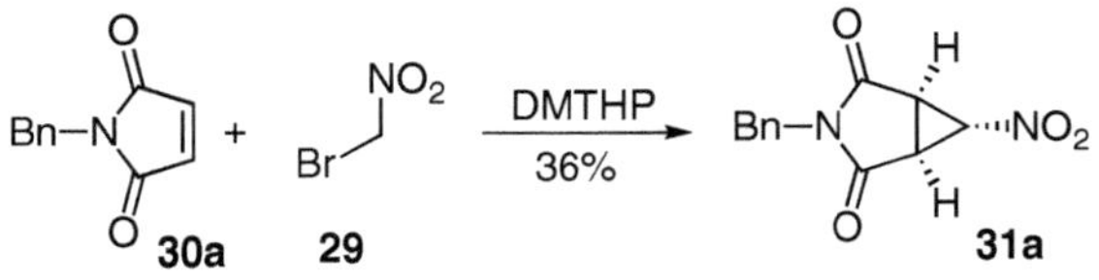

Scheme 11.9 Cyclopropanation of **30a** into **31a**.

This method was then applied to the synthesis of Trovafloxacin **32**, a novel broad spectrum antibacterial, via nitrocyclopropane **31a** (Scheme 11.10) [13].

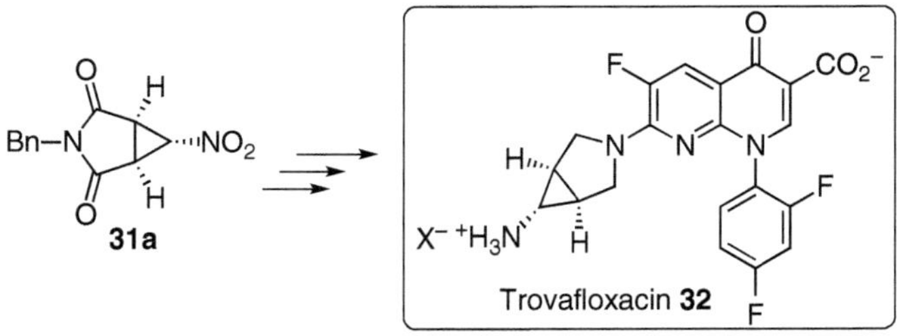

Scheme 11.10 Synthesis of Trovafloxacin **32**.

The aforementioned results suggested a general procedure for the synthesis of nitrocyclopropanes by reaction, under basic conditions, of **29** with alkenes **33** possessing two electron-withdrawing groups in α- and β-positions, respectively (Scheme 11.11) [14].

The success of the method depends from the correct addition of **29** to the basic solution of **33**. In fact, it is important to add bromonitromethane in several fractions

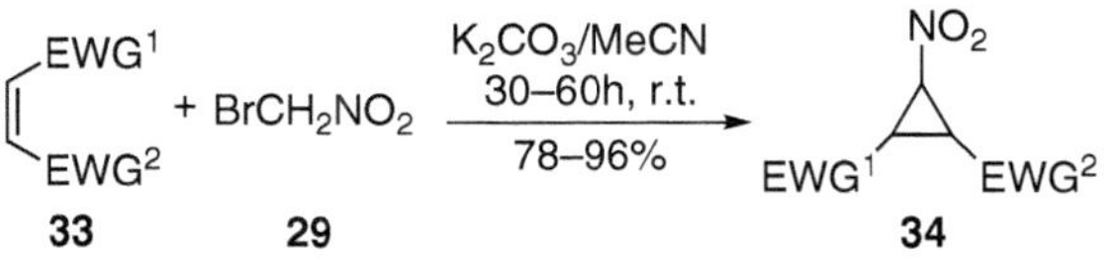

Scheme 11.11 Cyclopropanation of **33**.

Table 11.1 Synthesis of cyclopropanes **34**.

Entry	EWG^1	EWG^2	Yields of 34 (%)	dr ratio of 34
1	MeCO	MeCO	94	23 : 77
2	PhCO	MeCO	96	88 : 12
3	*p*-MeC_6H_3CO	MeCO	96	84 : 16
4	PhCO	PhCO	75	15 : 85
5	MeO_2C	MeCO	86	69 : 31
6	MeO_2C	MeO_2C	88	40 : 60
7	NC	NC	78	70 : 30

in order to avoid its decomposition observed under basic conditions. Overall yields of cyclopropanation into **34** are excellent (75–96%), while diastereoselectivity is modest (Table 11.1).

Further investigation was devoted to cyclopropanation of *N*-alkyl maleimides **30** with **29**, under the above reaction conditions (Scheme 11.12).

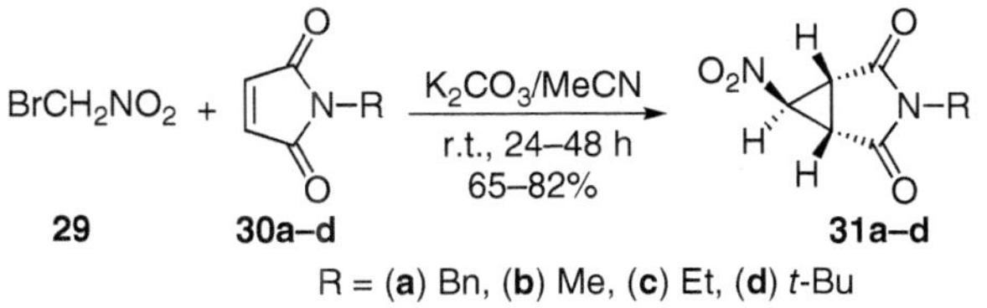

Scheme 11.12 Cyclopropanation of **30a–d** under K_2CO_3 catalysis.

Good yields are obtained and it is important to point out that compound **31a** (precursor of Trovafloxacin **32**) is prepared in 70% yield vs 36% of the original method (see Scheme 11.9).

A new regioisomeric synthesis of trifluoromethylated nitrocyclopropanes **37a,b** has been developed [15] by the Michael addition of bromonitromethane **29** with trifluoromethylcrotonate **35** (Scheme 11.13).

The authors reported even the use of **37a** as precursor of a number of functionalized CF_3-cyclopropanes, including amino acid. However, the immediate conversion of **37** into amino acid is difficult to obtain by direct reduction of nitro ester **37a** since the possible push–pull effect (Figure 11.3) due to the presence of an electron-withdrawing group such as ester.

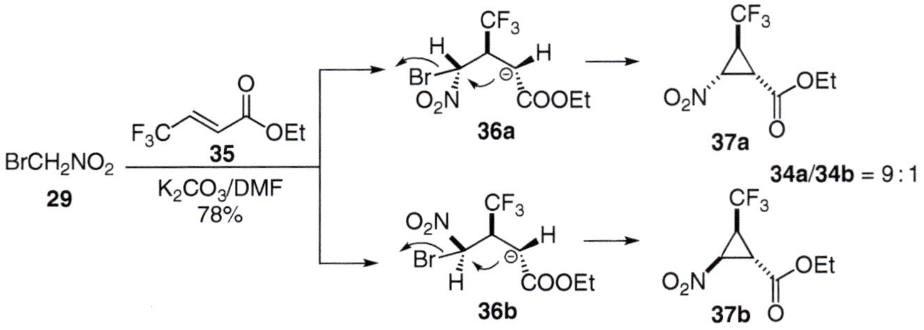

Scheme 11.13 Cyclopropanation of **35** into **37** under K_2CO_3 catalysis.

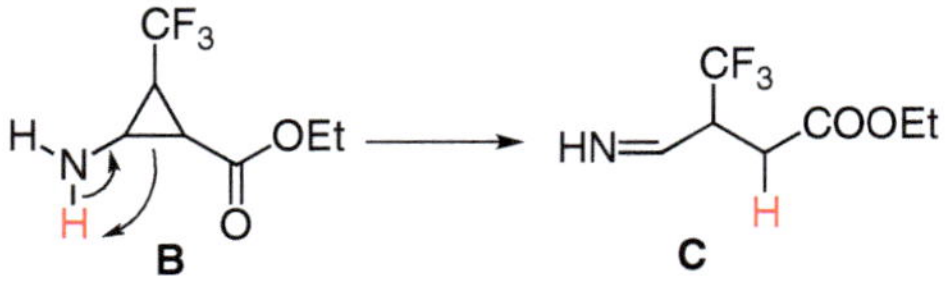

Figure 11.3 Push–pull effect.

Thus, in order to avoid this drawback, an alternative way to obtain amino acids **41** from **37a** has been developed [15], as reported in Scheme 11.14.

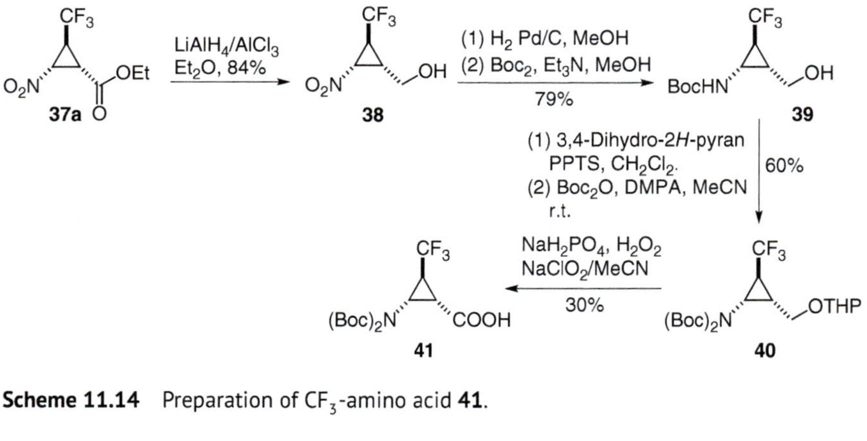

Scheme 11.14 Preparation of CF_3-amino acid **41**.

Firstly, the ester **37a** has been reduced to primary alcohol **38**. The nitro group is then reduced into amine and immediately protected into **39**. Successive protection of hydroxyl, and double Boc-protection of the amino group, affords **40** that, after oxidation (H_2O_2/$NaClO_2$), allows the target amino acid **41**.

Moreover, nitroalcohol **38** has demonstrated to be a very attractive building block also for the synthesis of CF_3-peptide **45**, in few steps (Scheme 11.15).

In fact, substitution of the hydroxyl with iodine **42** is strategic for the formation of azide **43**. Selective reduction of the latter by the Staudinger method and successive peptide coupling with Boc-valine-OH affords **44**. Finally, selective reduction of the nitro group into amine achieves **45** as 1 : 1 diastereomeric ratio.

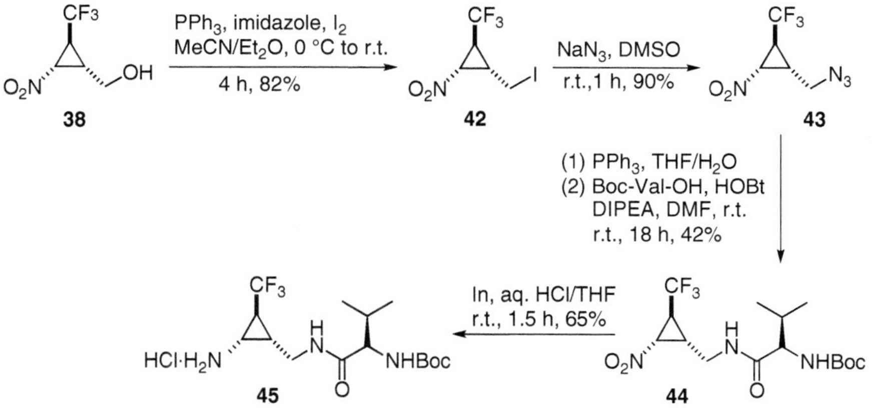

Scheme 11.15 Synthesis of CF_3-peptide **45**.

Stereoselective synthesis of polyfunctionalized nitrocyclopropanes has been developed [16] achieving an organocatalytic Michael addition of bromonitromethane **29** to β,γ-unsaturated α-keto esters **46** (Scheme 11.16).

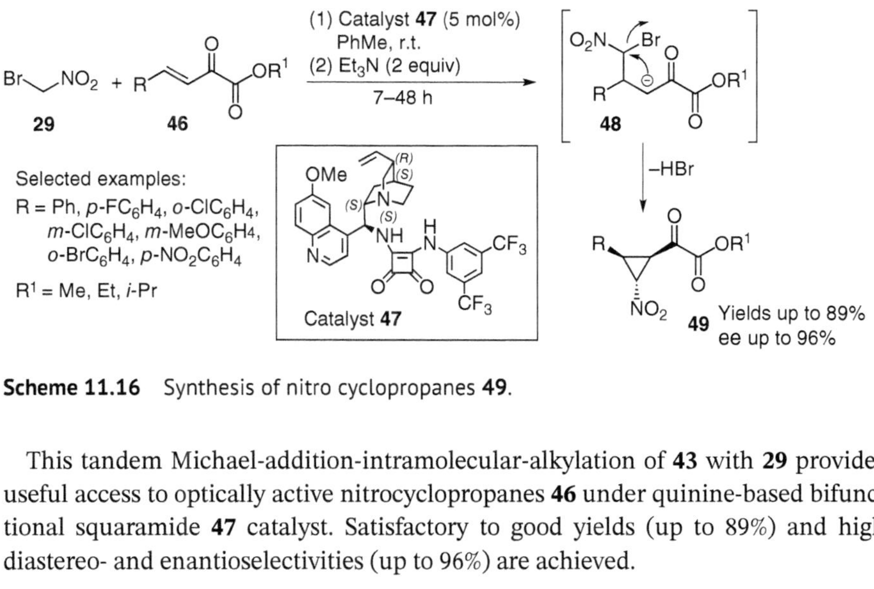

Scheme 11.16 Synthesis of nitro cyclopropanes **49**.

This tandem Michael-addition-intramolecular-alkylation of **43** with **29** provides useful access to optically active nitrocyclopropanes **46** under quinine-based bifunctional squaramide **47** catalyst. Satisfactory to good yields (up to 89%) and high diastereo- and enantioselectivities (up to 96%) are achieved.

11.1.2 Synthesis of Nitro Cyclopropanes from Conjugate Nitroalkenes

(i[1]) Nitroalkenes were proposed, two decades ago, as precursors of nitrocyclopropanes by 1,3-dipolar addition of diazomethane with nitroalkenes [17].

(ii[1]) Another research group discovered that optically active nitroalkenes allow cyclopropanation with sulfur ylide or dibromocarbene (Scheme 11.17) [18]. The obtained **52** have been then employed as strategic building blocks for the preparation of optical active cyclopropyl β-amino acids.

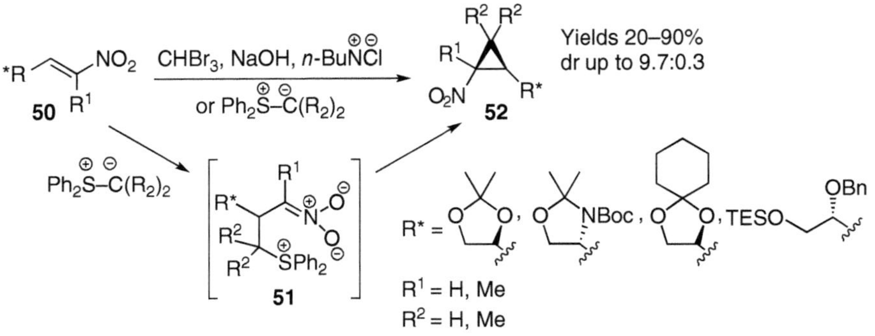

Scheme 11.17 Cyclopropanation of nitroalkanes **50** into **52**.

(iii[1]) A one-pot stereoselective synthesis of polyfunctionalized nitrocyclopropanes **56** was reported under the Michael addition of dimethylchloromalonate **54** to a large number of aromatic and aliphatic nitroolefins **53**, catalyzed with 1,8-diazabicyclo[5.4.0]undec-7-ene (DBU) and hexamethylphosphoramide (HMPA) as solvent, with one-pot cyclization of the adduct **55** into **56** in presence of DBU (Scheme 11.18) [19]. A single diastereomer (*anti* stereochemistry) is obtained in good yields.

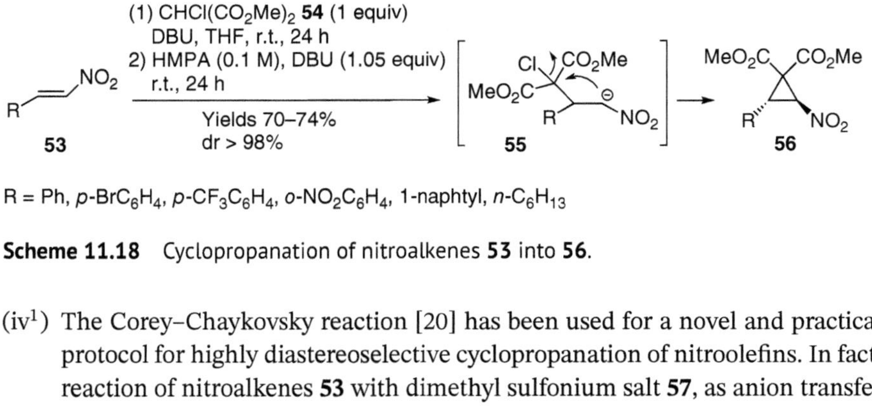

R = Ph, p-BrC_6H_4, p-$CF_3C_6H_4$, o-$NO_2C_6H_4$, 1-naphtyl, n-C_6H_{13}

Scheme 11.18 Cyclopropanation of nitroalkenes **53** into **56**.

(iv[1]) The Corey–Chaykovsky reaction [20] has been used for a novel and practical protocol for highly diastereoselective cyclopropanation of nitroolefins. In fact, reaction of nitroalkenes **53** with dimethyl sulfonium salt **57**, as anion transfer reagent, gives a large variety of highly functionalized nitrocyclopropanes **58** in good yields (up to 93%) and as single diastereomer (Scheme 11.19) [21].

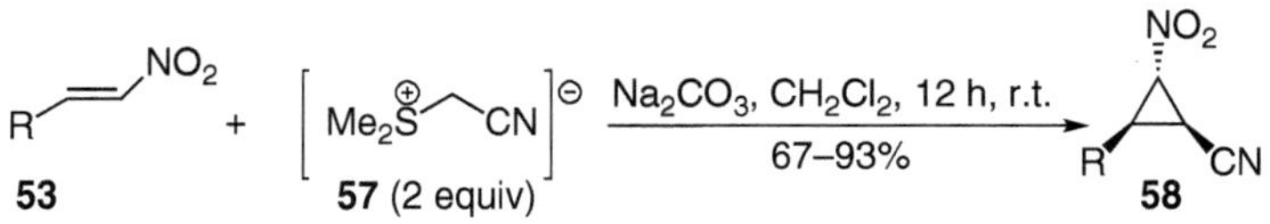

Selected examples:
R = c-C_6H_{11}, Ph, p-FC_6H_4, p-ClC_6H_4, p-$CH_3C_6H_4$, p-$MeOC_6H_4$, p-MeC_6H_4
m-$MeOC_6H_4$, o-$MeOC_6H_4$, o-$NO_2C_6H_4$, 3,4-$Cl_2C_6H_3$, 2-naphthyl,
2-furyl, 2-thiophenyl

Scheme 11.19 Cyclopropanation of nitroalkenes **53** into **58**.

Later, the same authors reported an improved Corey–Chaykovsky procedure [22] for a new efficient way by (**59**) sulfonyl ylide-mediated trifluoromethylation reaction of nitroolefins **53** to afford a large family of *cis*-configured trifluoromethyl nitrocyclopropanes **60** (Scheme 11.20). The generality of the procedure is demonstrated by "twenty-four" examples reported and obtained in good to excellent yields and good dr (up to 10 : 1). Moreover, these *cis*-cyclopropanes are important building blocks that possess unique and highly desirable configuration properties for the synthesis of pharmaceutical and agrochemical molecules.

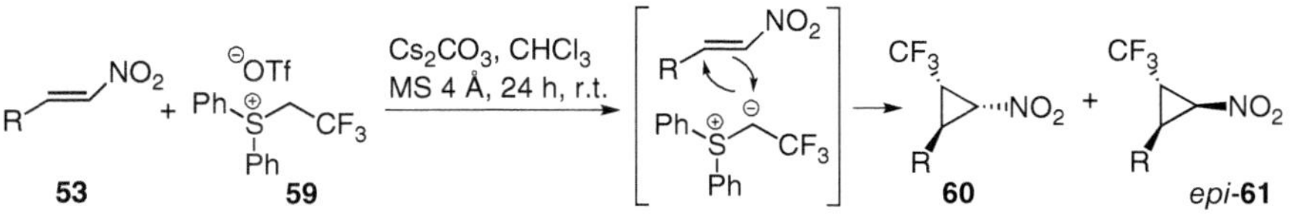

Selected examples of:
R = *c*-C_6H_{11}, Ph, 2-thiophenyl. *p*-PhC_6H_4, *o*-FC_6H_4, *p*-ClC_6H_4, *p*-$CF_3C_6H_4$, *p*-$MeOC_6H_4$, *o*-$NO_2C_6H_4$, 3,4-$Cl_2C_6H_3$, 3,5-$(OMe)_2C_6H_3$, 2-naphtyl, o-FC_6H_4, *m*-BrC_6H_4

60: Yields up to 94%
dr up to 10 : 1

Scheme 11.20 Cyclopropanation of nitroalkenes **53** into **60**.

Using a modified Corey–Chaykovsky procedure, the synthesis of a new histone demethylase LSD1 inhibitor, CBB3001 **65** (Scheme 11.21) [23] that potently inhibited LSD1 activity both in vitro and in vivo, has been reported.

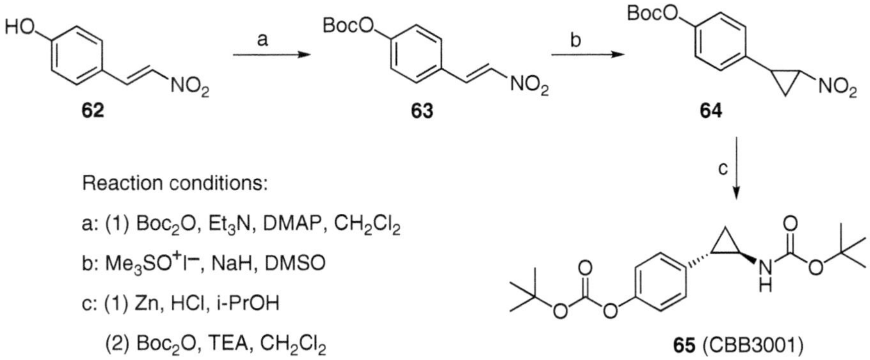

Reaction conditions:

a: (1) Boc_2O, Et_3N, DMAP, CH_2Cl_2

b: $Me_3SO^+I^-$, NaH, DMSO

c: (1) Zn, HCl, i-PrOH

(2) Boc_2O, TEA, CH_2Cl_2

Scheme 11.21 Synthesis of CBB3001 **65**.

The synthesis starts with the Boc-protection of the hydroxyl in nitroalkene **62**. The obtained derivative **63** is submitted to cyclopropanation using trimethylsulfoxonium iodide, instead of dimethyl sulfonium salt reported in the Corey–Chaykovsky method, with the formation of nitrocyclopropane **64**. Successive reduction of the nitro group and protection of the formed amine allows the target molecule **65** (CBB3001).

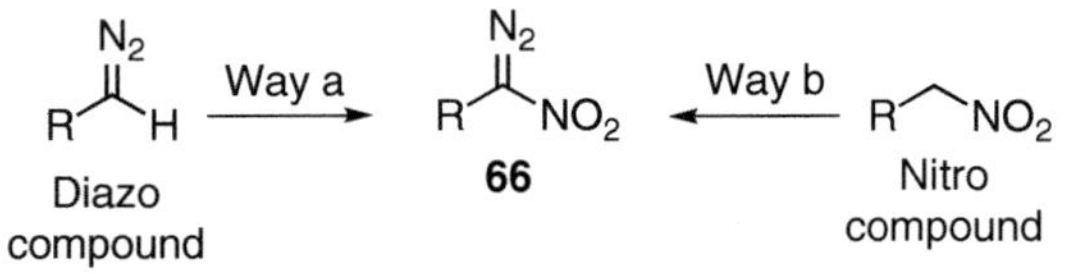

Figure 11.4 Preparation of nitrodiazo compounds **66**.

11.1.3 Synthesis of Nitro Cyclopropanes from Alkenes

Alkenes are useful and easily available class of molecules prone to afford nitrocyclopropanes with nitrodiazo compounds **66** (Figure 11.4) since they are ideal precursors of nitrocarbenes and potentially accessible via introduction of a nitro group into a diazo compounds (way a) [24], or introduction of a diazo group into nitro compounds (way b) [25].

(*i*²) Three decades ago, O'Bannon and Dailey discovered [24] that adding a solution of dinitrogen pentoxide to diazo compound at low temperature, in CCl_4, affords a variety of nitrodiaza compounds **66** (Scheme 11.22).

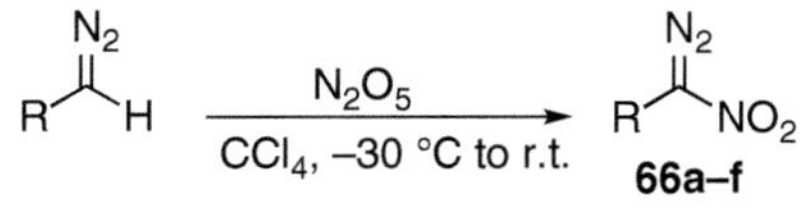

R = (**a**) CO_2Et, (**b**) CN, (**c**) ${}^tBuO_2$, (**d**) PhCO, (**e**) $TolSO_2$, (**f**) CF_3

Scheme 11.22 Nitration of diazo compounds into **66**.

These molecules **66** have then employed for cyclopropanation of a variety of alkenes **66** under Rhodium(II) acetate, achieving a large number of nitrocyclopropanes **67** in moderate to good yields (Scheme 11.23).

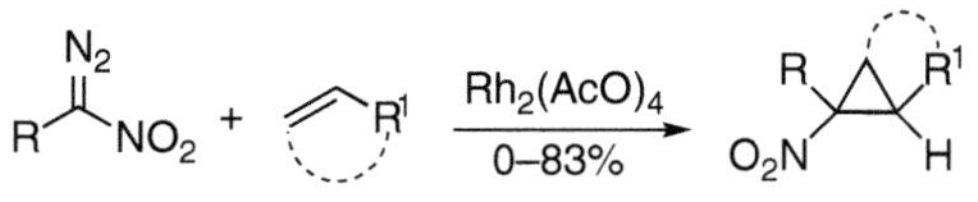

66 Alkenes 67

R = CO_2Et, CN, *t*-BuO_2, PhCO, $TolSO_2$, CF_3
Alkenes: styrene, isobutene, 1-hexene, *cis*-2-butene, *c*-C_6H_{11}, *trans*-2-butene, 2,3-dimethyl-1,2-butene

Scheme 11.23 Cyclopropanation of **66** into **67**.

The reaction works well with styrene, geminal, and *cis*-disubstituted alkenes, while *trans*- and tetra-substituted alkenes, generally, do not yield nitrocyclopropanes **67**.

The same author developed [26] an efficient method for decarboxylation of 1-nitrocyclopropanecarboxylates **67a** (R = CO_2Et) into nitrocyclopropanes **69** that can be easily reduced to cyclopropylamine **70**, very useful from many synthetic points of view (Scheme 11.24).

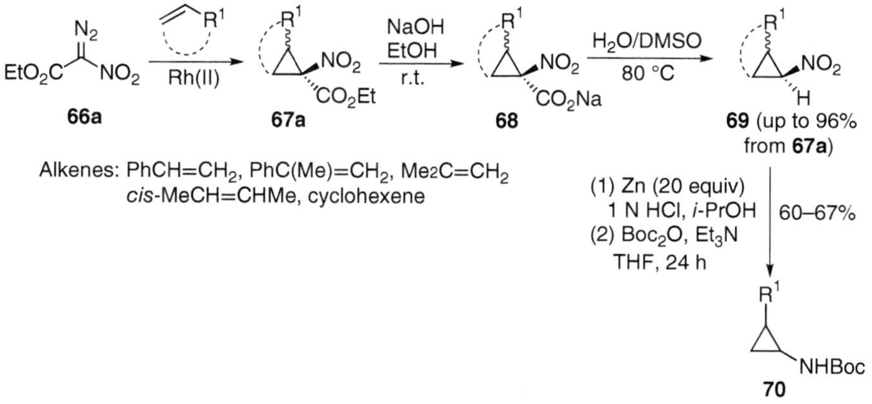

Scheme 11.24 Decarboxylation of **67a** into nitrocyclopropanes **69** and cyclopropyl amines **70**.

Thus, the ester **67a** can be converted into sodium carboxylates **68** with ethanolic NaOH solution at room temperature. Decarboxylation of these salts in wet DMSO solution occurs at 80 °C granting nitrocyclopropanes **69** in excellent yields. The latter can be easily reduced (Zn/HCl) into protected amines **70**.

(ii²) During studies devoted to the synthesis of cyclopropane α-amino acids **74**, a double procedure to obtain nitrocyclopropanes **73** ideal precursors of **74** (Scheme 11.25) has been reported [27].

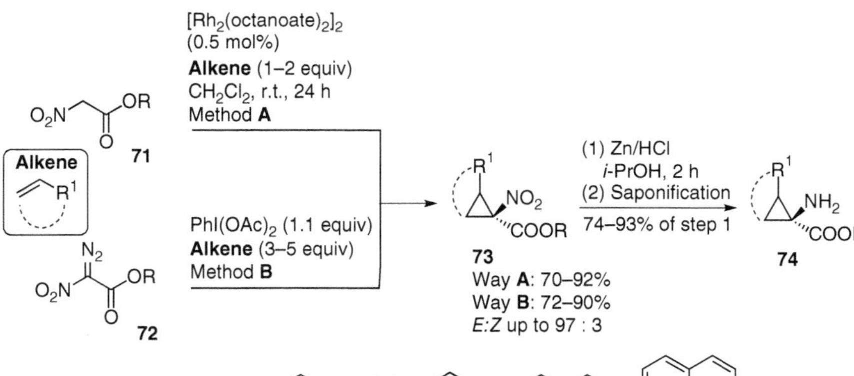

Scheme 11.25 Double ways (**A** and **B**) to obtain cyclopropyl amines **74**.

Method A starts from α-nitroester **71** and iodobenzene diacetate, while method B needs α-nitro diazoesters **72** with Rh(II) catalyst. Both methods allow access to a diverse series of nitrocyclopropane carboxylates **73**. In most cases, similar yields and diastereoselectivities are recovered for a variety of substrates. Thus, way **A** represents an alternative procedure in order to avoid the use of potentially dangerous dyazo-substrate **72**. Finally, compounds **73** can be easily converted, in good to excellent yields, into amino acids **74** in two steps:

(i) reduction with Zn/HCl aqueous solution in 2-propanol and (ii) saponification of ester functionality.

(iii²) The aforementioned cyclopropanation of alkenes, by diazo- or phenyliodonium ylide-derivatives of α-nitro esters **71a**, is performed with good yields and diastereocontrol but with poor enantiocontrol. For this reason a new procedure was then elaborated under Cu(I)–bisoxazoline complex catalyst **75**, in order to obtain an important improvement for asymmetric preparation of nitrocyclopropanes **76**, key precursors of amino acids **77** by simple Zn-mediated reduction, without loss of the high enantioselection, and successive saponification (Scheme 11.26) [28].

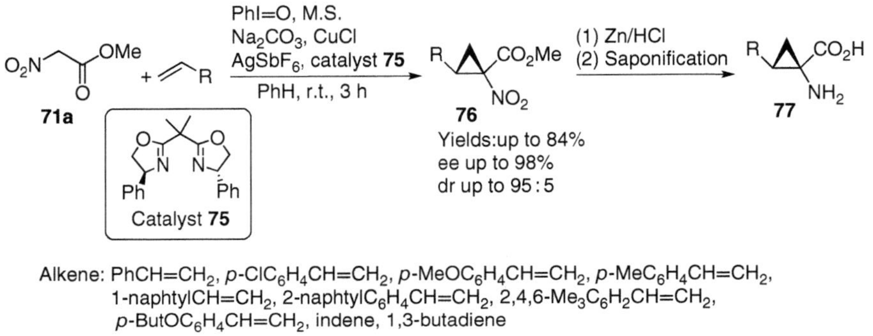

Scheme 11.26 Asymmetric cyclopropanation of alkenes **71a** into **76**.

Catalyst **75** is commercially available and needs the presence of molecular sieves (MS) to scavenge water. The use of Na_2CO_3 was found optimal for the complete consumption of the starting material.

The same authors, with the aim to optimize the enantiomeric ratios obtained with catalyst **75**, developed [29] the aforementioned procedure testing a variety of modified organocatalysts and found as more efficient the catalyst **78** (Figure 11.5) that possesses two couple of *gem*-dimethyl substituents at the α-position relative to oxygen.

As a further result, the authors developed a procedure in order to convert the *trans*-**76** into *cis*-**76**, via isomerization of sodium carboxylate **79**, as reported in Scheme 11.27.

Thus, formation of **79** can be achieved in very good yields treating with alcoholic NaOH. The isomerization step is realized stirring **79** in DMF under CO_2 atmosphere, in the presence of molecular sieves. The key step is the loss of CO_2 followed by isomerization and recombination with CO_2, affording the *cis*-isomer **76** via methylation [$(MeO)_2SO_2$] of the formed carboxyl.

(iv²) Belactosin A (**D**) is a naturally occurring molecule with moderate antitumor activity against cell-cycle progression at the G2/M phase, while Hormaomycin (**A**) has been discussed in the first page of this chapter. These molecules are structurally related since they contain 3-(*trans*-2-aminocyclopropyl) alanine (AcP)Ala **87** and 3-(*trans*-2-nitrocyclopropyl)alanine ((NcP)Ala) **88**.

Figure 11.5 Organocatalyst **78**.

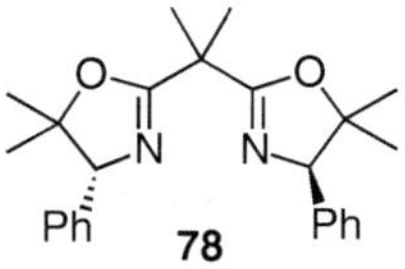

78

Diastereoselective intramolecular cyclopropanation of alkenes with α-nitrodiazoester **82** was reported (Scheme 11.28) as key step for the formal synthesis of Hormaomycin (**A**) and Belactosin A (**D**) (Figure 11.6) [30].

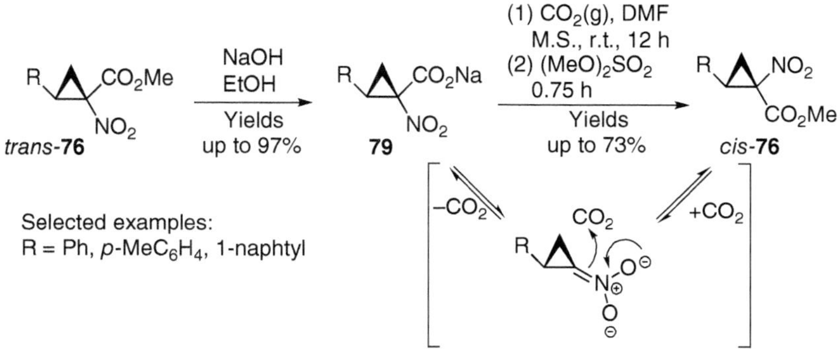

Scheme 11.27 *Trans–cis* isomerization of **76**.

The starting nitroacetate **81** is obtained by esterification of α-nitroacetic acid **80** with the appropriate polyfunctionalized alcohol. Successive formation of diazo derivative **82** opens the possibility to perform an intramolecular cyclopropanation an formation of bicyclic structure **83** prone to be cleaved into the corresponding amide **84** (**83** to **84** needs an intramolecular nucleophilic attack of nitrogen to the carbonyl, followed by oxidation of the obtained primary hydroxyl group into COO*t*-Bu and concomitant Boc-protection of the amino group). Successive ring cleavage of the heterocyclic ring **84**, under basic conditions, allows **86** through decarboxylation of **85**. Acidic hydrolysis of **86** furnishes (NcP)Ala **87** (24% overall yield) prone to achieves (AcP)Ala **88** (15% overall yield) via nitro group reduction.

11.1.4 Intramolecular Synthesis of Nitro Cyclopropanes from γ-Nitro Alcohols (the Mitsunobu Displacement)

γ-Nitro alcohols are a class of functionalized nitroalkanes in which the hydroxyl can acts as leaving group under basic condition, achieving an intramolecular nucleophilic substitution (Figure 11.7).

Following this concept, a new way to obtain nitrocyclopropanes from γ-nitro alcohols was reported by their treatment with a preformed complex of diethyl azodicarboxylate (DEAD) and PPh_3 (Table 11.2) [31]. Good yields and high diastereoselectivity are obtained. The predominant *trans*-disposition of the nitro group presumably reflects kinetic rather than thermodynamic factors. Even

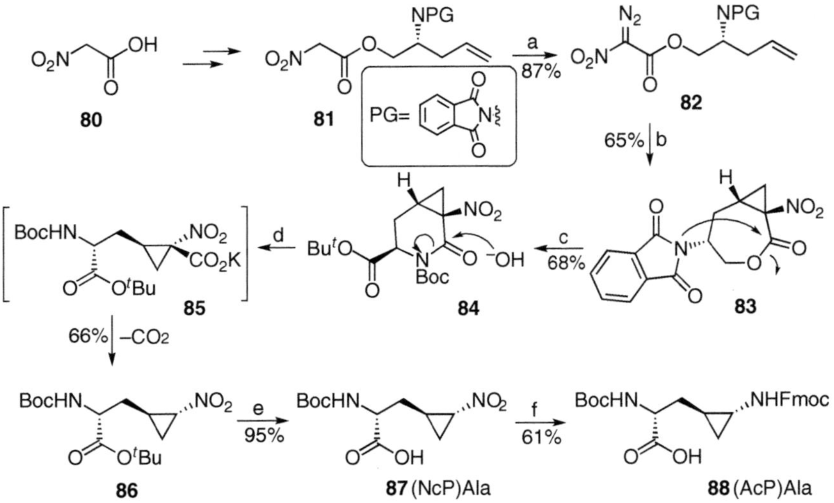

a: TfN_3, Py, MeCN, 0 °C.
b: $[Rh(OAc_2)_2]$, CH_2Cl_2, r.t..
c: (1) NaOH/EtOH, EtOH/CH_2Cl_2, r.t.; (2) Boc_2O/ DMAP, MeCN, r.t.; (3) TEMPO, TCAA, acetone, 0 °C to r.t.; (4) *t*-BuOH, H_2SO_4, $MgSO_4$, CH_2Cl_2, r.t..
d: KOH, DMSO,80 °C.
e: (1) HCl aq, Et_2O; (2) Boc_2O, Na_2CO_3 dioxane/H_2O.
f: (1) Zn/AcOH; (2) FmocOSu, Na_2CO_3, DMF.

Scheme 11.28 Synthesis of subunits **87** and **88**.

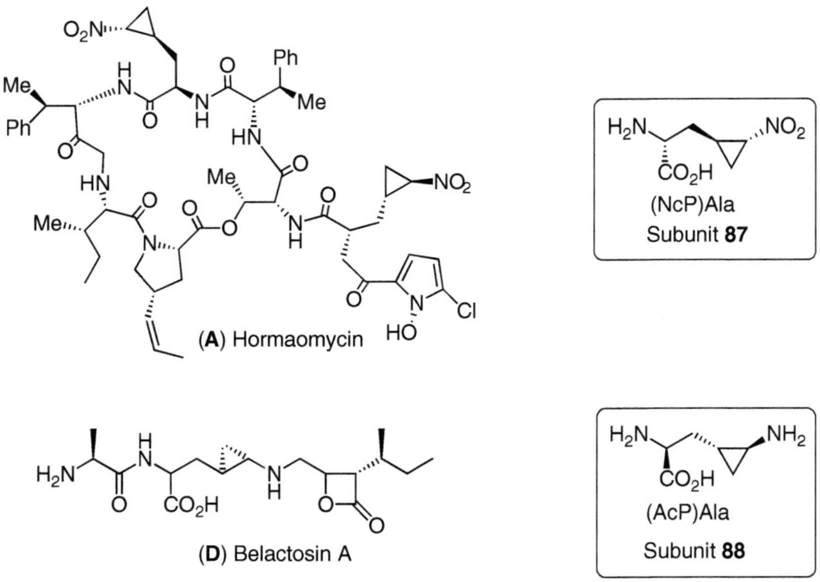

Figure 11.6 Hormaomycin (**A**), Belactosin A (**D**), and their subunits (NcP)Ala **87** and (AcP)Ala **88**.

Figure 11.7 Cyclopropanation of γ-nitroalcohols.

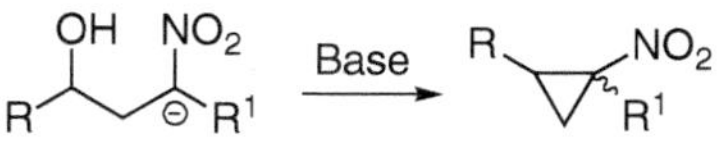

Table 11.2 Cyclopropanation of γ-nitro alcohols.

Entry	Nitroalcohol	Cyclopropane	trans/cis	Yield (%)
1	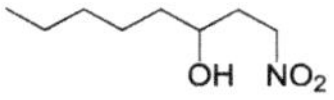	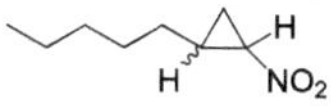	10 : 1	82
2	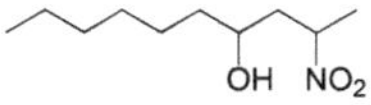	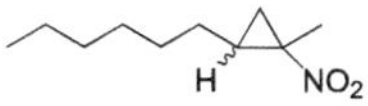	10 : 1	87
3	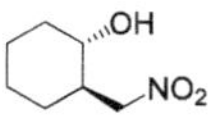	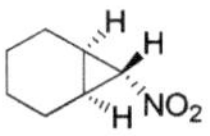	100 : 0	92
4	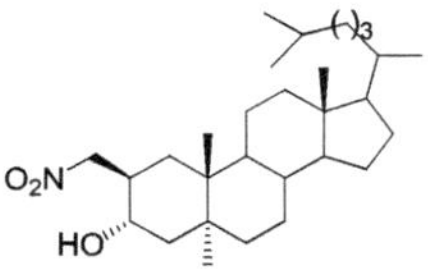	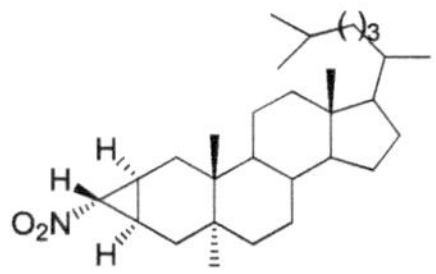	100 : 0	76
5	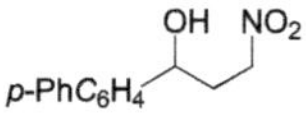	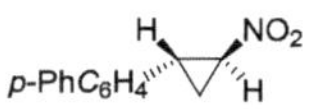	100 : 0	75
6	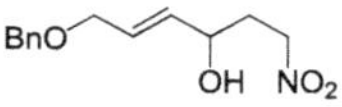	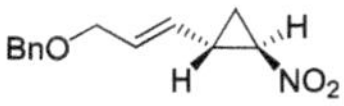	100 : 0	64
7	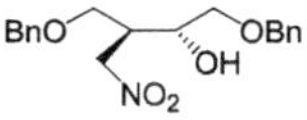	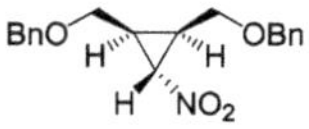	100 : 0	94
8	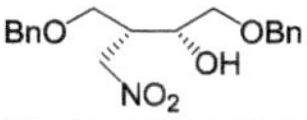	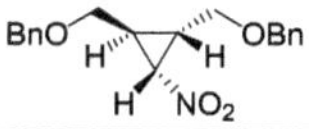	Not applicable	98

Reaction conditions: DEAD/Ph_3P, THF, r.t.

alcohols prone to dehydration (entries 5 and 6) allow cyclopropanes without any complication.

This source of nitrocyclopropanes has been applied to the synthesis of 3-(*trans*-2′-nitrocyclopropyl)alanine **95** (Scheme 11.29) [32], important structure present in Hormaomycin [(**A**), Figure 12.6)]. The synthetic sequence initiates with the readily available aldehyde **89** submitted to nitroaldol condensation with nitromethane and reduction ($NaBH_4$) of the obtained nitroalkene with the production of nitro acetal **90**. Cleavage of the latter into **91** with TsOH and selective protection of primary hydroxyl with triphenylmethyl chloride, in the presence of pyridine, gives **92**. The next step consists in the conversion of secondary hydroxyl into ether **93**, by treatment with MsCl, in which the ether functionality (OMs) acts a good leaving group.

Thus, under basic conditions, the obtained nitronate form produces γ-elimination with complete inversion of configuration at C-2, producing just the *trans*-configured

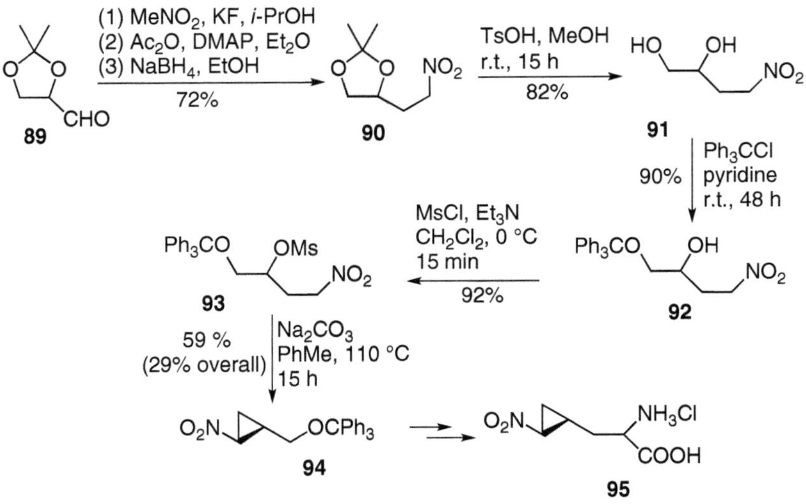

Scheme 11.29 Synthesis of **95**.

nitrocyclopropane **94** that can be easily converted into the target amino acid **95** in few steps.

11.2 Applications of Nitrocyclopropanes

Nitrocyclopropanes show important reactivities that can be very useful in organic synthesis and some of these will reported in this chapter.

11.2.1 Nitrocyclopropanes and Henry Reaction: Synthesis of Novel HIV-1 Protease Inhibitor.

The Henry reaction with nitro cyclopropanes is sensitive to the degree of substitution of these molecules. In fact, their propensity to undergo dimerization rather than react with an aldehyde under strong basic conditions was jet reported [33]. During a project devoted to the synthesis of novel HIV-1 protease inhibitors **101a,b**, Mitchell et al. [34] planned to start from diastereoselective nitroaldol reaction of nitrocyclopropane with aldehyde **96** to obtain nitroalkanol **97**, precursor of the targets **101a,b**, in good yield and diastereoselectivity (Scheme 11.30).

Thus, reduction (Zn/HCl) of **97** and reductive amination with isobutyraldehyde and sodium triacetoborohydride provides the secondary amine **98**.

Sulfonylation of the latter acquires **99a** that, by catalytic hydrogenolysis, yields **100a**. Finally, coupling **100a** with (3*R*,3a*S*,6a*R*)-hexahydrofuro[2,3-*b*]furan-3-yl (4-nitrophenyl) carbonate achieves the target molecule **101a**. Compound **101b** is prepared following the same steps starting from **99b** instead of **99a**.

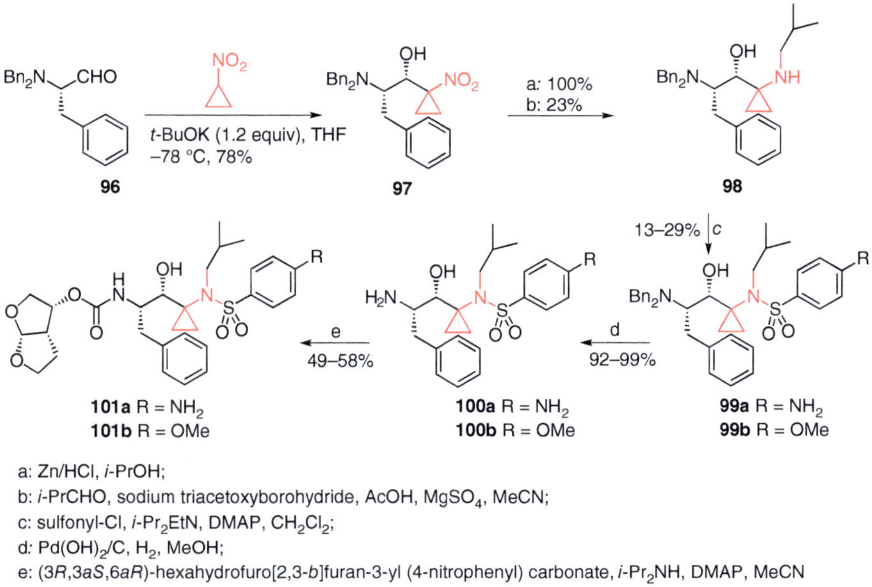

Scheme 11.30 Synthesis of novel HIV-1 protease inhibitors **101a,b**.

11.2.2 Cyclopropane Ring Expansion

Ring expansion is an interesting synthetic opportunity of nitrocyclopropanes that can be used mainly for the preparation of homo- and heterocyclic rings.

(i³) A tandem ring opening/cyclization of *trans*-nitrocyclopropanes **102** with aminopyridines **103**, without any assistance of catalyst and in water as solvent, affords pyrido[1,2-*a*[-pyrimidin-4-ones] **104** with the presence of a carboxylate group at C_3 (Scheme 11.31) [35].
A variety of compounds **104** are obtained in good yields, independently form the nature of substituents and functional groups present.
A plausible mechanism from the starting molecules **102** and **103** to the target **104**, via **I–V** structures, is reported in Scheme 11.32.

(ii³) Redox-neutral [3+2] cycloaddition of nitrocyclopropanes **105** with styrenes **106**, under visible-light photocatalysis, allows a large variety of polyfunctionalized nitrocyclopentanes **107** in high diastereoselectivities (Scheme 11.33) [36]. The reaction is carried out using [Ru(bpy)$_3$(PF$_6$)$_2$], in the presence of Et_3N/MeCN upon irradiation with an 18 W compact fluorescent lamp (CFL). The reaction is accelerated by adding 3 equiv of $LiBF_4$ as Lewis acid. The presence of $MgSO_4$ slightly promoted the reaction, possibly acting as a dray agent as well as Lewis acid.
The authors reported a possible mechanism under SmI_2, as a one-electron reductive reagent instead of visible-light photocatalysis, demonstrating that the newly development strategy underwent a one-electron reduction pathway to generate the corresponding nitro radical anion.

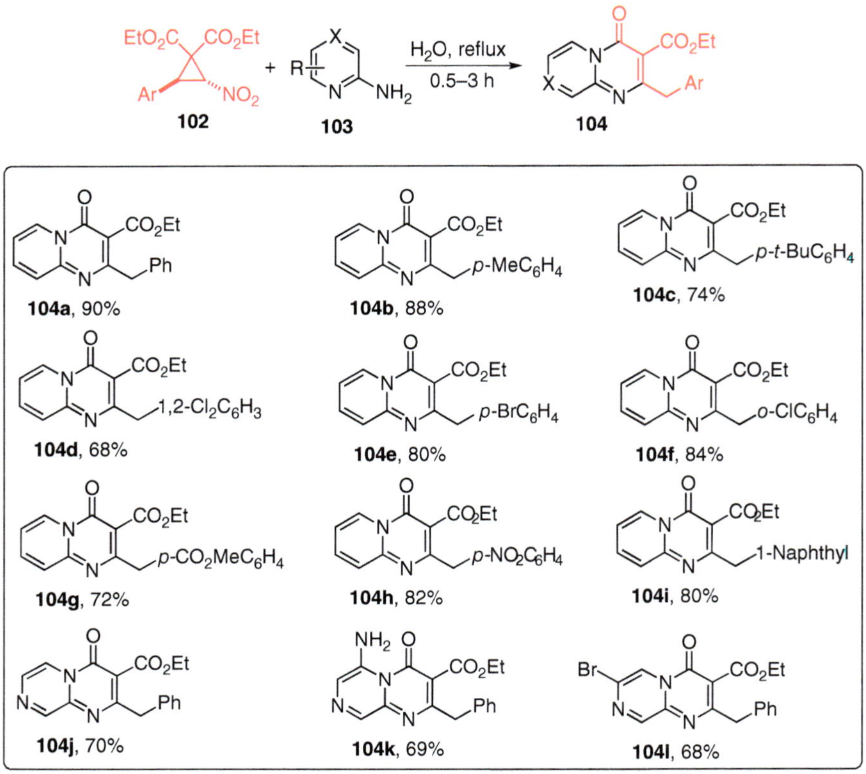

Scheme 11.31 Tandem ring opening/cyclization of cyclopropanes **102** into **104**.

(iii^3) A variety of nitrocyclopropanes **108**, differing from the nature of the functional group at C_2, gives intramolecular ring expansion, under visible-light photocatalytic conditions (catalyst **109**) affording enantioenriched dihydrofurans **111** or cyclopentenes **112**, depending from the nature of the electron-withdrawing group at C_4 (Scheme 11.34) [37].

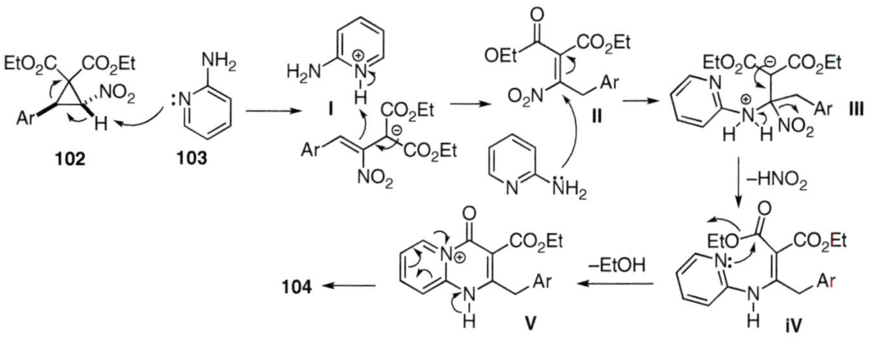

Scheme 11.32 Mechanism for the formation of **104**.

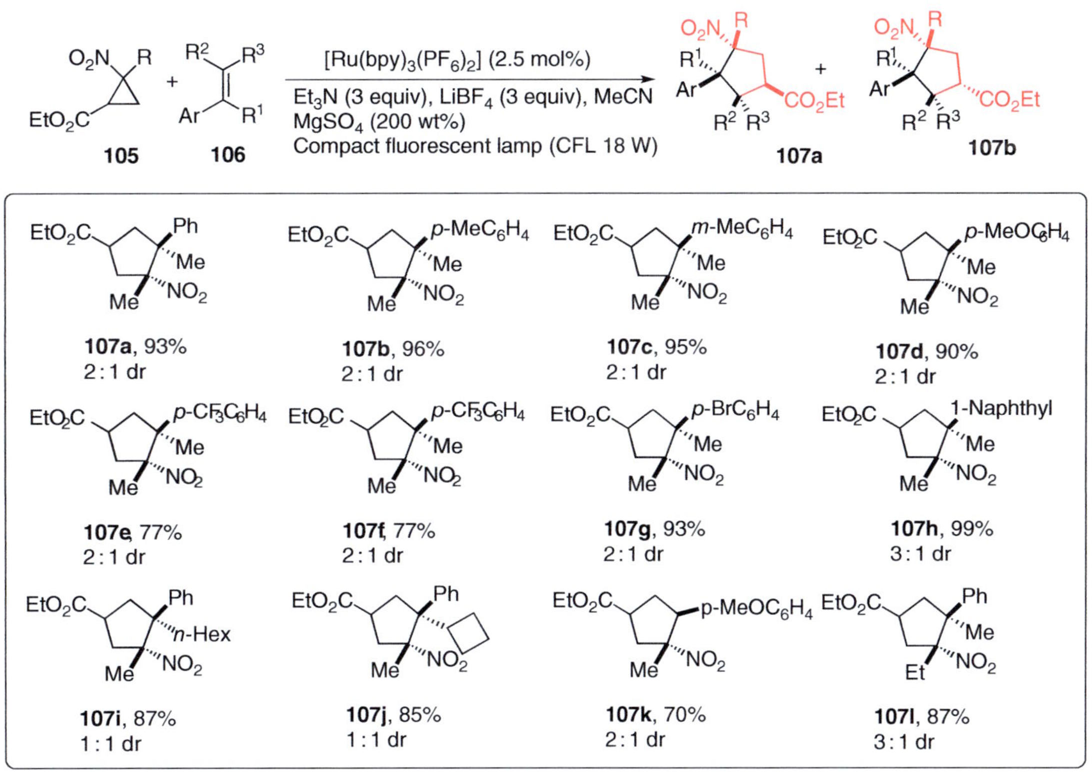

Scheme 11.33 [3+2]-Cycloaddition of nitrocyclopropanes **105** with styrenes **106**.

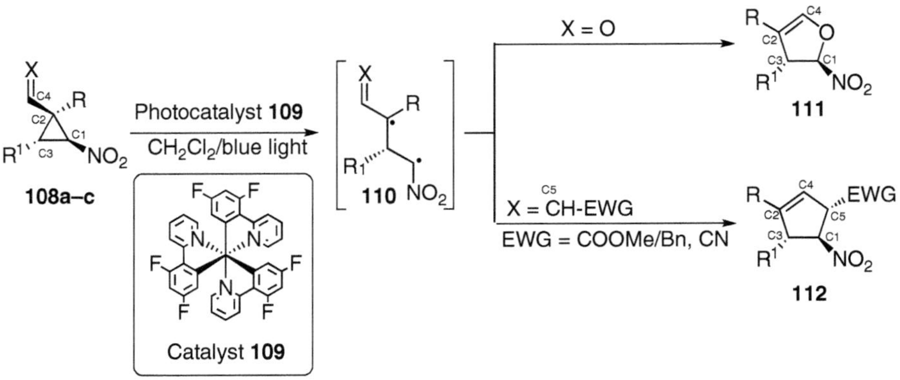

Scheme 11.34 Intramolecular reaction of nitrocyclopropanes **108** into digydrofurans **111** and cyclopentenes **112**.

The first results are obtained by intramolecular ring expansion of nitrocyclopropanes **108a** that, through the enol derivative of the aldehyde, allows excellent yields and enantioselectivities of dihydrofurans **111** since no loss of stereochemical information at C_3 is observed (Scheme 11.35).

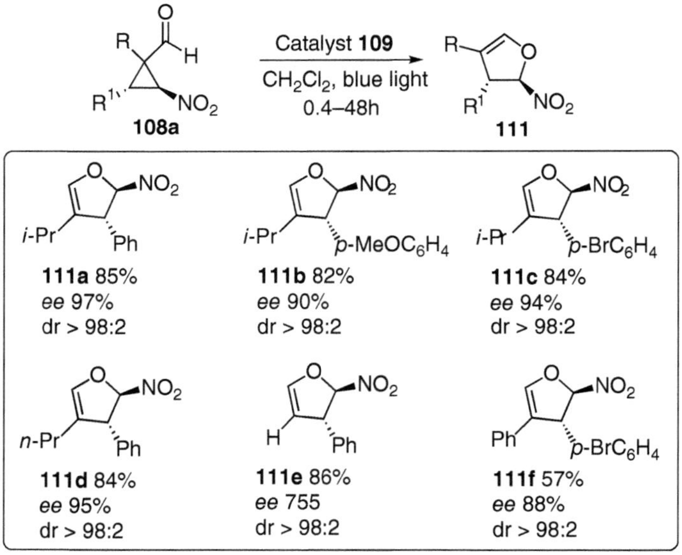

Scheme 11.35 Synthesis of dihydrofurans **111**. Selected examples

If the reaction is performed with the presence of ketone instead of aldehyde, the ring expansion is much slower (three days) with moderate yields.

To expand the scope of the protocol, the authors reported the conversion of **108b,c** into cyclopentene derivatives **112** (Scheme 11.36) applying the same reaction conditions, including the use of catalyst **109**.

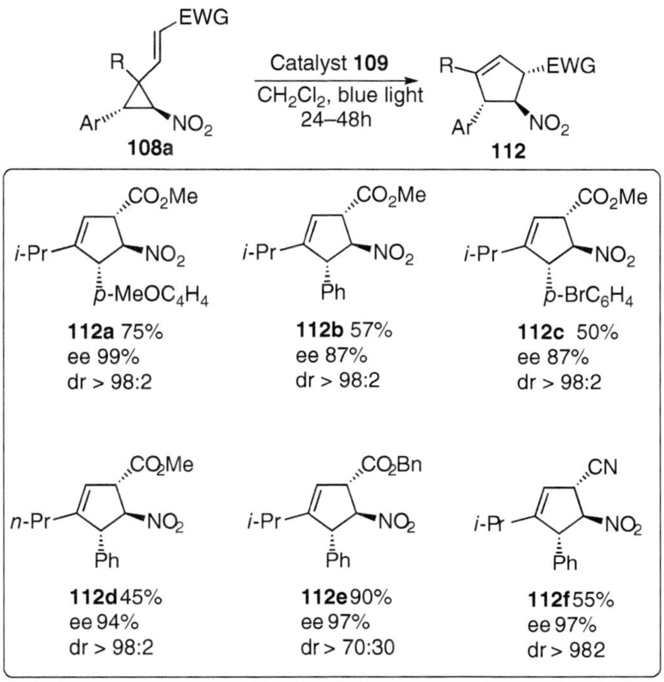

Scheme 11.36 Synthesis of cyclopentenes **112**. Selected examples.

Moderate to good yields and high ee and dr of pentacyclic polyfunctionalized alkenes **112** are achieved in 24–48 hours.

References

1 Rössner, F., Zeeck, A., and König, W.A. (1990). *Angew. Chem. Int. Ed. Engl.* 29: 64–65.

2 Agrawal, J.P. (1998). *Prog. Energy Combust. Sci.* 24: 1–30.

3 Alster, J., Iyer, S., and Sandus, O. (1990). Molecular architecture versus chemistry and physics of energetic materials. In: *Chemistry and Physics of Energetic Materials*, vol. 28 (ed. S.N. Bulusu), 641–652. Dordrecht, The Nederland: Kluwer Academic *Publishers.*

4 Patagoria, P.F., Lee, G.S., Mitchell, A.R., and Schmedt, R.D. (2002). *Termochimica Acta* 384: 187–204.

5 Sopova, A.S., Perekalin, V.V., Yurchenko, O., and Arnautova, G.M. (1969). *Zh. Org. Khim.* 5: 858–863. C A. 1969, 71, 38394.

6 Kocór, M. and Kroszczyński, W. (1976). *Synthesis*: 813–814.

7 Arai, S., Nakayama, K., Ishida, T., and Shioiri, T. (1999). *Tetrahedron Lett.* 40: 4215–4218.

8 Brandl, M., Kozhushkov, S.I., Loscha, K. et al. (2000). *Synlett*: 1741–1744.

9 Larionov, O.V., Savel'eva, T.F., Kochetkov, K.A. et al. (2003). *Eur. J. Org. Chem.*: 869–877.
10 Bezdudny, A.V., Klukovsky, D., Simurova, N. et al. (2011). *Synthesis* 1: 119–122.
11 (a) Rabasso, N. and Fadel, A. (2014). *Tetrahedron Lett.* 55: 6068–6071. (b) Perez, V., Rabaso, N., and Fadel, A. (2016). *Eur. J. Org. Chem.*: 320–324.
12 Braish, T.F., Castaldi, M., Chan, S. et al. (1996). *Synlett*: 1100–1102.
13 Norris, T., Braish, T.F., Butters, M. et al. (2000). *J. Chem. Soc., Perkin Trans. 1*: 1615–1622.
14 Ballini, R., Fiorini, D., and Palmieri, A. (2003). *Synlett*: 1704–1706.
15 Keita, M., De Bona, R., Dos Santos, M. et al. (2013). *Tetrahedron* 69: 3308–3315.
16 Yu, H., Wang, Q., Wang, Y. et al. (2013). *Chem. Asian J.* 8: 2859–2863.
17 Muray, E., Alvarez-Larena, A., Piniella, J.F. et al. (2000). *J. Org. Chem.* 65: 388–396.
18 (a) Galley, G., Hübner, J., Anklam, S. et al. (1996). *Tetrahedron Lett.* 37: 6307–6310. (b) Hübner, J., Liebscher, J., and Pätzel, M. (2002). *Tetrahedron* 58: 10485–10500.
19 McCooey, S.H., McCabe, T., and Connon, S.J. (2006). *J. Org. Chem.* 71: 7494–7497.
20 (a) Corey, E.J. and Chaykovsky, M. (1962). *J. Am. Chem. Soc.* 84: 3782–3783. (b) Corey, E.J. and Chaykovsky, M. (1965). *J. Am. Chem. Soc.* 87: 1353–1364.
21 Hommelsheim, R., Hock, K.J., Schumacher, C. et al. (2018). *Chem. Commun.* 54: 11439–11442.
22 Hock, K., Hommelsheim, R., Mertens, L. et al. (2017). *J. Org. Chem.* 82: 8220–8227.
23 Hoang, N., Zhang, X., Zhang, C. et al. (2018). *Bioorg. Med. Chem.* 26: 1523–1537.
24 O'Bannon, P.E. and Dailey, W.P. (1990). *Tetrahedron* 46: 7341–7358.
25 Charette, A.B., Wurz, R.P., and Ollevier, T. (2000). *J. Org. Chem.* 65: 9252–9254.
26 O'Bannon, P.E. and Dailey, W.P. (1990). *J. Org. Chem.* 55: 353–355.
27 Wurz, R.P. and Charette, A.B. (2004). *J. Org. Chem.* 69: 1262–1269.
28 Moreau, B. and Charette, A.B. (2005). *J. Am. Chem. Soc.* 127: 18014–18015.
29 Moreau, B., Alberico, D., Lindsay, V.N.G., and Charette, A.B. (2012). *Tetrahedron* 68: 3487–3496.
30 Vanier, S.F., Larouche, G., Wurz, R.P., and Charette, A.B. (2010). *Org. Lett.* 12: 672–675.
31 Yu, J., Falck, J.R., and Mioskowski, C. (1992). *J. Org. Chem.* 57: 3757–3759.
32 Zindel, J., Zeeck, A., König, W.A., and de Meijere, A. (1993). *Tetrahedron Lett.* 34: 1917–1920.
33 (a) Bordwell, F.G., Bartmess, J.E., and Hautala, J. (1978). *J. Org. Chem.* 43: 3113–3116. (b) Kai, Y., Knochel, P., Kwiatkowski, S. et al. (1982). *Helv. Chim. Acta* 65: 137–161.
34 Mitchell, M.L., Xu, L., Newby, Z.E., and Desai, M.C. (2017). *Tetrahedron Lett.* 58: 1123–1126.
35 Selvi, S. and Srinivasan, K. (2017). *Eur. J. Org. Chem.*: 5644–5648.
36 Wang, C., Ren, X., Xie, H., and Lu, Z. (2015). *Chem. Eur. J.* 21: 9676–9680.
37 Barrera, J.L., Martin, V.L., Rigotti, T. et al. (2017). *Angew. Chem. Int. Ed.* 56: 7826–7830.

12

Nitroalkanes as Source of Dicarbonyls

Dicarbonyls (Figure 12.1) are important key building blocks in organic synthesis for the preparation of a large variety of important chemicals including carbocycles, heterocycles, and several biologically active molecules. Nitroalkanes have demonstrated to be powerful precursors of these molecules due to their propensity to (i) generate the formation of C—C bond (the Michael or Henry reactions) and (ii) to be converted into carbonyls (the Nef reaction).

Thus, this chapter reports synthesis of 1,*n*-dicarbonyl and some representative applications.

12.1 1,2-Dicarbonyl Derivatives

A pioneering procedure was reported starting from nitroaldol (Henry) reaction of aryl aldehydes with nitromethane, under basic conditions, affording nitroalkanols **1**, then submitted to oxidative Nef reaction using molecular oxygen in presence of copper salts (Scheme 12.1) [1]. Under these conditions oxidation of both secondary alcohol into ketone and nitro to carboxyl, allows the one-pot formation of α-keto acids **2** in good yields and in few hours (three to five hours).

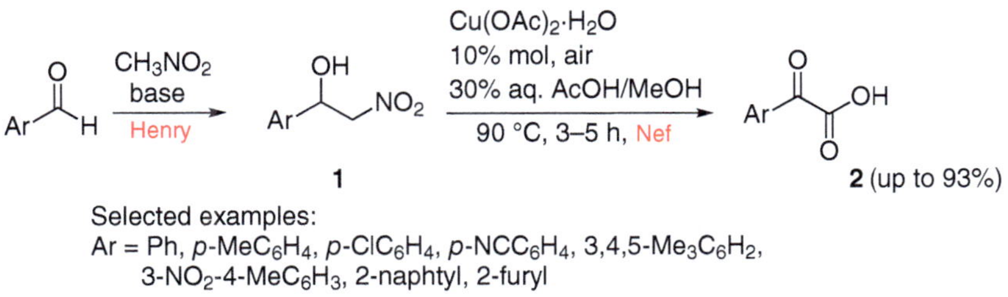

Scheme 12.1 Synthesis of α-ketoacids **2**.

The Henry reaction of 2-oxoaldehydes **4** with a number of nitroalkanes **3**, under basic and neat conditions, followed by microwave irradiation, achieves 1,2-diketones **8** (Scheme 12.2) in a one-pot procedure [2].

Nitroalkanes: Synthesis, Reactivity, and Applications, First Edition.
Roberto Ballini and Alessandro Palmieri.

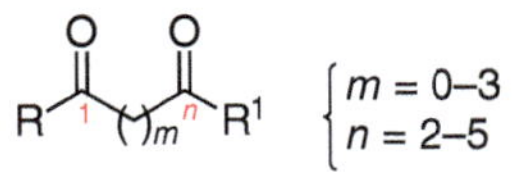

1,n-Dicarbonyl derivatives

Figure 12.1 1,*n*-Dicarbonyls.

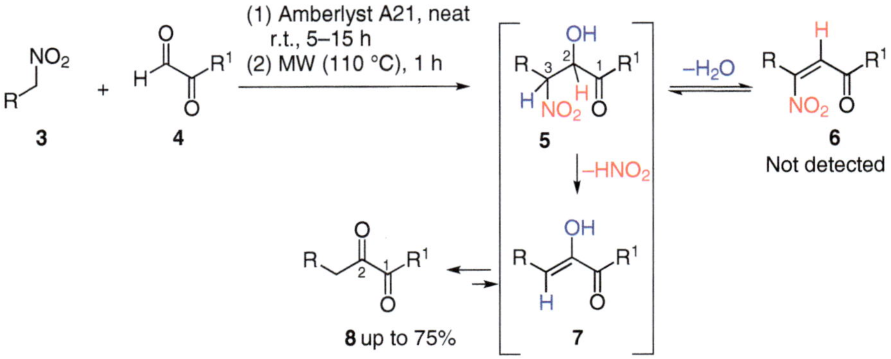

Selected examples:
R = *n*-Pr, Bu, $Ph(CH_2)_2$, *n*-C_9H_{19}, $Cl(CH_2)_3$, $NC(CH_2)_4$, $MeO_2C(CH_2)_4$
R^1 = Ph, *o*-FC_6H_4, *p*-$MeOC_6H_4$, 2-naphtyl

Scheme 12.2 Synthesis of 1,2-dicarbonyls **8**.

As it is possible to observe, the nitroalkanols **5** instead of the expected dehydration process with consequent formation of nitroalkenes **6** give nitrous acid elimination, under microwave irradiation at 110 °C, with the formation of the enols **7** ideal precursors of 1,2-dicarbonyls **8**.

The driving force of the process can be rationalized with the increased acidity of the proton in 2-position, due to the geminal presence of a carbonyl moiety which helps to the irreversible elimination of HNO_2 affording the enolic form **7** in good overall yields (52–76%). In addition, even the use of secondary nitroalkanes, such as 2-nitropropane and nitrocyclohexane produce **8** in satisfactory overall yields (>45%).

It is important to point out that the overall transformation (**3**+**4** into **8**) makes the nitroalkanes **3** as synthetic equivalent of carbanion synthon **A**, while the aldehyde **4** acts as the synthetic equivalent of the acyl cation synthon **B** (Figure 12.2).

Finally, this is one of the very rare cases where nitrous acid elimination is favored vs water elimination from β-nitroalcohol. In this sense, this report expands the extraordinary synthetic potential of the Henry reaction, discussed in Chapter 4.

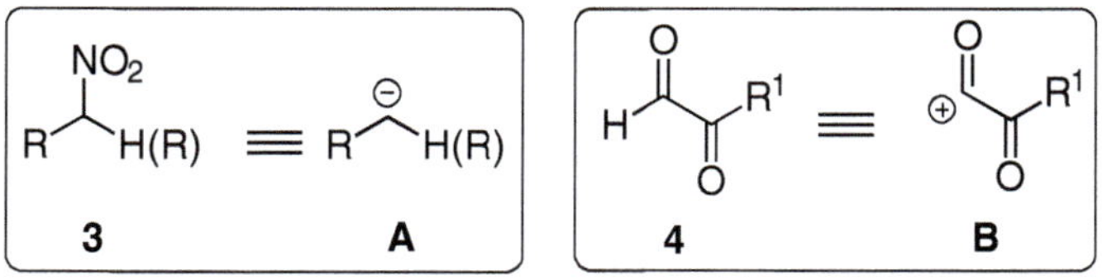

Figure 12.2 Synthetic equivalents of **3** and **4**.

12.2 1,3-Dicarbonyl Derivatives

Nitroaldol condensation is the key reaction for the preparation of 1,3-dicarbonyls from nitroalkanes. In fact (Scheme 12.3), reaction of β-nitroester **9** with aldehydes **10** at 50 °C and under basic alumina achieves the direct formation of nitroalkenes **11** that, by reductive (Raney-Ni, $NaH_2PO_2{\cdot}xH_2O$) Nef reaction convert into β-ketoesters **12** in satisfactory yields [3].

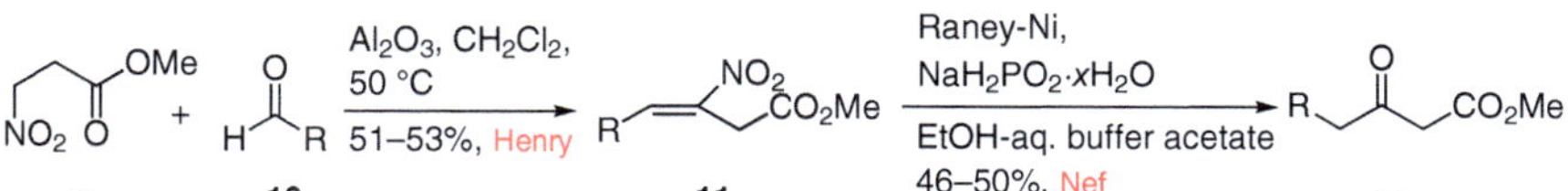

9 10 11 12

R = *n*-C_6H_{13}, *c*-C_6H_{11}, $Ph(CH_2)_2$

Scheme 12.3 Synthesis of β-ketoesters **12**.

1,3-Diketones have been prepared following the same approach of the latter procedure but with the right choice of the functionalized nitroalkane. In fact, starting from the nitroaldol condensation of aldehyde **10** with protected β-nitro ketone **13**, it possible to prepare mono-protected 1,3-diketones **15**, via nitroalkenes **14** submitted to the Nef reaction under reductive (Raney-Ni) conditions (Scheme 12.4) [4]. Successive ketal-deprotection produces 1,3-diketones **16** in excellent yields.

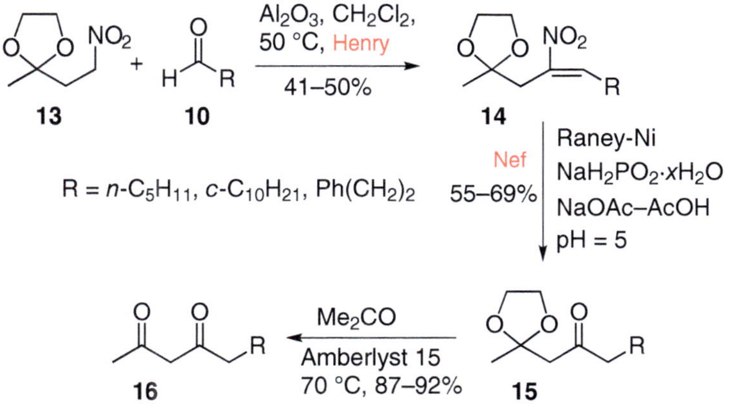

13 10 14 15 16

R = *n*-C_5H_{11}, *c*-$C_{10}H_{21}$, $Ph(CH_2)_2$

Scheme 12.4 Synthesis of 1,3-diketones **16**.

Nitroalkanes **3** and ethyl glyoxalate **17** are the key building blocks for one-pot synthesis of β-ketoesters **20** (Scheme 12.5) [5]. The process involves the Henry reaction of **3** with aldehyde **17** under basic condition (Amberlyst A21), with the formation of nitroalkanols **18** that by mesylation in presence of a base (Et_3N) are dehydrated into nitroalkenes **19** in good yields. Successive Nef reaction, under reductive conditions ($TiCl_3$), converts **19** into β-ketoesters **20**.

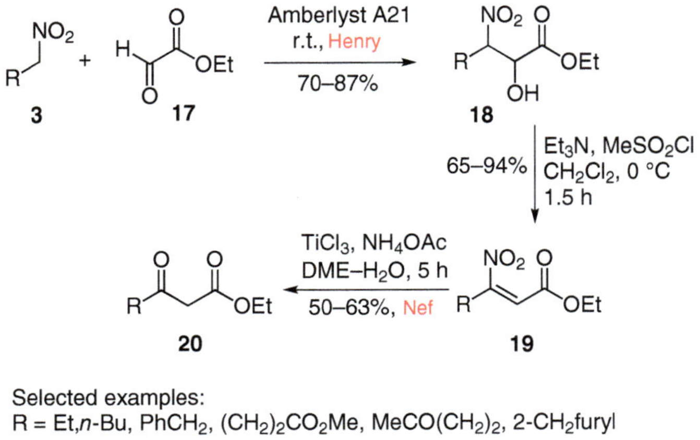

Scheme 12.5 Synthesis of β-ketoesters **20**.

12.3 1,4-Dicarbonyl Derivatives

1,4-Dicarbonyls structures are, probably, the most important and used dicarbonyls in organic synthesis and nitroalkanes have special aptitudes as strategic precursors for their preparation.

12.3.1 1,4-Diketones

A solid procedure for the preparation of 1,4-diketones is similar to those jet reported for the synthesis of 1,3-diones **16** from protected nitro ketone **13** (Scheme 12.4) [4]. In fact, as described in Scheme 12.6, changing nitroalkane **13** with **21**, the method furnishes monoprotected 1,4-dione **23** that, by acidic deprotection, allows 1,4-dicarbonyls **24**.

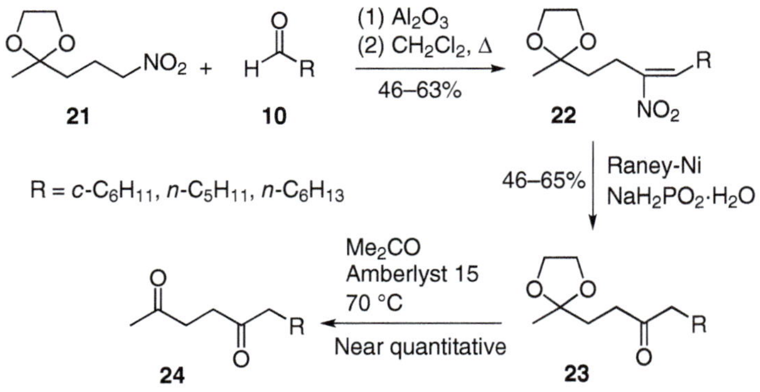

Scheme 12.6 Synthesis of 1,4-diketones **24**.

An important advantage, difficult to obtain following ways, is the availability of the monoprotected 1,4-dicarbonyls as strategic molecules for chemoselective synthetic procedures. Moreover, nitroalkane **21** can be easily prepared, as reported in Chapter 1.

A pioneering one-pot synthetic procedure for the preparation of 1,4-dicarbonyls **27** from nitroalkanes was reported three decade ago via Michael reaction of primary nitroalkanes **3** with enones **25**, under neat basic conditions, and in situ oxidative (H_2O_2) Nef reaction of the nitronate **26** (Scheme 12.7) [6]. Good overall yields are obtained of **27** (up to 90%) with a variety of nitroalkanes **3** and enones **25**. The one-pot process, very interesting from sustainability point of view, can be realized thanks to the first step in which basic alumina is used without the need of any solvent.

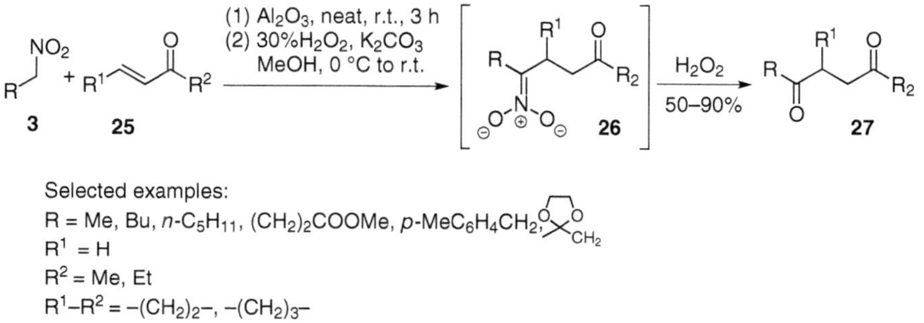

Scheme 12.7 One-pot synthesis of 1,4-diketones **27**.

In fact, the procedure consists in the conjugate addition of primary nitroalkanes **3** to enones **25** on basic alumina and under neat conditions. After mechanical stirring at r.t., in situ oxidation (30% H_2O_2) of the nitronate structure **26** affords 1,4-diketones **27** in satisfactory to excellent overall yields and in a one-pot process.

As extension, the authors applied this procedure to the one-pot synthesis of symmetrical 1,4,7-triketones **29,** as reported in Scheme 12.8, by double Michael reaction of CH_3NO_2 with 2 mol of enones **28**.

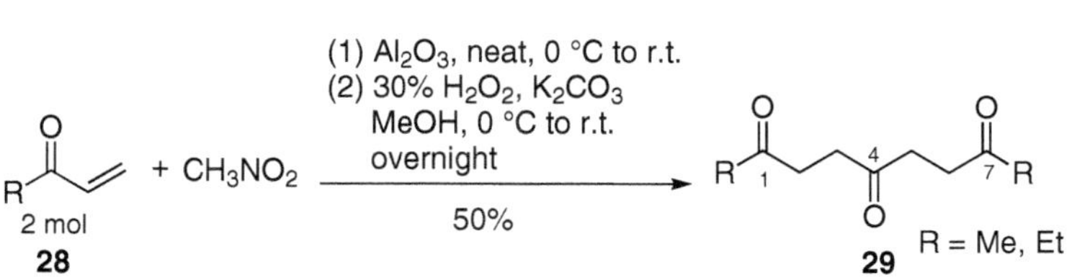

Scheme 12.8 One-pot synthesis of 1,4,7-triketones **29**.

The ability of 1,8-diazabicyclo[5.4.0]undec-7-ene (DBU) to promote the conjugate addition of primary nitroalkanes to enones as well as nitro to carbonyl conversion has been used in a tandem process that includes Michael and Nef reactions, with the one-pot formation of 1,4-diketones **30** (R^1 = alkyl) or γ-ketoesters (R^1 = OMe), as reported in Scheme 12.9 [7].

The strategic employment of 1-nitropent-4-ene **32** offers the opportunity to obtain 1,4-dicarbonyls **35** starting from its nitroaldol condensation with aldehydes under

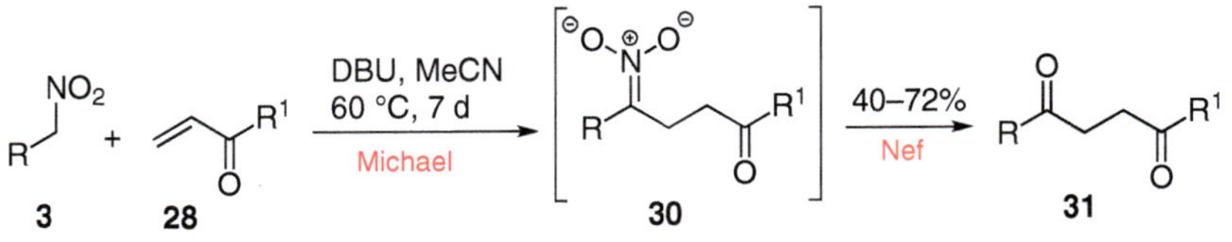

Scheme 12.9 One-pot synthesis of 1,4-diketones **31**.

refluxing basic Al_2O_3/CH_2Cl_2 (Scheme 12.10) [8]. The obtained nitroalkenes **33** are submitted to Nef transformation by reduction/oxidation process and formation of the enones **34**. A subsequent oxidation with Wacker method ($PdCl_2/CuCl_2/O_2$) allows 1,4-diones **35** in satisfactory yields.

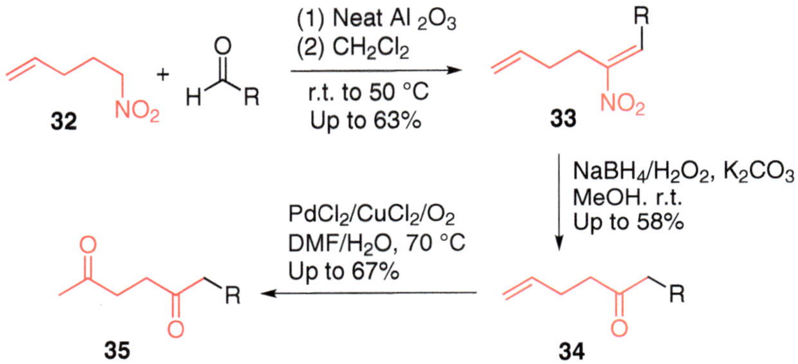

Scheme 12.10 One-pot synthesis of 1,4-diketones **35**.

Allylic acetates **36**, obtained by Baylis–Hillman procedure, react with nitroalkanes **3** (Scheme 12.11) [9] through a conjugate addition–elimination process that leads the formation of unsaturated nitro ketones **37**. Successive hydrolytic Nef reaction affords (*E*)-alkylidene-1,4-diketones **38** [9].

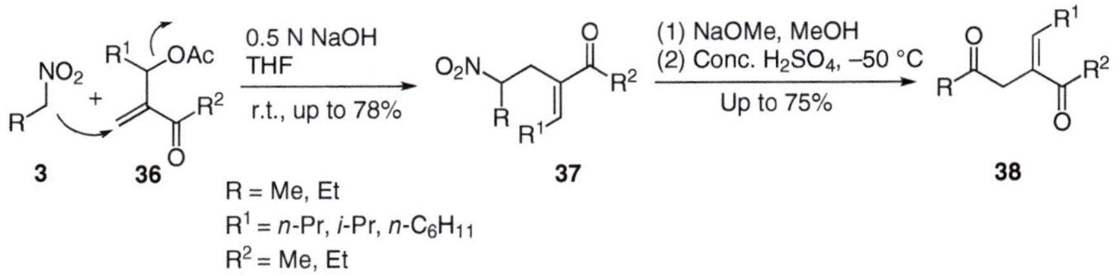

Scheme 12.11 One-pot synthesis of (*E*)-alylidene-1,4-diones **38**.

An important method for the synthesis of functionalized 1,4-diketones **42** was reported starting from 2-nitrocycloalkanones **39**, in three steps (Scheme 12.12) [10].

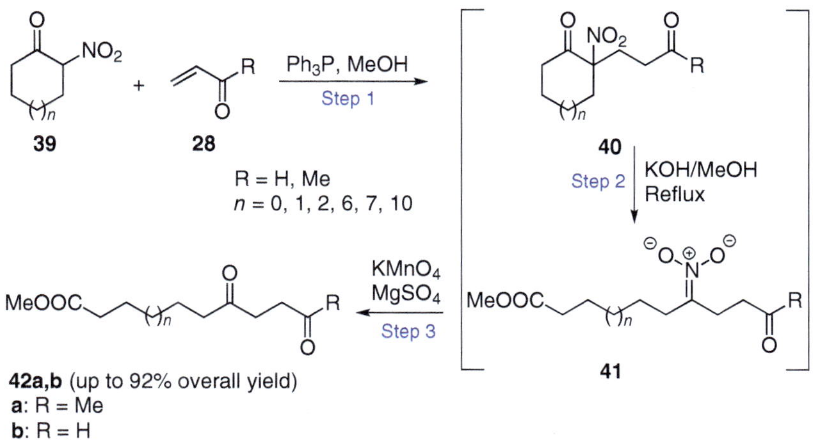

Scheme 12.12 One-pot synthesis of 1,4-diketones **42a** or γ-ketoaldehydes **42b**.

This procedure is highly representative because the high chemical versatility of nitroalkanes shows, in fact, Step 1 reports the formation of a new C—C bond via the Michael reaction and formation of the adduct **40** that, Step 2, is submitted to ring cleavage, favored by the electron withdrawing effect of the nitro group, with formation of the open-chain nitronate structures **41** that, under oxidative conditions *step 3*, give the Nef reaction, achieving functionalized 1,4-diketones **42a** (R = Me) or γ-ketoaldehydes **42b** (R = H).

(E)-Enediones are a particular and important class of 1,4-diketones because these are strategic intermediates in organic synthesis for the preparation of valuable targets such as pharmaceuticals, perfumes, pheromones, macrocycles, etc. Nitroalkanes have demonstrated to be valuable starting compounds even for these structures, as reported in Scheme 12.13.

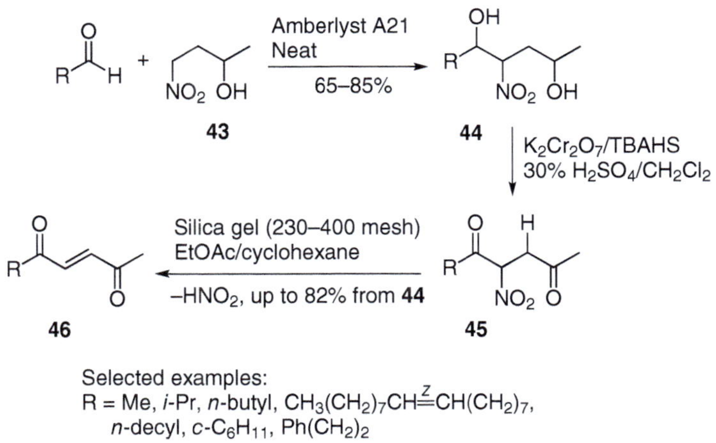

Scheme 12.13 One-pot synthesis of (*E*)-enediones **46**.

The process starts with the Henry reaction of a variety of aldehydes with nitroalkanol **43** [11], using Amberlyst A21 as base and under neat conditions. Successive oxidation of nitro diols **44** ($K_2Cr_2O_7$ under phase transfer conditions) affords nitro diketones **45**, prone to give nitrous acid elimination and formation of (*E*)-2-ene-1,4-diones **46** in good overall yields from **44**.

It is important to note that compound **46** with R = butyl (*E*-non-3-ene-2,5-dione) is the main component isolated from the fire bee derived from the cephalic secretion of workers of *Trigona tartaira* [12].

Thanks to the easily accessibility of 1,4-diones from nitroalkanes, a number of synthetic applications have been developed and here some of these are reported.

Ketoprofen **51** is one of the most prominent modern non-steroidal anti-inflammatory drug and its synthesis uses 1,4-diones **50** as key building block, prepared from phenylnitromethane **47** (Scheme 12.14) [13]. The process involves the Michael addition (92% yield) of **47** to the enone **48**, under basic alumina, without the need of any solvent. Consecutive Nef reaction, under basic/acidic conditions, achieves **50** in 74% overall yield, key precursor of **51** in few steps.

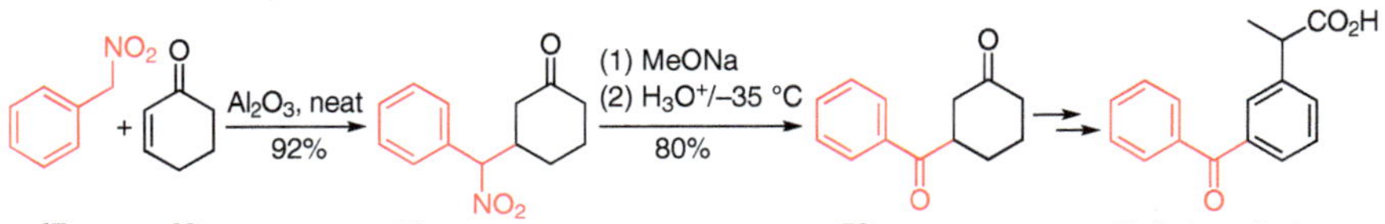

Scheme 12.14 Synthesis of 1,4-dione **50**, precursor of ketoprofen **51**.

The sesquiterpene "Cuparene" **56** is a natural product with important biological activity and its formal synthesis involves (Scheme 12.15) [14] the formation of cyclopentenone **55** from 1,4-dicarbonyls **54**.

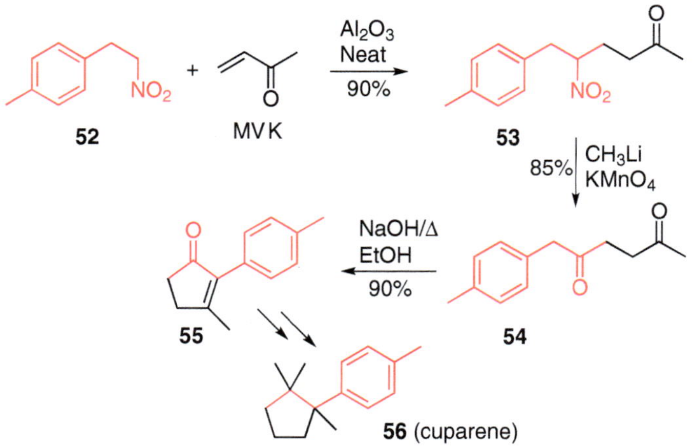

Scheme 12.15 Synthesis of cyclopentenone **55**, precursor of Cuparene **56**.

Thus, Michael reaction of aryl nitroalkane **52** with methyl vinyl ketone (MVK), in neat basic alumina, produces the adduct **53** that, under oxidative Nef reaction converts into 1,4-diketone **54**. An internal aldolic condensation affords **55** key precursor of "cuparene" **56** in few simple steps. The procedure offers a variety of advantages such as excellent yield (overall yield 69%), reduced need of solvents, and usage of simple reagents.

Allylrethrone **59** is an important component of insecticidal pyrethroid and strategic intermediate for the synthesis of allethrolone and pyrethrins. A one-flask and one-pot procedure for the preparation of Allylrethone **59** was developed form nitroalkane **32** and successive Michael and Nef reactions with formation of 1.4-diketone **58** that, by internal aldolic condensation, converts into **59** in 66% overall yield (Scheme 12.16) [15].

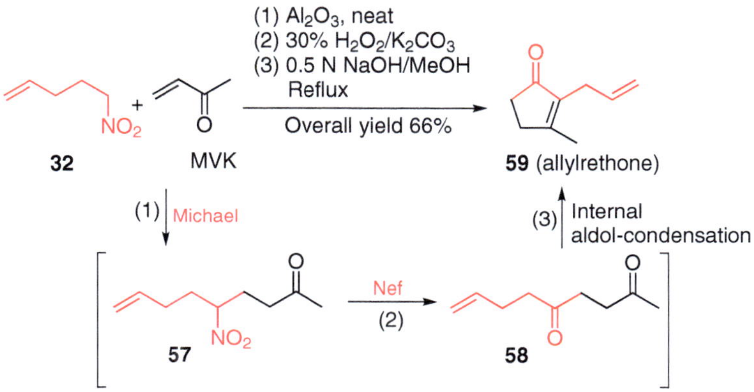

Scheme 12.16 One-pot synthesis of Allylrethrone **59**.

The key point for this one-pot process is the first step where the Michael reaction (formation of the adduct **57**) is performed in the absence of any solvents, so that the successive addition of reagents and solvent (MeOH) can be performed in the same flask without the need of any work-up. Very good overall **60** has been used as central intermediate in the synthesis of (*Z*)-jasmone **62b** and methyl jasmonate **63**, important components of essential oil of *Jasmine* flowers (Scheme 12.17) [16].

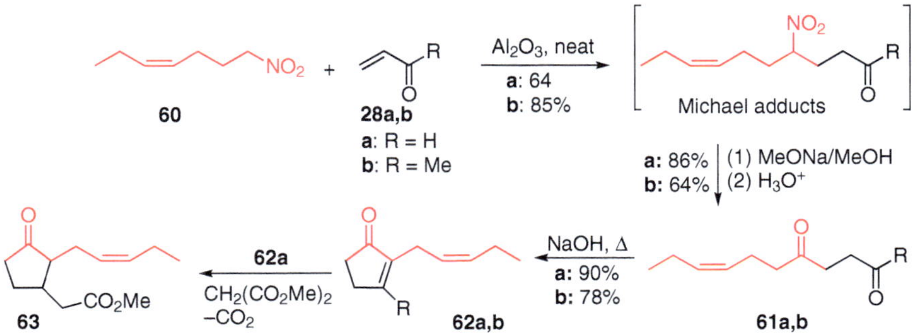

Scheme 12.17 Synthesis of (*Z*)-Jasmone **62b** and methyl Jasmonate **63**.

The Michael reaction of nitroalkane **60** with enones **28** affords an adduct that, by Nef reaction, converts into dicarbonyls **61**. Finally, intramolecular aldolic condensation yields (*Z*)-Jasmone **62b** and the cyclopentenone **62a** easily converted into methyl jasmonate **63**. Good yields are observed for each step.

12.3.2 γ-Ketoesters and γ-Ketoacids

γ-Ketoesters (or acids) are particular class of 1,4-dicarbonyl derivatives that often are prepared or are used in different ways respect to 1,4-diketones.

Following the same procedure jet reported in Scheme 12.3 (synthesis of β-ketoesters **12** from β-nitroester **9**), it is possible to obtain γ-ketoesters **66** by nitro esters **64** via reductive Nef conversion of nitroalkenes **65** [3], while oxidative Nef reaction produces γ-ketoacids **67** (Scheme 12.18) [8].

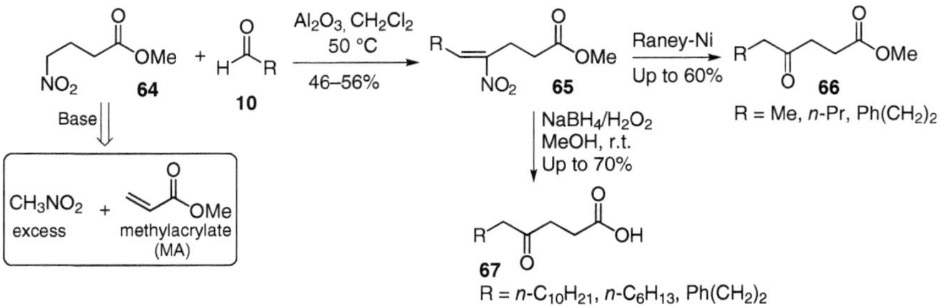

Scheme 12.18 Synthesis of γ-ketoesters **66** or γ-ketoacids **67**.

Compound **66a** (R = Me) can be reduced using baker's yeast, affording (*R*)-(+)-γ-caprolactone, the pheromone of *Trogoderma glabrum* (Scheme 12.19).

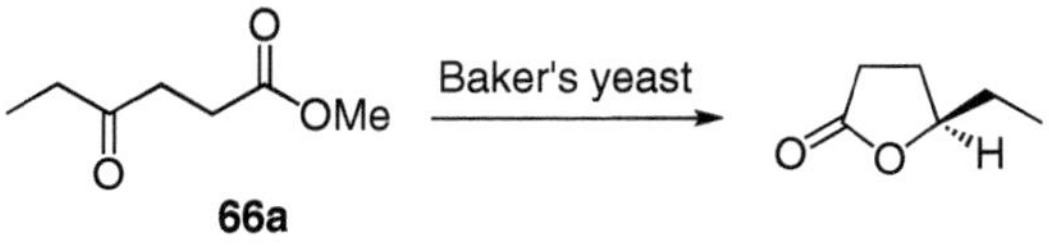

Scheme 12.19 Synthesis of (*R*)-(+)-γ-caprolactone.

An interesting procedure, for γ-ketoacids preparation, starts (Scheme 12.20) [17] with the Michael addition of nitroalkanes to methyl acrylate (MA) on Amberlyst A21 surface (neat conditions), granting the formation of γ-nitroesters **68**. Successive Nef reaction, using permanganate, furnishes 4-oxoalkanoic esters **69** in good yields.

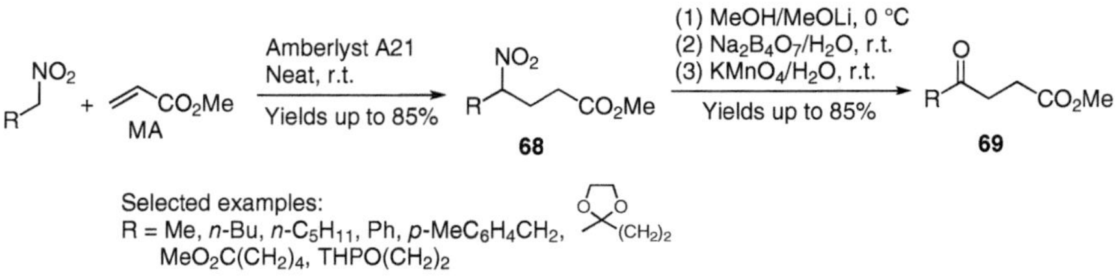

Scheme 12.20 Synthesis of γ-ketoesters **69** via Michael reaction.

The mildness and the efficiency of the procedure are demonstrated by the high yields and the survivals of a variety of other functionalities.

The Michael reaction of alkyl acrylates **70** (1 equiv) with CH_3NO_2 (1 equiv) [18] in the presence of Cs_2CO_3 as base and CuCN as catalyst, furnishes, in a one-pot way, (Scheme 12.21) keto diesters **72**, as major product, in presence of small amount of Michael adducts **71** (*ratio* up to 84 : 10) [18]. The in situ Nef reaction is strongly favored by the presence of CuCN. Anyway, the products **71** and **72** can be simply separated using flash chromatography while the residue **71** can be easily converted into **72** with a further nitro/carbonyl conversion.

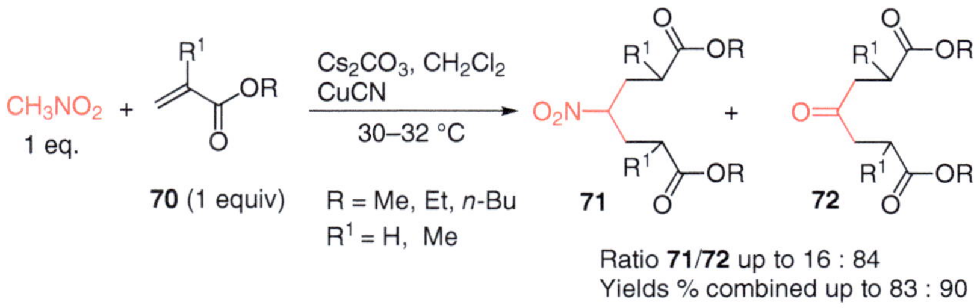

Scheme 12.21 One-pot synthesis of γ-ketodiesters **72**.

α-Nitro cycloalkanones **39** can be strategic for the synthesis of γ-ketodiesters **75** (Scheme 12.22) [10]. Good overall yields (up to 70%) are reported.

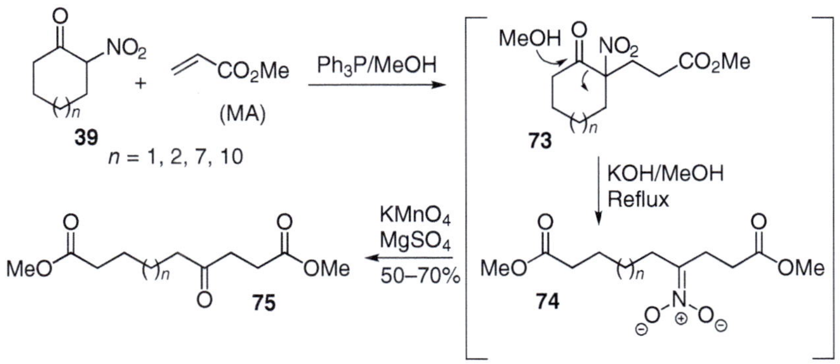

Scheme 12.22 One-pot synthesis of γ-ketodiesters **75**.

Conjugate addition of primary nitroalkanes to ethyl 2-(bromomethyl)acrylate **76** allows nitro adduct **77** through a tandem nucleophilic addition–elimination process, followed by Nef reaction and formation of α-methylene γ-ketoesters **78** (Scheme 12.23) [19]. The latter are prone to lactonization into *exo*-methylene butyrolactones **79** in good overall yields by $NaBH_4$ reduction. These classes of *exo*-methylene lactones are very important moiety present in many biologically active natural products and in pharmaceuticals.

γ-Ketoacids are synthetically strategic intermediates for many transformations and their easy availability is very important. The Michael addition of nitroalkanes

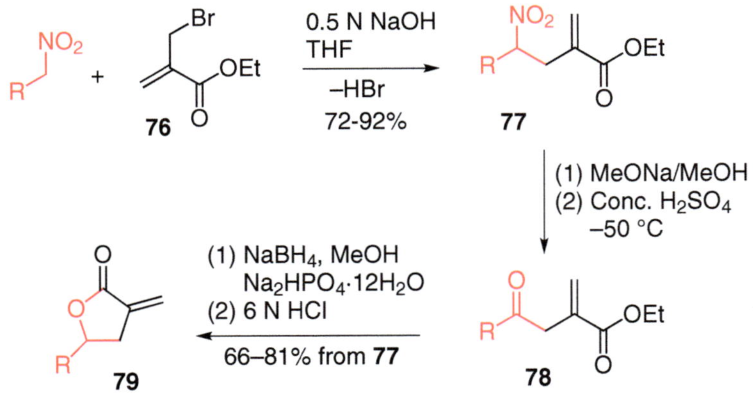

Selected examples:
R = n-C_5H_{11}, n-C_9H_{19}, $MeOCO(CH_2)_4$, $HO(CH_2)_3$, $CH_2{=}CH(CH_2)_8$, $PhCH_2$,

Scheme 12.23 Synthesis of *exo*-methylene butyrolactones **79**.

to acrolein, under basic alumina and in absence of solvents, allows nitroaldehydes **80** that, under oxidative (H_2O_2/K_2CO_3) Nef reaction, convert (way a) into keto acids **81** (Scheme 12.24) [20].

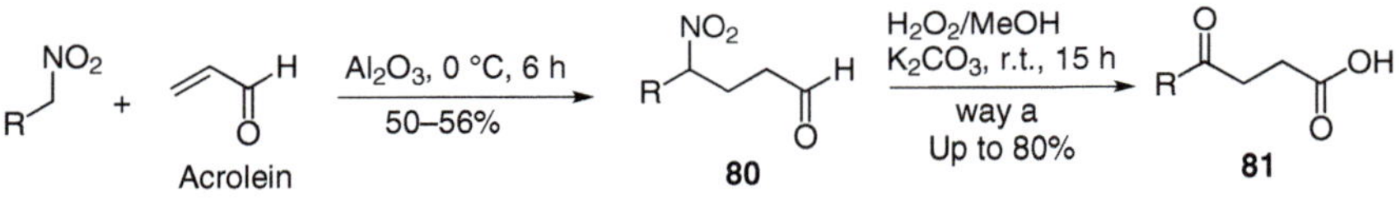

Selected examples:
R = Me, Et, n-Bu, $MeOCO(CH_2)_4$

Scheme 12.24 Synthesis of γ-ketoacids **81** or γ-ketoaldehydes **83**.

Alternatively (way b) (i) ketal-protection of the aldehyde, (ii) H_2O_2-Nef reaction, and (iii) deprotection of ketal with Me_2CO/TsOH achieve the formation of γ-ketoaldehydes **83**.

The Michael reaction of nitroalkanes with methyl propiolate **84** is precursors of synthesis of α,β-unsaturated γ-ketoesters **86** by Nef transformation of the adducts **85** (Scheme 12.25) [21].

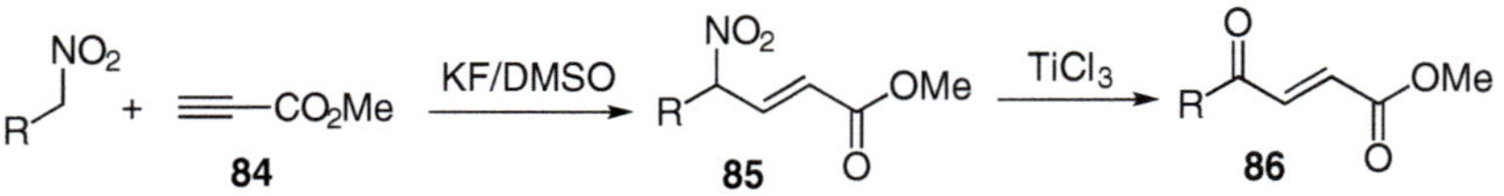

Scheme 12.25 Synthesis of γ-keto α,β-unsaturated esters **86**.

12.4 1,5-Dicarbonyl Derivatives

1,5-Diketones can be obtained following the same strategy of Schemes 12.4 and 12.5, in which 1,3- or 1,4-diketones are prepared from ketal-protected nitroketone **13** and **21**, respectively. Thus, starting from nitroketal **88**, easily prepared from the corresponding bromo derivative **87** (Sub-Scheme **A**), furnishes 1,5-diketones **91** (Scheme 12.26).

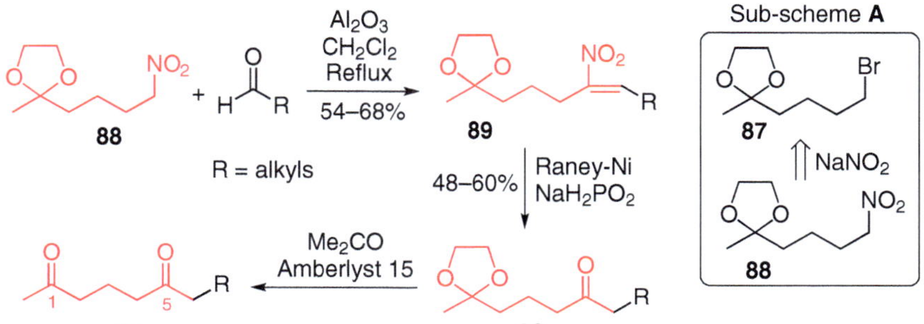

Scheme 12.26 Synthesis of 1,5-diketones **91**.

Primary α-nitroketones **92** reacts with enones **28**, under Ph_3P as base, affording nitrodiones **93**. Successive radical denitration by Bu_3SnH/azoisobutyrronitrile (AIBN)/C_6H_6 produces 1,5-diketones **94** in good yields (Scheme 12.27) [22].

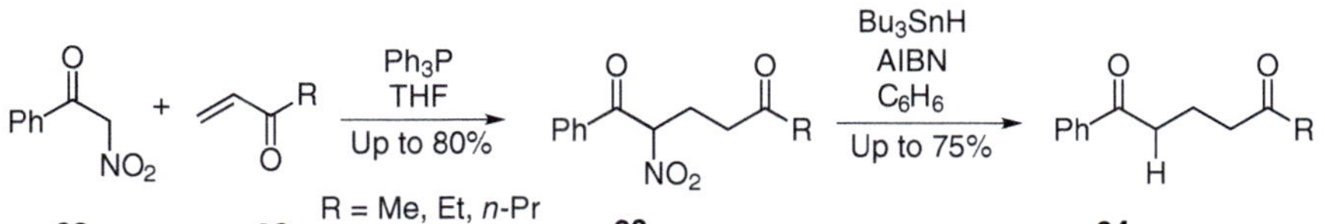

Scheme 12.27 Synthesis of 1,5-diketones **94**.

Of particular interest is the dione **94** with R = Et, because it is an important natural product isolated from fungus *Phellinus tremulae*.

δ-Ketoesters, an important class of 1,5-dicarbonyl derivatives, can be obtained following the same procedure previously reported in Schemes 12.3 and 12.6 [3].

Thus (Scheme 12.28), methyl 5-nitropentenoate **95** (easily prepared as reported in Sub-Scheme **B**) is submitted to nitroaldol condensation with aldehydes and formation of nitroalkenes **96** that, by reductive Nef reaction, convert into δ-ketoesters **97** in good yields.

Compound **97a** ($R = n\text{-}C_{10}H_{21}$) was submitted to baker's yeast reduction with formation of (*R*)-(+)-δ-Hexadecanolide **98** (Sub-Scheme **C**) queen substance of oriental hornet *Vespa orientalis*.

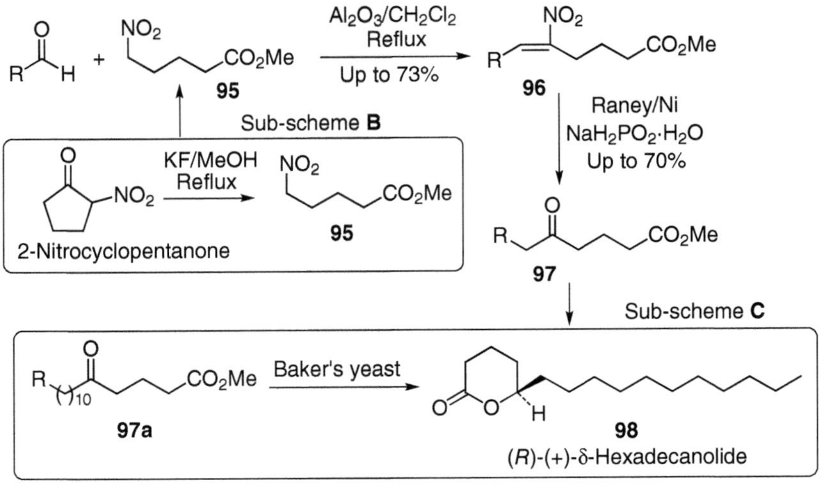

Scheme 12.28 Synthesis of δ-ketoester **97**.

References

1 Nikalje, M.D., Sayyed, A.I., Dewkar, G.K., and Sudalai, A. (2000). *Tetrahedron Lett.* 41: 959–961.
2 Palmieri, A., Gabrielli, S., Sampaolesi, S., and Ballini, R. (2015). *RSC Adv.* 5: 36652–36655.
3 Ballini, R. and Bosica, G. (1993). *J. Chem. Res. (S)*: 435.
4 Ballini, R. and Bartoli, G. (1993). *Synthesis*: 965–967.
5 Ballini, R., Fiorini, D., and Palmieri, A. (2004). *Tetrahedron Lett.* 45: 7027–7029.
6 Ballini, R., Petrini, M., Marcantoni, E., and Rosini, G. (1988). *Synthesis*: 231–233.
7 Ballini, R., Barboni, L., Bosica, G., and Fiorini, D. (2002). *Synthesis*: 2725–2728.
8 Ballini, R. and Bosica, G. (1994). *Synthesis*: 723–725.
9 Chamakh, A., M'hirsi, M., Villiéras, J. et al. (2000). *Synthesis*: 295–299.
10 Ballini, R., Bosica, G., and Gigli, F. (1998). *Tetrahedron* 54: 7573–7580.
11 Ballini, R. and Bosica, G. (1994). *J. Org. Chem.* 59: 5466–5467.
12 Kerr, W.E. and Costa Cruz, C. (1961). *Rev. Bras. Biol.* 21: 1–16.
13 Ballini, R. and Bosica, G. (1995). *Org. Prep. Proced. Int.* 27: 561–564.
14 Ballini, R. and Petrini, M. (1987). *Synth. Commun.* 17: 543–545.
15 Ballini, R. (1993). *Synthesis*: 687–688.
16 Ballini, R. and Petrini, M. (1989). *Synth. Commun.* 19: 575–583.
17 Ballini, R., Petrini, M., and Rosini, G. (1987). *Synthesis*: 711–713.
18 Farooq, S., Sangwan, P.L., Aleti, R.R. et al. (2012). *Tetrahedron Lett.* 53: 3305–3309.
19 Ballini, R., Bosica, G., and Livi, D. (2001). *Synthesis*: 1519–1522.
20 Ballini, R. and Petrini, M. (1986). *Synthesis*: 1024–1026.
21 Kalita, D., Khan, A.T., Saikia, A.T. et al. (1998). *Synthesis*: 975–976.
22 Ballini, R. and Bosica, G. (1994). *J. Nat. Prod.* 57: 1462–1463.

13

Nitroalkanes as Source of Spiroketals

Spiroketals moieties are main structures (i.e. pheromones) or sub-structures of many natural products [1] such as steroidal Sapogenin and marine organism (Figure 13.1).

Thus, it is clear that the most common spiroketal units **I–V** are those reported in Figure 13.2.

So, due to the great importance of these molecules, their efficient access is of great interest and this chapter reports the key role of nitroalkanes as starting molecules, capable of promoting the synthesis of a variety of spiroketals [2].

13.1 1,6-Dioxaspiro[4.4]nonanes

A double, successive Michael reactions of nitromethane **2** with two different conjugated enones **1** and **4** (Scheme 13.1) were used as key strategy [3] for the preparation of racemic 2-ethyl-1,6-dioxaspiro[4.4]nonane **7**, known as Chalcogran, the main component of the aggregation pheromone of the bark beetle *Pityogenes chalcographus*.

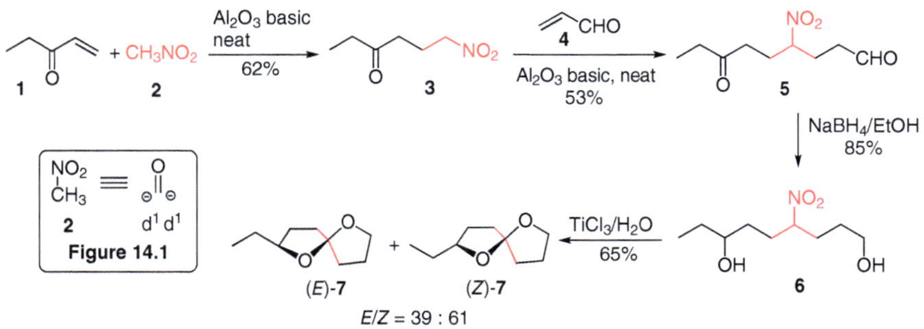

Scheme 13.1 Synthesis of Chalcogran **7**.

Thus, the Michael addition of **2** to 1-penten-3-one **1**, under neat basic alumina, affords the adduct **3**, then submitted to a further Michael reaction with acrolein **4**, under the same reaction conditions. The obtained nitro-dicarbonyl **5** is selectively reduced ($NaBH_4$) to the nitro-diol **6**. Finally, reductive Nef reaction ($TiCl_3$) of **6** is

Nitroalkanes: Synthesis, Reactivity, and Applications, First Edition.
Roberto Ballini and Alessandro Palmieri.

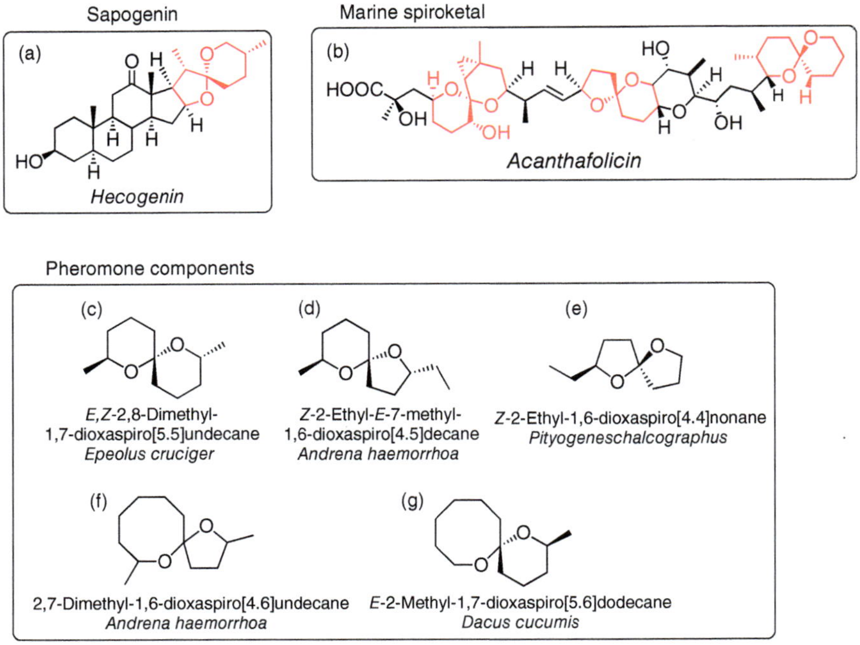

Figure 13.1 Some representative examples of spiroketals.

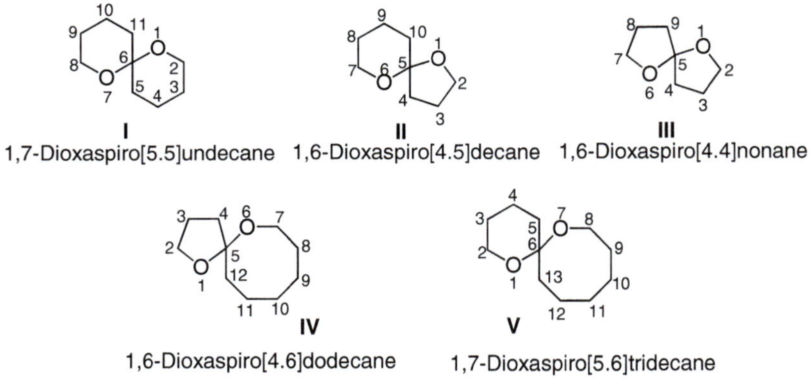

Figure 13.2 Most common spiroketal structures **I–V**.

accomplished by spontaneous spiroketalization with the formation of Chalogran **7** as a mixture of stereoisomers, in 20% overall yield.

It is important to observe, examining the synthetic role of nitromethane **2**, that it acts (see Figure 13.1, inserted in Scheme 13.1) as d^1d^1 multiple coupling reagent [4].

Symmetric 2,7-dialkyl-1,6-dioxaspiro[4.4]nonanes can be prepared from Michael process, under both Amberlyst A21 and solvent-free conditions, involving two moles of enones **1** and 1 mol of nitromethane (Scheme 13.2) [5]. The obtained adducts **8** are reduced ($NaBH_4$), without any work-up or purification steps, affording **9** in good yields (up to 80%).

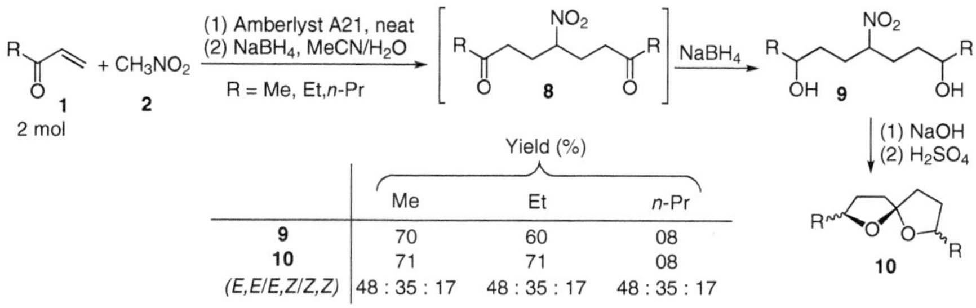

Scheme 13.2 Synthesis of symmetric 2,7-dialkyl-1,6-dioxaspiro[4.4]nonanes 10.

Standard Nef reaction (NaOH/H_2SO_4) achieves the one pot formation of spiroketals **10**, components of pheromonic mixture in insect of *Andrena* species.

An alternative enantioselective reduction of diketones **8** with baker's yeast or diisopinocampheyl-chloroborane (DIP-Cl™) provides a straightforward entry to optically active nitro diols **11** (Scheme 13.3) [6, 7].

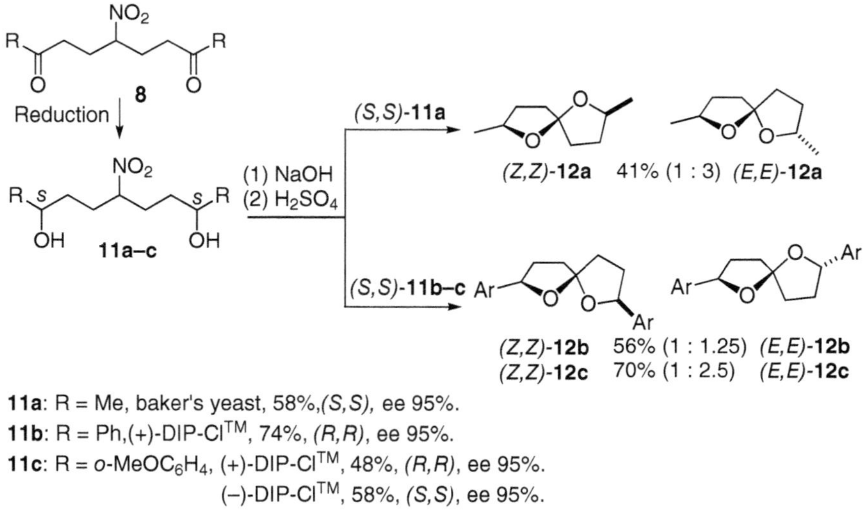

Scheme 13.3 Enantioselective synthesis of **12**.

Successive Nef reaction provides spiroketals **12a–c** in satisfactory to good yields and moderate stereoselectivity.

13.2 1,6-Dioxaspiro[4.5]undecanes

1,6-Doxaspiro[4.5]skeleton needs a diverse synthetic approach respect to [4.4]structure because the oxane ring can be formed just by ring closure involving a 1,5-relationship between the carbon atoms bearing hydroxyl and nitro groups. Thus, (*E*)-7-methyl-1,6-dioxaspiro[4.5]decane **18**, odor's component of wasp *Paravespula vulgaris*, was efficiently synthesized starting from commercially available

1-bromo-5-hexene **13** that, upon Wacker-type oxidation (Scheme 13.4) [8] furnishes **14**, key precursor of nitroalkane **15** by nucleophilic displacement of the bromine atom using $NaNO_2$. Successive Michael reaction of **15** with acrolein **4**, under neat conditions, affords the adduct **16** that, under chemoselective reduction ($NaBH_4$), allows nitrodiol **17**. Finally, Nef reaction completes the synthesis of spiroketal (*E*)-**18** in high selectivity and good overall yield (50% from **15**).

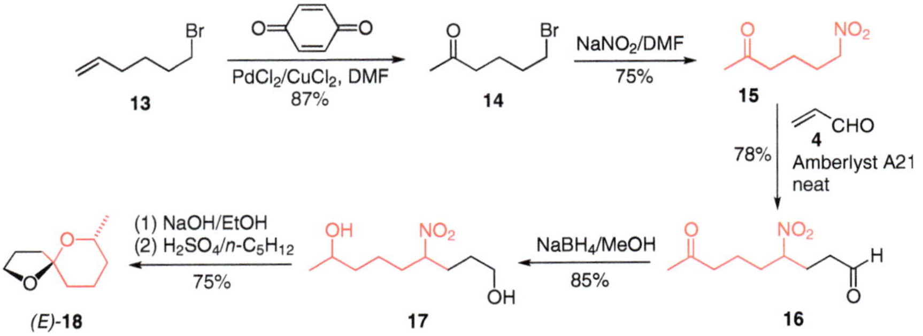

Scheme 13.4 Enantioselective synthesis of (*E*)-**18**.

2-Methyl-1,6-dioxaspito[4.5]decane **22**, one more pheromonic component found in *P. vulgaris*, was synthesized starting from 2-nitrocyclopentanone **19** (Scheme 13.5). This method is highly representative of the "chameleonic" behavior of nitroalkanes. In fact,

Step 1 (Michael reaction): The nitro group promotes the generation of a C—C bond by Michael reaction of 19 (carbon 2) with methylvinyl ketone (MVK) under basic alumina, affording the adduct 20.

Step 2 (retro-Claisen reaction): The electron withdrawing effect of the nitro group favors the ring cleavage of cyclic structure 20, with the formation of open-chain nitro ketones **21** that, by $NaBH_4$ reduction, furnishes nitrodiol 21.

Step 3 (Nef reaction): Nitro to carbonyl in carbon-2 of **21** achieves the simultaneous spiroketalization into **22** as *E*/*Z* mixture (*ratio* 2 : 1).

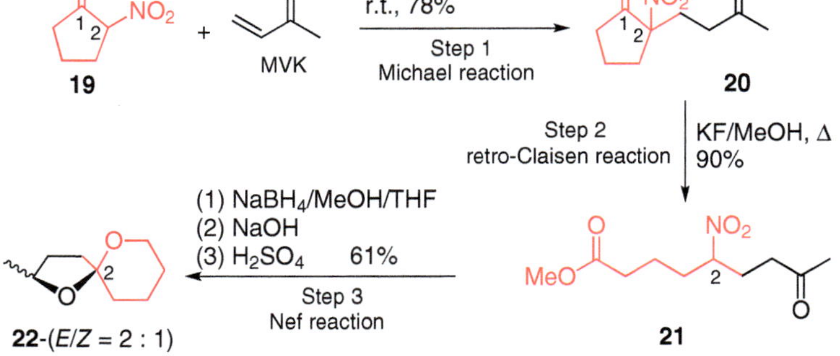

Scheme 13.5 Synthesis of spiro[4.5]decane **22**.

γ-Rubromycin **29** is a pigment isolated from *Streptomyces collinus*, particularly active against the reverse transcriptase of HIV-1 that can be characterized as bis-benzanellated 1,6-dioxaspiro[4.5]decane structure that presents some challenging aspects from a synthetic standpoint.

Nitroaldol reaction of aldehyde **23** (Scheme 13.6) [9] with nitromethane, promoted by cetyltrimethylammonium bromide (CTABr), produces quantitatively nitroalkanol **24** that upon dehydration and double bond reduction leads to nitroalkane **25**. Nitroaldol condensation of the latter with arylaldehyde **26** gives the nitroalkene **27** and represents the key step for the convergent synthesis of the carbon unit of the target molecule. Successive Pd/reduction of **27** and acidic (HCl) treatment produces the Nef reaction with simultaneous formation of spiroketal **28**, immediate precursor of the target compound **29** in few steps.

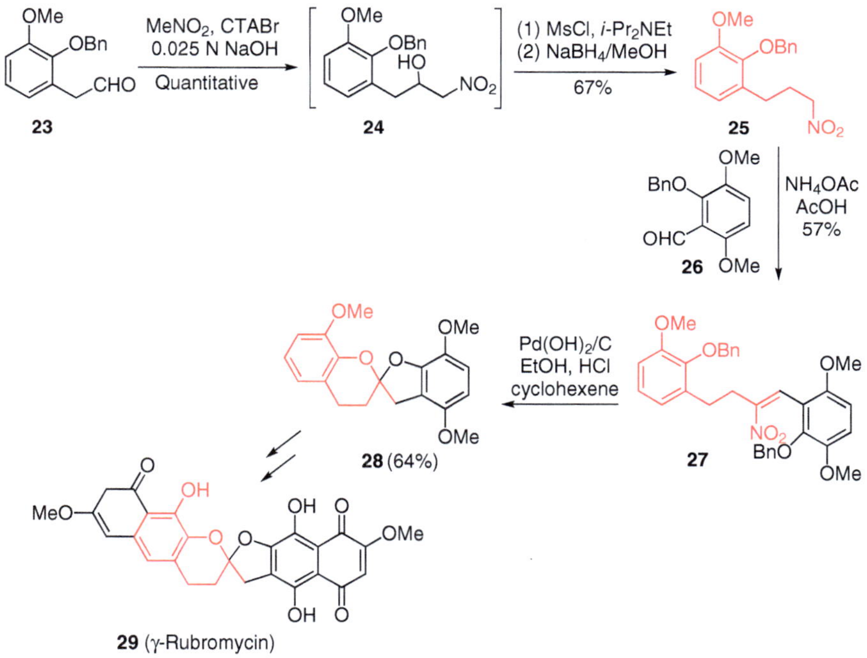

Scheme 13.6 Synthesis of γ-Rubromycin **29**.

13.3 1,6-Dioxaspiro[4.6]undecanes

Spiro[4.6] group are relatively rare, however this structure is present in pheromones produced by *Andrena haemorrhoa* and *Andrena wilkella*. Thus, a new one-pot synthetic approach was reported modifying the procedure previous reported in Scheme 13.5.

In fact (Scheme 13.7) [10], quantitative Michael reaction of nitrocyclohexanone **30** with enones **1**, under neat basic alumina, allows the adducts **31** that under reductive ($NaBH_4$/MeCN/H_2O) conditions yield spiroketals **33** in a one-pot way (60–75% overall).

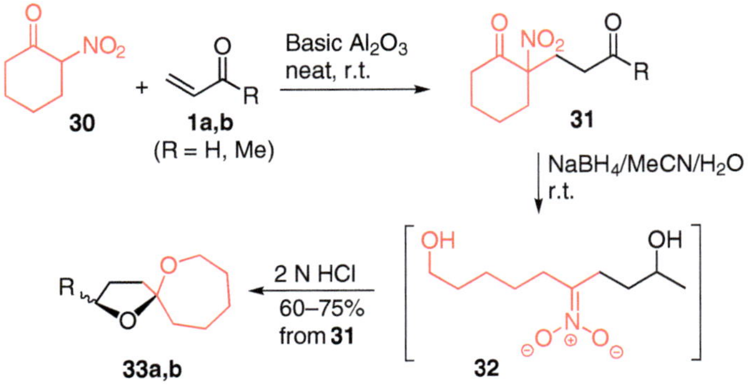

Scheme 13.7 Synthesis of spiro[4.6]undecanes **33**.

The process can be explained with the double action of $NaBH_4$ as reducing agent (carbonyls to hydroxyls) and as base (nitro to nitronate) with the formation of the intermediate **32** that, after acidic work-up (HCl), converts the nitronate group into a carbonyl (Nef reaction) and consequent direct formation of the spiroketals **33**.

A short, enantioselective synthesis of 2,7-dialkyl-1,6-dioxaspiro[4.6]undecanes, pheromones produced by *A. haemorrhoa* and *A. wilkella* was developed from ring cleavage of nitrocyclohexanols **34** with $CuSO_4$ in refluxing benzene (Scheme 13.8) [11]. The formed nitro ketones **35** were protected as ketals affording **36** that, after Michael reaction with MVK in Amberlyst A21 achieves the adducts **37**. Ketal deprotection and bioreduction with baker's yeast afford the stereoselective formation of diols **30**, prone to be converted into spiroketals **40a,b** with the (*S,S*)-stereochemistry, by Nef reaction.

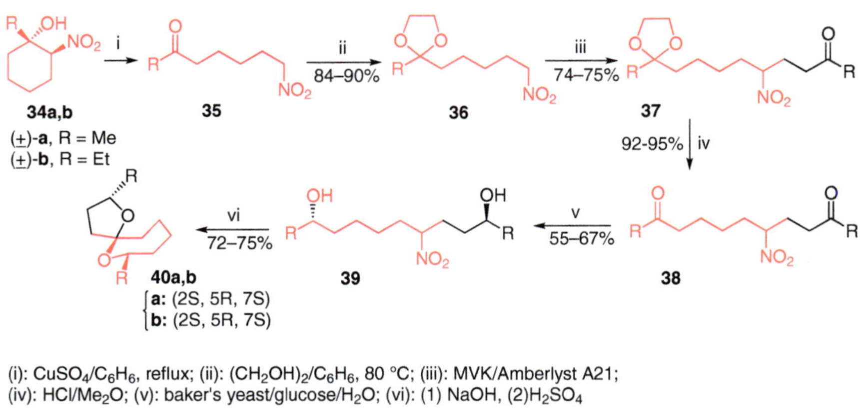

Scheme 13.8 Synthesis of 2,7-dialkyl-1,6-dioxaspiro[4.6]undecanes **40**.

13.4 1,7-Dioxaspiro[5.5]undecanes and 1,7-Dioxaspiro[5.6]dodecanes

Nitroaldol condensation of hydroxyl nitroalkane **41** with protected aldehyde **42**, under basic alumina in refluxing DCM, gives nitroalkenes **43** that, after $NaBH_4$ reduction, furnishes the nitronate **44** predisposed to be converted into a carbonyl under acidic (HCl) work-up, followed by spontaneous spiroketalization into **45a** "spiro[5.5]undecanes" and **45b** "spiro[5.6]dodecanes," the main components of the sex pheromones of the olive fruit fly *Dacus oleae* **45a** and *A. haemorrhoa* **45b**, respectively (Scheme 13.9) [12].

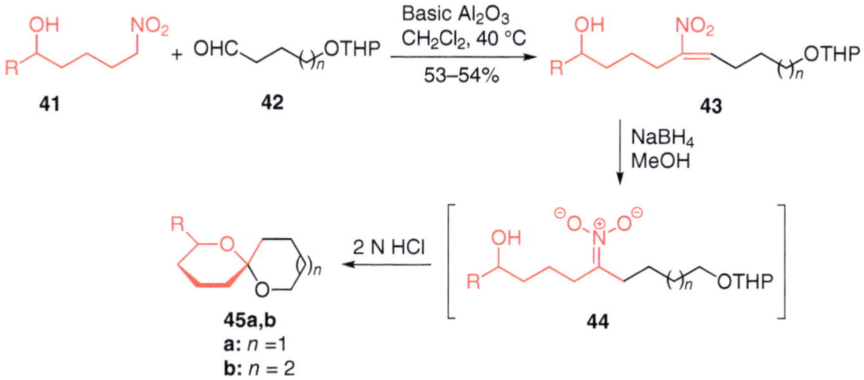

Scheme 13.9 Synthesis of 1,7-dioxaspiro[5.5]undecanes **45a** and dioxaspiro[5.6]dodecanes **45b**.

All the reported synthetic processes are based on the nitro to carbonyl conversion (Nef reaction) for the spiro-ketalization step. Anyway, in the synthesis of dioxaspiro[5,6]dodecane **52**, a pheromone component of *A. haemorrhoa*, the spiro-carbon is preformed and the nitro group plays a different role. As reported in Scheme 13.10 [13], nitroalkane **46** is submitted to C-acylation with *N*-acetylimidazole **47** under basic conditions, affording the α-nitro ketone **48** in 62% yield.

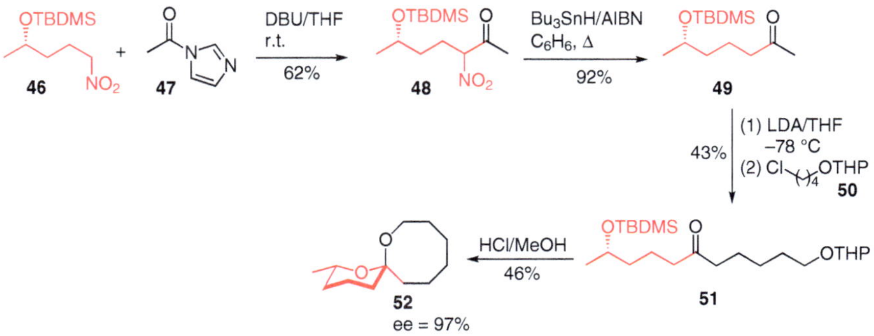

Scheme 13.10 Synthesis of 2-methyl-1,7-dioxaspiro[5.6]dodecane **52**.

Denitration of **48** under radical conditions (Bu_3SnH/azoisobutyronitrile [AIBN]) provides the ketone **49** that, after introduction of a further carbon framework **50** containing a second hydroxyl group, yields **51**, precursor of spiroketal **52** in 46% yield, after acidic hydrolysis of the protecting groups (OTBDM and OTHP).

References

1 Perron, F. and Albizati, K.F. (1989). *Chem. Rev.* 89: 1617–1661.
2 Ballini, R., Petrini, M., and Rosini, G. (2008). *Molecules* 13: 319–330.
3 Rosini, G., Ballini, R., Petrini, M., and Marotta, E. (1986). *Angew. Chem. Int. Ed.* 25: 941–942.
4 Seebach, D. and Knochel, P. (1984). *Helv. Chim. Acta* 67: 261–283.
5 Ballini, R., Bosica, G., and Uselli, A. (1994). *J. Heterocyclic Chem.* 31: 259–260.
6 Occhiato, E.G., Guarna, A., De Sarlo, F., and Scarpi, D. (1995). *Tetrahedron: Asymmetry* 6: 2971–2976.
7 Occhiato, E.G., Scarpi, D., Menchi, G., and Guarna, A. (1996). *Tetrahedron: Asymmetry* 7: 1929–1942.
8 Rosini, G., Ballini, R., and Marotta, E. (1989). *Tetrahedron* 45: 5935–5942.
9 (a) Capecchi, T., de Koning, C.B., and Michael, J.P. (1998). *Tetrahedron Lett.* 39: 5429–5432; (b) Capecchi, T., de Koning, C.B., and Michael, J.P. (2000). *J. Chem. Soc., Perkin Trans. 1*: 2681–2688.
10 Ballini, R., Petrini, M., and Rosini, G. (1990). *Tetrahedron* 46: 7531–7538.
11 Bez, G., Sarma, M., Saikia, A.K., and Barua, N.C. (2000). *Synthesis* 4: 537–540.
12 Ballini, R. and Petrini, M. (1992). *J. Chem. Soc., Perkin Trans. 1*: 3159–3160.
13 Nakamura, K., Kitayama, T., Inoue, Y., and Ohno, A. (1990). *Tetrahedron* 46: 7471–7481.

Index

Nitroalkanes: Synthesis, Reactivity, and Applications, First Edition.
Roberto Ballini and Alessandro Palmieri.

b